Allocation in the Europe
Emissions Trading Schen

T0253904

A critical issue in dealing with climate change is deciding who has a right to emit carbon dioxide, particularly when those emissions are limited. *Allocation in the European Emissions Trading Scheme* provides the first in-depth description and analysis of the process by which rights to emit carbon dioxide were created and distributed in the European Emissions Trading Scheme. This is the world's first large-scale experiment with an emission trading system for carbon dioxide and is likely to be copied by others if there is to be a global regime for limiting greenhouse gas emissions. The book consists of contributions by participants in the allocation process in ten representative member states and at the European Commission. The problems encountered by these ten representative countries, the solutions found and the choices they made, will be of interest to all who are concerned with climate policy and the use of emissions trading.

A. DENNY ELLERMAN is Senior Lecturer at the Sloan School of Management, Massachusetts Institute of Technology.

BARBARA K. BUCHNER is Senior Researcher at the Fondazione Eni Enrico Mattei (FEEM).

CARLO CARRARO is Professor of Econometrics and Environmental Economics at the University of Venice and Director of Research at the Fondazione Eni Enrico Mattei (FEEM).

Allocation in the European Emissions Trading Scheme

Rights, Rents and Fairness

Edited by

A. DENNY ELLERMAN
Massachusetts Institute of Technology

BARBARA K. BUCHNER
Fondazione Eni Enrico Mattei

CARLO CARRARO
University of Venice

CAMBRIDGE UNIVERSITY PRESS

CAMBRIDGE UNIVERSITY PRESS
Cambridge, New York, Melbourne, Madrid, Cape Town, Singapore,
São Paulo, Delhi, Dubai, Tokyo, Mexico City

Cambridge University Press
The Edinburgh Building, Cambridge CB2 8RU, UK

Published in the United States of America by Cambridge University Press, New York

www.cambridge.org
Information on this title: www.cambridge.org/9780521182621

© Cambridge University Press 2007

First published 2007
Reprinted 2008
First paperback edition 2010

A catalogue record for this publication is available from the British Library

ISBN 978-0-521-87568-4 Hardback
ISBN 978-0-521-18262-1 Paperback

If factors of production are thought of as rights, it becomes easier to understand that the right to do something which has a harmful effect...is also a factor of production.

Ronald Coase,
The Problem of Social Cost

Contents

Figures

Boxes

Tables

Contributors

DANIELE AGOSTINI Department for Global Environment, International and Regional Conventions, Italian Ministry for the Environment

CONOR BARRY CDM – Secretariat, UN Framework Convention on Climate Change, Registration and Issuance Unit

ISTVAN BART DG Environment, European Commission

BARBARA K. BUCHNER Fondazione Eni Enrico Mattei (FEEM)

CARLO CARRARO Department of Economics, University of Venice and Fondazione Eni Enrico Mattei (FEEM)

TOMAS CHMELIK Climate Change Department, Czech Ministry of Environment

PABLO DEL RÍO Department of Economics, University of Castilla–La Mancha

A. DENNY ELLERMAN Center for Energy and Environmental Policy Research, Sloan School of Management, Massachusetts Institute of Technology

DAVID HARRISON National Economic Research Associates (NERA) Economic Consulting

BOLESLAW JANKOWSKI Badania Systemowe 'EnergSys' Sp. z o.o.

FELIX CHRISTIAN MATTHES Öko-Institut (Institute for Applied Ecology)

SIGURD LAUGE PEDERSEN Danish Energy Authority

DANIEL RADOV National Economic Research Associates (NERA) Economic Consulting

FRANZJOSEF SCHAFHAUSEN German Federal Environment Ministry

PETER ZAPFEL DG Environment European Commission

LARS ZETTERBERG Swedish Environmental Research Institute (IVL)

Foreword

Environmental policy-making in Europe has undergone a profound change over the last two decades: new policies have been initiated primarily at the EU rather than Member State level, and the influence of economic thinking has become quite important. There is probably no other policy area where this shift is clearer than on climate change, and the development of the EU ETS is undoubtedly the most remarkable example of both developments.

It was indispensable for all those involved in the development of climate change policy to be able to rely on solid academic work in both Europe and the world at large. It is very welcome for me as a policy-maker that this work is not limited only to conceptual thinking, but also takes in the evaluation of practical choices made in mundane matters such as allocation. This book is therefore a timely contribution for all those involved in the design of further trading schemes around the world for the first Kyoto Protocol period 2008–2012 and beyond.

All observers realised that the learning-by-doing phase of the EU ETS in 2005 would be challenging from several perspectives. Perhaps the most important was the absence of a comprehensive database concerning the emissions of the 11,500 installations covered by the EU ETS. Many pragmatic solutions had to be invented on the spot by the national officials in the capitals and in the Commission's offices. In the meantime, the first compliance cycle came to an end on 15 May 2006, and this has generated a wealth of verified real-life emissions data for the year 2005 at the level of each installation. The process of designing allocation plans for the trading period 2008–2012 will benefit a lot from this empirical base, as all those involved have moved up the NAP learning curve in a significant manner. And this applies not only to policy-makers, but also to companies and academics.

The new piece of analysis in front of us brings also a number of insights which will influence the policy process much beyond the NAP process. The legislators may have underestimated the significant distributional battle that every allocation process entails, between Member States, between sectors, and within sectors. Sorting these out in a satisfactory manner was a precondition for the solid implementation of the EU ETS. Moreover the small print of the allocation plans may also have a considerable impact on the effectiveness of the system. It is of utmost importance to acquire a thorough insight into all aspects of this allocation experience, not least in the light of the review of the EU ETS Directive. It would be surprising if the decentralised nature of the allocation process, the structure of twelve general criteria and the limited role of auctioning would not become major issues in the preparation of the EU ETS for the post-2012 period.

Jos Delbeke
DG Environment
European Commission
July 2006

Acknowledgements

This book originated from a EU project on 'Lessons Learnt from the National Allocation of Allowances', sponsored by the European Commission's Directorate General for Environment. We would therefore like to thank the European Commission for financial support, and Jos Delbeke and Peter Zapfel for their intellectual engagement and guidance.

In order to prepare this book, two workshops were organised in which the contents of the different chapters were discussed. We are therefore grateful to the staff of the Fondazione Eni Enrico Mattei, and in particular to Rita Murelli and Barbara Racah, who helped to organise the workshops and who provided useful assistance throughout the preparation of this book.

Several discussions and comments from friends and colleagues have improved our understanding of the EU ETS allocation process. We thank our colleagues at the Fondazione Eni Enrico Mattei, at the Massachusetts Institute of Technology Center for Energy and Environmental Policy Research and at the Joint Program on the Science and Policy of Global Change for helpful remarks and suggestions. Comments from participants at seminars in Cambridge, MA and in Venice and from those colleagues who attended the special session on the design of climate policy at the 3rd World Congress of Environmental Economists in Kyoto are gratefully acknowledged.

We would also like to thank the various contributors to this book for their efforts, their commitment and their patience during the completion of the project. And, finally, we wish to thank Chris Harrison of Cambridge University Press for his encouragement throughout the process that led to this book.

Glossary and abbreviations

AAU	Assigned Amount Unit
Accession countries	This term is usually used to refer to the ten new EU Member States from Central and Eastern Europe, which joined the European Union on 1 May 2005 (Cyprus, Czech Republic, Estonia, Hungary, Latvia, Lithuania, Malta, Poland, Slovakia and Slovenia)
Allowance	According to the EU ETS Directive Article 3(a), 'allowance' means an allowance to emit one tonne of carbon dioxide equivalent during a specified period, which shall be valid only for the purposes of meeting the requirements of this Directive and shall be transferable in accordance with the provisions of this Directive
Annex B	The list of countries taking on legally binding commitments along with a listing of their actual commitments as defined in the Kyoto Protocol
Annex I	Industrialised countries that, as parties to the UNFCCC, have pledged to reduce their greenhouse gas emissions by the year 2000 to 1990 levels as per Article 4.2 of the Kyoto Protocol are listed in Annex I. Annex I Parties consist of countries belonging to the OECD, the Economies-in-Transition and Turkey
Banking	Saving emission permits for future use in anticipation that these will accrue value over time

Baseline	Reference scenario which is used for comparisons with other scenarios
Base year	Year that serves as base for calculation of emission targets. In the UNFCCC and the Kyoto Protocol it is 1990 for Annex I countries (except Bulgaria 1988, Hungary average 1985-87, Poland 1988, Romania 1989). A base year may also be used as a reference for establishing an emissions baseline, or standard by which to measure verifiable changes in carbon stocks for the purpose of determining net changes of GHG emissions from anthropogenic land-use change and forestry activities
Basket	The six gases CO_2, CH_4, N_2O, HFC, PCF and SF_6 form a basket in which the Kyoto commitments are denominated
BAT	Best available technology
BAU	Business as usual
Benchmark	A measurable variable used as a baseline or reference in evaluating the performance of projects or actions
Bubble	The idea that emissions reductions anywhere within a specific area count toward compliance. The possibility of forming a 'bubble' represents one of the flexible mechanisms included in the Kyoto Protocol
BSA	The Burden Sharing Agreement or so-called European Bubble allows the EU to reallocate its overall Kyoto target of −8% among the different EU Member States. In June 1998, a political agreement has been reached on the allocation of emission reduction efforts within the EU. The reduction commitments of each EU Member State can be found in the Communication from the Commission to

	the Council and the European Parliament (COM(99) 230 final, 19 May 1999)
Cap	Absolute emissions limit
Carbon dioxide equivalent	The concentration of CO_2 that would cause the same amount of radiative forcing as the given mixture of CO_2 and other greenhouse gases
Carbon sequestration	The uptake and storage of carbon; trees and plants, for example, absorb carbon dioxide, release the oxygen and store the carbon
Carbon sink	Any reservoir that takes up carbon released from some other part of the carbon cycle; for example, the atmosphere, oceans and forests are major carbon sinks because much of the CO_2 produced elsewhere on the Earth ends up in these bodies
CCGT	Combined cycle gas turbine
CDM	Clean Development Mechanism: in Article 12 of the Kyoto Protocol, the parties established the CDM for the purposes of assisting developing countries in achieving sustainable development and helping Annex I parties meet their emissions targets; carbon currency: certified emission reduction units (CERs)
CER	Certified emission reduction unit
CHP	Combined heat and power
CO_2	Carbon dioxide: the main greenhouse gas affected directly by human activities
COP	Conference of the Parties: the supreme body of the UNFCCC
Directive	In our context, the reference to the Directive usually denotes the Directive 2003/87/EC of the European Parliament and of the Council of 13 October 2003 establishing a scheme for greenhouse gas emission allowance trading within the Community

	and amending Council Directive 96/61/EC, *OJ* L 275, 32–45, 25 October 2003
EC	European Commission
ECJ	European Court of Justice
Economies in transition	The countries listed in Annex I or Annex B that are undergoing the process of transition to a market economy but that are also classified along with the EU, Japan and the US as Annex I parties to the UNFCCC
Emissions leakage	A concept often used by policy-makers in reference to the problem that emissions abatement achieved in one location may be offset by increased emissions in unregulated locations
Emission permit	In general a tradable entitlement to emit a specified amount of a substance. In the context of the EU ETS, the operators of included installations need to hold a 'greenhouse gas emissions permit' issued by a competent authority in accordance with Articles 5 and 6 of the EU ETS Directive
ERU	Emissions reduction unit
ETS	Short for EU ETS
EU	European Union
EUA	European Union Allowances
EU ETS	The European Union Emissions Trading Scheme, specified by the Directive 2003/87/EC, and launched in January 2005
GHG	Greenhouse gas: any trace gas that does not absorb incoming solar radiation but does absorb longwave radiation emitted or reflected from the Earth's surface. The most important greenhouse gases are water vapour, carbon dioxide, nitrous oxide, methane and chlorofluorocarbons (CFCs).
Hot air	The amount by which an Eastern European country's Kyoto Protocol target exceeds its

	probable emissions in 2012 even without any abatement actions. The reason for this excess emission reductions is the economic collapse which these countries suffered after the base year 1990
Installation	According to the EU ETS Directive Article 3(e), 'installation' means a stationary technical unit where one or more activities listed in Annex I are carried out and any other directly associated activities which have a technical connection with the activities carried out on that site and which could have an effect on emissions and pollution
International Emissions Trading	Emission trading as defined in Article 17 of the Kyoto Protocol; carbon currency: Assigned Amount Units (AAUs)
IPPC	Integrated Pollution Prevention and Control
JI	Joint Implementation: GHG mitigation projects between developed countries as defined in Article 6 of the Kyoto Protocol; carbon currency: emission reduction units (ERUs)
Kyoto Mechanisms	Generic term for the flexible mechanisms of the Kyoto Protocol: bubbles, JI, CDM and international emissions trading
KPD	Known planned development
MS	Member State of the European Union
MW_{th}	Megawatt thermal
NAP	National Allocation Plan. For each period of the EU ETS, each Member State is supposed to develop a national plan stating the total quantity of allowances that it intends to allocate for that period and how it proposes to allocate them
Non-Annex I country	Any country that does not belong to Annex I of the UNFCCC, i.e. the developing countries and some countries in transition

NER	New Entrant Reserve set aside for any installation carrying out one or more of the activities indicated in the Directive, which has obtained a GHG permit or an update of its GHG emissions permit because of a change in the nature or functioning or an extension of the installation
NO_x	Nitrogen oxides
Operator	According to the EU ETS Directive Article 3(f), 'operator' means any person who operates or controls an installation or, where this is provided for in national legislation, to whom decisive economic power over the technical functioning of the installation has been delegated
SO_2	Sulphur dioxide: a gas representing a form of air pollution It results from the combustion of fuels that contain sulphur, most prevalently from coal
UNFCCC	United Nations Framework Convention on Climate Change: a multilateral agreement that lays the basis for international climate negotiations
VOC	Volatile organic compounds

PART I
The EU ETS allocation process

1 | The EU ETS allocation process: an overview

A. DENNY ELLERMAN, BARBARA
K. BUCHNER AND CARLO CARRARO

1 Motivation

From 1 January 2005, a price has been imposed on emissions of carbon
dioxide (CO_2) for a significant segment of the economies of twenty-
five sovereign nations constituting a major geographical region of the
world. This day marked the start of the European Union's Emissions
Trading Scheme (EU ETS), a cap-and-trade programme inspired by the
Kyoto Protocol, but more importantly, the prototype of the multinational collaboration that will be required if a global climate change
regime is to be established.

This achievement is remarkable in view of the differing circumstances
of these nations, their varying obligations under the Kyoto Protocol
and their uneven commitment to climate policy. Even more remarkable is the ambition of the project in its scope, timing and novelty.
Commitments under the Kyoto Protocol were known as of the end of
1997, but as late as 2000, when a formal discussion on including a
cap-and-trade system in the EU's Climate Change Programme began
(European Commission 2000), it would have been visionary to expect
that more than a few of the nations most committed to climate change
policy would adopt such a policy, not to mention those less committed
to effective action, those hostile to emissions trading or those not facing binding targets under the Kyoto Protocol. Moreover, the decision
to have a pre-Kyoto trial period from 2005 to 2007 compressed the
timing to beyond what seemed feasible. A typical view of the time (and
for some time after) was succinctly stated in Point Carbon's issue of
The Carbon Market Analyst of September 2001: 'We believe that the
chances of having a Community-wide trading scheme in place by 2005
is a low-probability scenario.'

Political agreement among the then fifteen member governments of
the EU (EU15) was not reached until the summer of 2003, which left

barely a year and a half to transpose the Directive, allocate allowances and develop the requisite infrastructure.

The resulting market is not only the world's largest environmental market, surpassing by far the scope of the US SO_2 allowance trading program,[1] but it is the embryo from which a future global regime may emerge. As such, this European endeavour will be much watched and the experience gained from it holds lessons for those who wish to follow, as well as for those who contemplate an eventual global regime to deal with climate change concerns.

The subject of this book concerns only one – very important – aspect of the EU ETS: the process of creating and distributing allowances by ten representative EU Member States. A trading system cannot exist without something to trade and among the first requirements of any emissions trading system is the creation and distribution of rights to emit, usually termed allowances, and in this case, European Union Allowances, or EUAs. These allowances give the holder the right to emit (1 t) of CO_2 emissions within some time period. They are tradable throughout the EU and they are required to be held and surrendered annually by the owners of all emitting sources to the appropriate national regulatory authority in an amount equal to emissions during the year. In the case of the EU, each Member State was responsible for developing a National Allocation Plan (NAP), subject to review by the European Commission (EC). This NAP both determined the total number of EUAs and distributed these allowances to the entities that would eventually be required to surrender them to the appropriate regulatory authority.

While allowances have been created and distributed in other trading systems, mostly in the United States, the EU ETS is the first trading system to allocate rights for CO_2. Accordingly, one motivating interest of this volume is to examine whether allocating carbon rights is different from allocating rights to the more conventional regulated emissions that have characterised earlier trading systems. But the more important motivation is that, if there is to be a global trading regime for limiting greenhouse gas (GHG) emissions, other nations will find it necessary to create and to allocate CO_2 allowances and they will naturally look

---

[1] While both the EU ETS and the US SO_2 program enjoy continental dimensions, the former includes approximately 11,500 installations and annual emissions of 2.2 billion metric tons of CO_2, which compares to the latter's 3,000 installations and 9 million short tons of SO_2.

to this precedent for instruction. For them, the recent experience of the EU offers rich lessons. The goal of the editors and contributors to this volume is to make this experience more accessible and to draw appropriate observations and lessons from it.

2 Methodology

The process of writing this book has involved more than the compilation of eleven individual contributions with editorial introduction and conclusion. In early 2005, the Fondazione Eni Enrico Mattei assembled a small group of individuals who were closely involved in the EU allocation process at the Member State level to participate in a structured discussion and analysis of their experiences. In order to keep the process manageable, the number of Member States was limited to ten, selected to constitute a representative selection of large and small states, ones with and without problems in meeting their Kyoto obligations and those that were both early and late in submitting NAPs to Brussels. These participants agreed to describe and to comment on the NAP process in their countries and those ten written contributions constitute the core of the book. The eleventh contribution is written by an official of the EC to provide a perspective from the centre on the process of guidance, coordination, review and approval. While all of these contributions are written by persons actively involved in the allocation process often in official positions, all are writing as individuals and not as representatives of their respective governments. In addition, several experts from academia, media commentators and carbon market intermediaries were asked to join the discussions to provide their perspectives in both oral and written form. Table 1.1 lists the contributors, external experts and editors. Appendix I includes a brief biography of each.

To ensure a unified product, two coordinating mechanisms were adopted. First, a common outline (Appendix II) was developed to provide a more uniform presentation of the experience of each Member State and to facilitate inter-country comparisons by readers. Second, a structured interaction between contributors, experts and editors was followed to bring out the common features and to highlight the important differences among countries. This process consisted of two highly focused workshops in the course of 2005 to provide the opportunity for informed discussion and a common knowledge base among

Table 1.1 *Participants in the FEEM project*

Country	Name	Affiliation
Contributors		
European Commission	Peter Zapfel	DG Environment
United Kingdom	Daniel Radov	NERA Economic Consulting
	David Harrison	
Germany	Felix Christian Matthes	Öko-Institut/Institute for Applied Ecology
	Franzjosef Schafhausen	German Federal Environment Ministry
Denmark	Sigurd Lauge Pedersen	Danish Energy Authority
Sweden	Lars Zetterberg	The Swedish Environmental Research Institute (IVL)
Ireland	Conor Barry	Irish EPA[b]
Spain	Pablo del Río	University of Castilla–La Mancha
Italy	Daniele Agostini	Adviser, Italian Ministry for the Environment
Hungary	Istvan Bart	Hungarian Ministry of Economy and Transport[a]
Czech Republic	Tomas Chmelik	Czech Ministry of Environment
Poland	Boleslaw Jankowski	Badania Systemowe 'EnergSys' Sp. z o.o.
External experts		
Norway	Atle Christer Christiansen	Point Carbon
France	Christine Cros	French Ministry of Environment
Ireland	Frank Convery	University of Dublin
United Kingdom	Per Lekander	UBS Investment Research
The Netherlands	Hans Warmenhoven	Spin Consult
Editors		
Italy	Barbara Buchner	Fondazione Eni Enrico Mattei
Italy	Carlo Carraro	Fondazione Eni Enrico Mattei
USA	A. Denny Ellerman	Massachusetts Institute of Technology

[a] Currently: European Commission, DG Environment.
[b] Currently: Secretariat of the United Nations Framework Convention on Climate Change, CDM – Registration and Issuance Unit.

participants, and repeated email and telephone contact during the writing and revising of this book. The purposively structured interaction among participants and the common outline ensured the discussion of the key features of the allocation process in all contributions.

The result of this process is the present volume consisting of this introduction, a contribution describing the NAP process as seen from Brussels, the ten individual Member State contributions and the editors' concluding chapter. The final product provides a unified analysis of how these nations came to decide limits on CO_2 emissions for a significant segment of their economies and to distribute the resulting allowances to the affected CO_2-emitting facilities. As a last part of this introduction, a brief guide to the individual contributions is provided to highlight the key features of each.

3 Guide to the contributions

The initial contribution presents the perspective from the *EU Commission*, as seen by *Peter Zapfel*, in his reflections on what he aptly terms a brief but lively chapter in EU climate policy. This contribution provides a summary of the key provisions of the Emissions Trading Directive, the timeline of the key steps and the institutional setting in which the allocation process took place. More importantly, it describes the challenges faced and the key factors that contributed to the success of the endeavour. Many of the difficulties encountered during the allocation process were expected because of the novelty of the enterprise, but some came as a surprise, for example, the resistance that industry showed to the use of emissions trading, the tendency in some NAPs to substitute bureaucratic rules for the market and the degree of soft harmonisation that was eventually achieved. Readers are particularly directed to his insightful comments on the Commission's role as the guardian of the environmental integrity and its heavy reliance on education, back-channel communication, and informal coordination to ensure the ultimately successful outcome of 25 approved NAPs and thus a functioning ETS.

The *United Kingdom* has viewed itself as a leader in emissions trading and it was the first Member State to publish a draft NAP in January 2004, which was of course much examined by others. As explained by *David Harrison* and *Daniel Radov*, the NAP process started early in the UK and there was extensive stakeholder consultation. This did not make the challenge of allocation any easier, however. The development of the NAP in the UK first gave shape to many of the features that would be common to most NAPs: no auctioning, a two-stage allocation to sectors then installations, a reliance on past emissions for installation-level allocation, and new entrant and closure provisions. A broad spectrum of pre-existing climate policies in the UK – including renewable targets, voluntary agreements and a prior emissions trading scheme – overlapped with the EU ETS and had to be integrated into NAP development. Along with other complications, this led to significant use of the opt-out provision. The authors also provide insights into the UK's dispute with the European Commission over the UK's total allocation, a dispute that ultimately went to the European Court of Justice (ECJ) for resolution.

Germany is the largest Member State in terms of emissions and sources and it was another country that began the process of drawing up a NAP very early. In recounting the German experience, *Felix Matthes* and *Franzjosef Schafhausen* explain the factors influencing the determination of the total cap and how other energy policies, such as the nuclear phase-out, strongly influenced allocation choices. Germany is unusual in deciding upon a one-stage process that avoided sector allocations and developed instead a compliance factor model that led to more or less uniform allocations relative to recent emissions after accounting for a number of special provisions. Germany tried to use benchmarking for allocation, but found it infeasible for reasons that are well explained in this contribution. Readers will also find a good discussion of the controversy surrounding new entrant and closure provisions and their relation to other policies. Finally, Germany is notable in its planned use of *ex-post* adjustments, which became and remains a matter for litigation with the Commission before the ECJ.

After these first two prominent countries in the European allocation process, we turn to a set of three smaller countries, two of which, Denmark and Sweden, like the UK, have some experience with emissions trading and governments strongly supportive of climate policy, and a third, Ireland, the government of which was committed to being an early mover in part because Ireland held the presidency of the EU Council in the first half of 2004 when the NAPs were due to be submitted.

Denmark stands out in a number of respects. It is the only Member State for which an existing national emissions trading programme covered a significant share of the EU ETS CO_2 emissions (75%). As explained by *Sigurd Lauge Pedersen*, this earlier experience simplified the NAP process and made it considerably less contentious than in other Member States. As a result, Denmark was one of the few Member States to submit the NAP on schedule, and to have it approved with a minimum of debate; and it was the only Member State to have an operating registry in January 2005. Denmark is also unique in being the only Member State to auction the full 5% of the total allowances and it is one of the few to use benchmarking for part of the allocation to installations. Finally, it stands out among North European Member States for having a large gap to close between projected emissions and its national commitment under the Kyoto Protocol as modified by the European Burden Sharing Agreement (BSA).

Sweden is a good example of a country with high energy intensity (due to its location and industrial structure) but low carbon intensity (due to its reliance on nuclear and hydro power). Sweden has no real problem in meeting its Kyoto/BSA target. As explained by *Lars Zetterberg*, emissions trading was, like in Denmark and the UK, a generally accepted approach to CO_2 emissions reduction. One main problem was obtaining reliable installation-level data, despite having previously collected what were believed to be good energy data for other purposes. Also like Denmark and the UK, Sweden had adopted pre-existing measures, notably a carbon tax, that had to be integrated with the trading system. Sweden is also unusual in having a significant number of opt-in units and in allowing closed facilities to retain their allowance endowments for the rest of the trading period.

Ireland is one of the fastest-growing economies of the EU and, like Denmark, it faces a significant projected shortfall in meeting its Kyoto goal. It was also one of the few that submitted on time. But, as explained by *Conor Barry*, the really unique aspect of the Irish NAP experience was the relatively technical nature of the process, and the structure of consultation with industry. While the Irish NAP is perhaps the least politically influenced and most 'rational' of those included in this volume, the government reserved a final say on some key choices and in fact overruled some recommendations from the technical group. Despite its different process, the important choices were very similar to those taken in other countries where the process was more overtly political. The NAP process in Ireland was also facilitated by a carbon tax that had been adopted to go into effect for the non-covered sectors at the same time as the ETS, but which in the end was not implemented.

Having covered five 'northern' Member States whose governments were all supportive of action on climate change and had adopted meaningful actions to limit GHG emissions prior to the NAP process, we now turn to five southern and eastern Member States for which the same cannot be said. Spain and Italy are countries where climate policy did not have the same salience it did in the 'northern' nations that are presented first. Interestingly, the main effect of the lack of early commitment to climate policy was delay in submitting NAPs to Brussels, not significantly different choices.

Spain is perhaps the best example of an EU Member State that is far from being on track to meet its Kyoto/BSA goal because of a high rate of economic and emissions growth. As described by *Pablo del Río*, the NAP process in Spain was one of a country without a meaningful commitment to climate policy coming to terms with the implementation of an effective instrument to reduce CO_2 emissions. A remarkable feature of the Spanish NAP process is the degree of interaction and openness in developing the NAP, the government's role as manager and arbiter, and an outcome of general acceptance among all the interested parties. Although the choices in Spain largely mirror those taken elsewhere, there are some unusual features, for instance pooling for non-electric installations and a greater than usual use of benchmarks.

Italy is the Member State that had perhaps the most confrontational relationship with Brussels during the NAP process. As explained by *Daniele Agostini* the dispute stemmed from the different visions the two had of the role of the EU ETS within national climate change policy: whether the ETS' role was to be 'policy setting' or 'policy compliance', that is, whether it was a framework within which the existing Italian GHG Reduction Plan would operate or a specific instrument which itself had to be adopted within the framework of Italy's GHG Reduction Plan emissions scenarios in order to ensure compliance with Italy's Kyoto/BSA target. The issue came to a head, as in several other Member States, over the total to be allocated when the Commission demanded a ten per cent reduction in the total proposed by Italy. Italy's NAP is also distinguished by the innovative approaches taken to allocation at the installation level, which reflect in part the uniqueness of Italy's electricity sector, which has no nuclear generation and very little coal generation, and the generally low energy intensity of Italy's economy.

In many ways, the most interesting Member State NAPs were those from the ten accession countries, which, with the exception of Slovenia, have no problem meeting their Kyoto targets. How to handle 'hot air' became a complicating issue in deciding the total cap, but many of the problems and solutions found elsewhere show up in the three examples described here: Hungary, the Czech Republic and Poland.

Hungary was the only one of these three to propose a total that was not rejected by the Commission and, as noted by *Istvan Bart*, the government was roundly criticised by domestic stakeholders for not attempting to distribute more allowances to participants. As in the other accession countries, opinion was divided between whether the ETS was an opportunity or an imposition. Hungary is unusual in having chosen a one-stage approach to allocation like Germany and to the extent that individual negotiation could be used to make installation-level allocations. This unusual approach was made possible by the relatively small number of installations and it helped to overcome the data problems that afflicted nearly every Member State. A good discussion of the desirability and infeasibility of benchmarking, at least in the 2005–2007 period, is also contained in this contribution.

In contrast to Hungary, the *Czech Republic* was forced to accept a fifteen per cent reduction to its proposed total allocation, because the Commission did not accept the arguments offered in support of the total. As explained by *Tomas Chmelik*, the Czech process was characterised by strong interaction with industry and extensive use of projections as a means of arriving at a compromise total that seemed to be a reasonable forecast of business-as-usual (BAU) emissions. As in other Member States using a two-stage approach to allocation, the more controversial stage was the allocation to sectors, not to individual installations. In the Czech Republic, this issue had to be resolved by the government. This contribution is also noteworthy in describing clearly the useful aspects of projections and in illustrating the extent to which recent emissions were the reference point for individual unit allocations.

Poland is the last of the Member State NAPs that are described, not only because it is the largest of the ten accession countries, but also because the disagreements with the Commission were so clearly etched and the difficulties of implementation so great. As *Boleslaw*

Jankowski describes it, the EU ETS was an 'ill-fitting suit' that was not made for the accession countries who did not face a problem of compliance with the Kyoto Protocol and for whom other priorities were more important than climate policy. Like the Czech Republic, Poland proposed a total that was reduced significantly by Brussels. This reduction was seen by many as creating a negative balance of benefits and costs due to the high transaction costs of implementing the scheme. Accordingly, emissions trading offered no cost savings in CO_2 emission reduction since no cost would be incurred in complying with the Kyoto Protocol in the absence of the ETS. Although the Polish government accepted the cut as the price of entry into the EU, it had a very hard time gaining industry acceptance of the reduced allocations that were implied. For this reason, as well as the delays inherent in a change of government, the Polish NAP and registry were the last to be actually implemented. Notwithstanding the disagreements, delays and different circumstances, the choices made in the Polish NAP closely track those made in other Member States.

References

European Commission 2000. 'Green paper on greenhouse gas emissions trading within the European Union', COM(2000) 87, March 2000.

Point Carbon 2001. 'Towards EU-wide emissions trading? Politics, design and prices', *The Carbon Market Analyst*, 25 September 2001.

2 | A brief but lively chapter in EU climate policy: the Commission's perspective

PETER ZAPFEL[1]

1 Introduction

The EU greenhouse gas emission trading scheme constitutes a path-breaking new chapter in EU environmental law and policy. It is the first continent-wide cap-and-trade scheme that has been put in place and several aspects of the implementation of the scheme created a formidable challenge for authorities at European and national level. Never before has an EU environmental policy created an economic asset whose annual value runs into the tens of billions of Euros and has set up a process of shared tasks among the European and national levels to organise the distribution of these valuable assets to private economic actors.

In this chapter we look at the first allocation round from the European perspective. We do this by outlining the allocation process and rules (Section 2), discussing the challenges faced in developing the first allocation plans (Section 3), describing the role of Commission guidance (Section 4) and reviewing the Commission's assessment of plans (Section 5). In Section 6 we describe the key factors that influenced and shaped the first round allocation process, followed by conclusions in the final section.

---

[1] The author would like to thank participants of two workshops held over the course of 2005 by the Fondazione Eni Enrico Mattei for valuable comments and discussion on a presentation that served as the background for this paper and the editors for comments on a draft version. The views expressed in this paper are those of the author and do not necessarily coincide with those of the European Commission.

2 The allocation process and rules in Directive 2003/87/EC

Allocation is governed by Articles 9 to 11 and Annex III[2] of Directive 2003/87/EC (European Community 2003).[3]

Article 9(1) provides that each Member State has to draw up a National Allocation Plan in advance of each trading period. This provision implies a wide delegation of tasks to fix the details of allocation arrangements to the national level. In this allocation plan each Member State has to decide how many allowances it intends to allocate in total for the forthcoming trading period and how it proposes to share out these allowances to covered installations. The plan has to be drawn up in an objective and transparent manner and respect a set of eleven criteria defined in Annex III to the Directive. The Commission is furthermore mandated to draw up guidance on the implementation of the criteria.[4] The obligations on each Member State to take into account comments from the public in elaborating the plan and to publish the final plan at the same time as it is notified to the Commission and other Member States ensures transparency of the process leading to and the content of the allocation plan.

Article 9(2) foresees that the notified allocation plan has to be considered by the Committee set up under Directive 2003/87/EC. Committee meetings are chaired by the Commission and attended by Member States. Records of Committee meetings are not available to the general public.

Article 9(3) mandates the Commission to assess a notified allocation plan and empowers it to prevent the implementation of a plan at national level, if it finds the plan to be incompatible with the criteria in Annex III of the Directive or the provisions of Article 10. The Commission is given three months from the notification of a plan to carry out the assessment. It is furthermore obliged to justify any rejection decision. In the event that a plan is rejected the Commission has to accept amendments proposed by the Member State before the plan can be implemented at national level.

Article 10 regulates the allocation method and foresees that in the first trading period at least 95% of the allowances have to be allocated

---

[2] See Appendix IV for the text of Articles 9 to 11 and Annex III.

[3] For details of the negotiation of the allocation provisions in Directive 2003/87/EC see Zapfel (2005).

[4] The Commission guidance document is discussed in the following section.

free of charge. This implies that each Member State has the discretionary power to sell or auction up to 5% of the total number of allowances it allocates. In the second trading period the Directive allows for a minimum of 90% to be allocated free of charge. No allocation method is defined for further trading periods.

Article 11(1) regulates the allocation process following the approval of the plan by the Commission. It foresees the implementation of the plan by taking a final national allocation decision at least three months before the beginning of the first trading period.

Article 11(2) relates to further trading periods and has the same content except that the final allocation decision has to be made twelve months prior to the start of the trading period.

Article 11(3) contains a reference to the state aid articles of the Treaty[5] and foresees that the allocation has to take into account the need to provide access to allowances for new entrants, without specifying any further details.

Article 11(4) finally provides that the allowances allocated for the multi-year trading period shall not be issued, i.e. credited in the registry account of the installation, in total in the first year of the trading period, but rather in annual proportions.

Annex III lists eleven criteria to be respected or taken into account in the allocation plan. Some of the criteria have a mandatory character (e.g. the need for consistency with a Member State's Kyoto target and the obligation to include a list of covered installations with allocated amounts per installation), while others are of an optional nature (e.g. accommodation of early action). The common character of the criteria is that they are of a principled and general rather than operational nature. This means that Member States have considerable freedom to implement the criteria, while the Commission has no clear guidance for the assessment of plans.

The main characteristics of the allocation rules are the *ex-ante* nature, the periodic decision-making, a wide delegation of tasks to the national level (decentralised process) and strong central control by the Commission.

---

[5] For information about state aid policy, which is based on Articles 87 and 88 of the Treaty, consult http://europa.eu.int/comm/competition/index_en.html.

3 The allocation challenges

The first allocation round has been shaped by challenges that are likely to materialise again in later allocation rounds but also others that are likely not to play as much of a role in the future. The allocation challenges to be mastered at the national level to develop the first round allocation plans consisted of:

- Identifying the installations covered
- Gathering and processing of relevant data
- Fixing the national cap and deciding on the path to the Kyoto target
- Elaborating allocation formulae at sector and installation level
- Designing rules for new entrants and installation closure
- Overcoming know-how gaps in authorities and among stakeholders
- Organising public consultation and securing political acceptability
- Tight time schedule.

Identifying the installations covered

A very time-consuming task in the first allocation round was the identification of installations covered by the scheme. In the preparation of the allocation plans it emerged that the legal interpretation of which installations are captured by Annex I of the Directive differed across Member States, in particular regarding the question of what constitutes a combustion installation.

Gathering and processing of relevant data

A second task that turned out to be extremely time-consuming was the issue of gathering and processing of relevant data from covered installations. A complicating factor in this task was that some data collection exercises were commenced at a stage when the EU-wide (data) monitoring and reporting guidelines, the binding rules that have to be applied to annually monitor and report actual emissions data as of 2005, were still under development. Another challenge in this regard was that some Member States started data gathering at a time when no legal basis for requesting data from installation operators was in place yet, as the Directive had not yet been implemented in national law. Some Member States did in fact face delays in initiating data collection because of the lack of a legal basis or had to redo the data collection after the law was in place. The issue of relevance of data has also factored in this task. As data collection is a very demanding task there was pressure to collect only data that would in the end be used in the allocation formulae. Hence Member States had to decide early on for example the base period for sector and/or installation level allocations. The limits in data availability did in fact limit the great number of theoretical choices for allocation formulae. A final factor that made the data issue a heavy task was the decision of several Member States to classify emissions data into combustion or process-related.

Fixing the national cap and deciding on the path to the Kyoto target

The most important allocation decision from a macro perspective is the total number of allowances to be created, i.e. the setting of the cap. This is because the sum of these 25 national decisions defines the overall scarcity in allowances and the environmental quality of the instrument. The Directive offers considerable flexibility for Member States in setting national caps. In the first allocation round two types of situations could arise in the cap-setting exercise. A Member State that was likely on current and projected trends not to reach the overall Kyoto target in 2008 to 2012 had to decide on a (national) path to achieving the Kyoto target. As the trading scheme covers a share[6] of a Member State's total emissions, cap-setting in that case needed to take into account also other measures to reduce emissions, including the publicly funded purchase of Kyoto mechanism units.[7] Alternatively a Member State that was well on track to achieve or even over-achieve the Kyoto target or was not subject to a Kyoto target at all[8] was faced

---

[6] The share of emissions covered by the Directive varies across Member States, with a range of 20% to 60%. The major determinant for the share is the fuel mix in the power-generation sector such that Member States with a large share of carbon-free power (nuclear and/or hydro) have a low share covered, while those with a large share of coal in the fuel mix are closer to the high end of the range.

[7] Under the terms of the Kyoto Protocol a Party to it, i.e. a country, can in addition to undertaking emission cuts on its own territory outsource some reductions to other countries and gain credit in the form of various carbon currencies, most prominently certified emission reduction units (CERs) in the context of the Clean Development Mechanism (CDM), emission reduction units (ERUs) in the context of Joint Implementation (JI) and Assigned Amount Units (AAUs) in the context of International Emission Trading.

[8] The latter being the case for Malta and Cyprus.

with fewer constraints and was mainly limited in not setting a cap above projected emissions.

Elaborating allocation formulae at sector and installation level

The core of the allocation work (to be distinguished from the cap-setting) is to break down the total number of allowances to be allocated into allocated amounts for each individual installation, including the funding of a new entrants reserve. The Directive does not give much guidance for these tasks. It leaves it up to the discretion of a Member State whether to break down the total into sectoral budgets as an intermediate step to the installation level (one-step vs. two-step allocation process). It leaves also the sectoral definitions in the hands of national decision-making. There are pros and cons for both the one-step and two-step allocation process. In the end most Member States chose a two-step process in order to be able to account for specific circumstances of sectors. However, the two-step allocation process brought with it a lot more issues to be discussed and decided than a one-step process. In the elaboration of allocation formulae at installation level most plans are in principle based on average annual emissions in a base period stretching back some three or four years. Adjustments to the approach were made by, for example omitting the lowest and/or highest annual emissions figures over the base period. The base period emissions were subject to various factors and agreeing these factors was the subject of lively debate in national capitals. Most plans have a variant of a so-called compliance factor, which is a discount factor to bring the amount of allowances resulting from the installation level formulae into line with the total cap. Installation-level formulae also had to be discounted to fund new entrants reserves.

Designing rules for new entrants and installation closure

Another major challenge in designing the first plans relates to new entrants and closure rules. Again on this aspect the Directive does not give much guidance and leaves it to national discretion whether to build a new entrants reserve and how to do so as well as whether to continue to allocate allowances to installations closed during the trading period.

While the Directive contains a definition of a new entrant[9] it does not define closure. A plan without new entrants and closure rules, i.e. new installations do not get allocated allowances free of charge but rather have to satisfy their needs by market purchases and closed installations continue to receive allowances until the end of the trading period, is much simpler from an administrative point of view. In the end all Member States chose to set aside a new entrants reserve and almost all Member States opted also for closure provisions. The fact that these high-order decisions were made at a rather late stage in most Member States meant that the time available for developing the necessary details in the design of the rules was very tight. A decision to build a new entrants reserve meant that a Member State had *inter alia* to decide how many allowances to set aside in the reserve, how to fund it (e.g. proportional contributions from sectoral budgets), allocation formulae for eligible new entrants, rule upon exhaustion of the reserve (some form of replenishment or not) and treatment of allowances left in the reserve at the end of the trading period.

Overcoming know-how gaps in authorities and among stakeholders

The challenges discussed so far are predominantly of a technical nature. The EU greenhouse gas emission trading scheme is the first application of this environmental policy instrument at EU level and few Member States had previously implemented or considered the use of the instrument at national level.[10] Therefore neither authorities charged with the task to develop the allocation plan nor stakeholders involved had much experience with an allocation process.[11] This relative inexperience made the process design (see next point) even more challenging. In many cases know-how gaps had to be addressed as a first step towards

---

[9] Article 3(h) states that 'new entrant' means any installation carrying out one or more of the activities indicated in Annex I, which has obtained a greenhouse gas emissions permit or an update of its greenhouse gas emissions permit because of a change in the nature or functioning or an extension of the installation, subsequent to the notification to the Commission of the national allocation plan.

[10] In the field of greenhouse gases Denmark and the United Kingdom had developed and implemented national trading schemes.

[11] For a discussion of some of the misperceptions about emission trading that existed in the early phase of the public debate in Europe that led to the adoption of Directive 2003/87/EC see Zapfel and Vainio (2002).

making informed decisions on technical issues. Where this was not possible technical issues had to be resolved against the backdrop of prevailing gaps.

Organising public consultation and securing political acceptability

The Directive mandates wide-ranging transparency and public consultation. Stakeholders have to be consulted both prior to notification of a plan to the Commission and after a decision by the latter. In addition to the formal stakeholder consultation, governments had to decide about the allocation of various tasks (data collection, cap-setting, allocation to sectors and installations etc.) to individual ministries or authorities, the design and set-up of interministerial working groups and external groups with stakeholder participation, and the need for technical support of allocation work by means of external expertise (consultancy services etc.). In some cases (e.g. Spain and the Czech Republic) controversy over the designation of the lead ministry delayed technical work and in particular decision-making considerably. Finding the right balance of stakeholder involvement in the technical work was also difficult. Progress on the establishment of a design could have and has been hampered both by too high a degree as well as too low a degree of stakeholder participation.

Tight time schedule

As elaborated above the establishment of the first allocation plan is complex both in terms of technical and process issues. An additional factor that characterised the first allocation process was the very tight time schedule. The Directive was politically agreed in July 2003 and formally entered into force on 25 October 2003. This means that the first allocation plans were due less than six months after the formal entry into force of the Directive.

4 The role of Commission guidance

Because of these many challenges the allocation process embodied the Commission took an active role in the run-up to the allocation process to assist and guide Member States. Prior to the adoption of the

Directive the Commission had conducted and published two studies (Harrison and Radov 2002, as well as PriceWaterhouseCoopers and ECN 2003) that not only supported the negotiation on the allocation process and rules, but also offered assistance in the implementation process. Harrison and Radov have *inter alia* synthesised the allocation rules in all pre-existing cap-and-trade schemes.

Following political agreement on implementing a trading scheme[12] the Commission prepared in April 2003 a non-paper (European Commission 2003a) for a technical working group with Member States representatives that elaborated in a practical manner the steps to be undertaken to prepare an allocation plan. This paper, while being an unofficial document to assist Member States, was quite instrumental in defining the set-up of many national allocation debates, as for example terms like 'top-down' and 'bottom-up' approaches to allocation became standard vocabulary in many capitals.

In accordance with Article 9(1) of the Directive the Commission was mandated to develop guidance on the implementation of the criteria listed in Annex III by the end of 2003. This guidance is non-binding for Member States, to which it essentially is addressed, while it binds the Commission in its assessment of plans. In this allocation guidance (European Commission 2003b) the Commission established a hierarchy of and linkages between the eleven allocation criteria by means of categorisation.[13] It distinguished between mandatory and optional criteria as well as outlining whether the criteria had or could be applied in determining the total cap, the allocation at sector level or the allocation to individual installations. In order to improve the accessibility, transparency and comparability of plans, the Commission developed and recommended to Member States the use of a common format for the presentation of plans.

The guidance could not and did not change the rather general character of the allocation criteria and the Commission could therefore not make the application of the criteria much more operational. In some instances, e.g. on criterion 6 on new entrants and criterion 7 on early action, it elaborated alternative options for implementing

---

[12] This political consensus was reached in early December 2002 after both the European Parliament and the Council had concluded the first reading of the proposed Directive, seeking some technical amendments but supporting in principle the adoption of a Directive.

[13] See table 1 on p. 3 of European Commission (2003b).

them. In some instances the guidance contains recommendations on the preferred way to apply a criterion. With regard to criterion 4 the Commission strived to simplify the implementation by means of recommending a threshold below which the effect of other Community legislation should not be taken into account. With regard to two aspects that turned out to be important elements in the assessment of the total cap, i.e. the intended purchase of Kyoto mechanism units with public funds and projected needs for covered installations, the guidance stressed the importance but did not go as far as outlining operational conditions to be fulfilled. On the intended use of Kyoto mechanisms units the guidance stated: 'In the national allocation plan, a Member State must substantiate any such intentions to use the Kyoto mechanisms. The Commission will base its assessment notably on the state of advancement of relevant legislation or implementing provisions at the national level.'[14] Operational criteria to assess the substantiation were established in the first assessment of plans (see next section).

While the allocation process was largely delegated to the national level, the crucial role of the Commission to assess the plans meant that Commission guidance, be it in a formal or informal way, was actively sought along the way of establishing a plan both by national authorities and stakeholders involved in national processes. This was to convince the Commission about the merits of features that were considered in the national debate, but also to test the waters and smooth the process of the Commission's assessment of the plan. At an advanced stage of the national debate attempts were also made to involve the Commission as an arbiter between conflicting interests. The Commission has been actively involved to assist in technical questions but it did not offer any upfront assessment and did not express a view on controversial aspects prior to the notification of a plan.

5 The assessment of plans

In accordance with the Directive Member States had to notify the first allocation plan by 1 May 2004. For those Member States from Central and Eastern Europe, which joined the European Union on 1 May 2005, the obligation to notify the allocation plan arose on the day of accession, as none of the new members had requested in the accession

---

[14] Extract from paragraph 32, p. 7 of European Commission (2003b).

negotiations a temporary derogation from the Directive. Table 2.1 indicates the effective notification dates of the first allocation plans. Only seven Member States (Austria, Denmark, Germany, Ireland, Lithuania, Slovak Republic and Slovenia) notified a plan close to the official due date. On 7 July 2004, the date of the adoption of the Commission decisions on the first plans, nine plans were still outstanding. The last plan was received by the Commission on 3 January 2005, i.e. some nine months after the due date. The Commission launched infringement proceedings against some Member States in order to speed up national decision-making processes.

The Directive further specified that the Commission had to take a decision on a plan within three months of notification, otherwise the plan would be approved tacitly. Table 2.1 shows the assessment dates and the time between the initial notification and assessment. The staggered submission of plans has also resulted in a staggered approval schedule. During 2004 the Commission passed three batches of decisions (7 July – eight plans, 20 October – eight plans and 27 December – five plans). In 2005 four more plans were approved, in March, April, May and June respectively. The approval of the Greek plan on 20 June 2005 marked the end of the assessment phase that had lasted almost fifteen months. As for the length of the assessment process for individual plans, only eight plans were assessed within the three-month period specified by the Directive. The Commission assessed the Italian plan for a period of ten months. An assessment period beyond three months does not mean that the Commission has violated its obligation under the Directive, but reflects the fact that the majority of plans notified to the Commission did not contain all the information necessary for the Commission to carry out a complete assessment. Hence the initial notification was frequently followed up by further notifications and in such cases the three-month period applied from the last notification completing the allocation plan. A number of plans were notified lacking allocations at installation level or even a list of installations covered.

Prior to the conclusion of the assessment a plan was considered by the Committee set up under Directive 2003/87/EC. This Committee process was an opportunity for other Member States to express views on the content of a plan and highlight issues for the Commission to consider in its assessment. The opinion expressed by the Committee does not formally bind the Commission. However, the discussion in

Table 2.1 *Overview of initial notification and assessment dates of National Allocation Plans*

Member State[a]	Plan initially notified[b]	Assessment concluded	Months to assess[c]
Austria	2 April 2004	7 July 2004	>3
Belgium	30 June 2004	20 October 2004	<4
Cyprus*	3 November 2004	27 December 2004	<2
Czech Republic*	12 October 2004	12 April 2005	6
Denmark	31 March 2004	7 July 2004	>3
Estonia*	23 June 2004	20 October 2004	<4
Finland	13 April 2004	20 October 2004	>6
France	26 July 2004	20 October 2004	<3
Germany	1 April 2004	7 July 2004	>3
Greece	3 January 2005	20 June 2005	>5
Hungary*	9 November 2004	27 December 2004	<2
Ireland	1 April 2004	7 July 2004	>3
Italy	26 July 2004	25 May 2005	10
Latvia*	10 May 2004	20 October 2004	>5
Lithuania*	6 May 2004	27 December 2004	>8
Luxembourg	15 April 2004	20 October 2004	>6
Malta*	3 November 2004	27 December 2004	<2
Netherlands	26 April 2004	7 July 2004	>2
Poland*	22 September 2004	8 March 2005	<6
Portugal	25 June 2004	20 October 2004	<4
Slovak Republic*	4 May 2004	20 October 2004	<6
Slovenia*	3 May 2004	7 July 2004	>2
Spain	7 July 2004	27 December 2004	<6
Sweden	29 April 2004	7 July 2004	<3
United Kingdom	10 May 2004	7 July 2004	<3

[a] For Member States marked with an asterisk (*) the due date was 1 May 2004; all other Member States were obliged to notify by 31 March 2004.

[b] The initial notification dates indicated in this table are the registration dates by the Commission rather than the dates the plans were adopted or released at national level.

[c] < up to 15 days less than the month indicated; > up to 15 days more than the month indicated.

Source: European Commission (2004a, 2004b, 2004c, 2004d, 2005a, 2005b, 2005c, 2005d, 2005e).

Committee provided an indication for each Member State about how strong concerns were in other Member States about particular features of its plan and may have boosted the preparedness of a Member State to agree amendments with the Commission.

The staggered submission of plans had important implications for the assessment process. On the one hand it meant that the Commission had to establish an assessment policy without having received several allocation plans. On the other hand plans that were finalised at the national level after the adoption of the first batch of Commission decisions on 7 July 2004 could be adapted so as to be in line with anticipated Commission requirements. Some Member States may in fact have delayed notification of plans after the first batch of decisions not merely for technical reasons, but also to see what standard the Commission would apply.[15] In view of the general nature of the Annex III allocation criteria and the absence of complete indications in the Commission guidance document, an assessment policy had in fact to be established largely with the first decisions. In order to explain the decisions and the assessment policy the Commission adopted communications accompanying the first and second batch of Commission decisions (European Commission 2004b, 2004d).

Turning to the substance of the assessment the Commission required changes mainly with regard to three aspects: the total cap, the installations listed in the plan and the intention to adjust allocations *ex post*. Table 2.2 provides summary information on the plans as approved by the Commission and indicates which plans were approved with a lower than proposed total cap and after the removal of *ex-post* adjustments. The plans of France, Italy and Spain were only approved on condition of or after more installations than listed in the initially notified plan were included in the trading scheme.

A lowering of the total cap was required in more than half of the plans. In total the assessment resulted in some 290 million allowances fewer than intended in the initially notified plans to be allocated in the first trading period. The Commission justified the necessity to lower caps mainly with the inconsistency of the proposed path to a Member State's Kyoto target – a violation of criterion 1 in Annex III of the

---

[15] An example in this regard is Hungary. The first paper with allocation principles that was published for consultation contained – most likely inspired by the notified German allocation plan – a number of *ex-post* adjustments at installation level. The plan notified in November 2004 no longer contained these features.

Table 2.2 *Summary information on the first allocation process*[a]

Member State	Allowances (in million)[b]	Share in EU allowances (%)	Installations covered	Cap lowered	Ex-post disallowed
Austria	99.0	1.5	205	Yes	Yes
Belgium	188.8	2.9	363	Yes	Yes
Czech Republic	292.8	4.4	435	Yes	Yes
Cyprus	16.98	0.3	13	No	No
Denmark	100.5	1.5	378	No	No
Estonia	56.85	0.9	43	Yes	No
Finland	136.5	2.1	535	No	No
France	469.5	7.1	1,172	Yes	Yes
Germany	1,497.0	22.8	1,849	No	Yes
Greece	223.2	3.4	141	No	Yes
Hungary	93.8	1.4	261	No	Yes
Ireland	67.0	1.0	143	Yes	Yes
Italy	697.5	10.6	1,240	Yes	Yes
Latvia	13.7	0.2	95	Yes	No
Lithuania	36.8	0.6	93	Yes	Yes
Luxembourg	10.07	0.2	19	Yes	Yes
Malta	8.83	0.1	2	No	No
Netherlands	285.9	4.3	333	Yes	Yes
Poland	717.3	10.9	1,166	Yes	Yes
Portugal	114.5	1.7	239	Yes	Yes
Slovak Republic	91.5	1.4	209	Yes	No
Slovenia	26.3	0.4	98	No	No
Spain	523.3	8.0	819	No	No
Sweden	68.7	1.1	499	No	No
United Kingdom	736.0	11.2	1,078	No[c]	No
Total	6,572.4	100.0	11,428	n.a.	n.a.

[a] Figures do not take into account any opt-ins and opt-outs of installations in accordance with Article 24 and 27 of Directive 2003/87/EC.

[b] One allowance corresponds to 1 t of CO_2.

[c] While the plan notified was approved without lowering the cap, the United Kingdom submitted a new plan on 16 November 2004 to the Commission that foresaw an increase in the total cap by almost 20 million allowances to 756 million for the period. This increase after the Commission's approval of the plan on 7 July 2004 was rejected by the Commission on 12 April 2005.

Source: European Commission (2004a, 2004b, 2004c, 2004d, 2005a, 2005b, 2005c, 2005d, 2005e).

Directive – and with the intended allocation above projected needs of the covered installations – a violation of criterion 2. Several plans were found to be on an inconsistent path to the Kyoto target because of insufficient substantiation of the intended purchase of Kyoto mechanism units with public funds. In assessing the first batch of plans the Commission developed a catalogue of criteria[16] to evaluate the substantiation of intended purchase of Kyoto mechanism units which was applied to all plans that relied on this component. A number of plans, predominantly from Member States that joined the EU in May 2004, were found to propose a total cap above projected needs. The Commission relied in its assessments mainly on projections of total greenhouse gas emissions that Member States produce and report on a regular basis to the Commission and the Secretariat to the UN Framework Convention on Climate Change.

Ex-post adjustments in many variants were also disallowed in more than half of the plans. The national allocation decision based on the plan as approved by the Commission has a final and irreversible character. This implies that neither the total number of allowances nor the quantities indicated in the notified plan and the final allocation decision may be revised *ex post*. A number of notified allocation plans had various provisions that would possibly result in *ex-post* adjustments of allowances allocated to individual installations. A few examples: in the Dutch plan it was intended that any allowances not allocated to newly constructed installations out of the new entrants reserve would be distributed pro rata to installations listed in the plan towards the end of the trading period. In several plans it was intended that the allowances allocated to an installation would be curtailed, in case the installation was found to emit less than a certain percentage of its baseline emissions or allocation in a given year.[17] These examples demonstrate that *ex-post* adjustments were intended both upwards (increasing the allocation of an installation) and downwards (reducing the allocation of an installation). Most intended *ex-post* adjustments were rule-based, i.e. the circumstances under which they would be applied and the exact character were laid down in abstract terms in the plan.

In view of the number and nature of the changes to notified plans the Commission has required, it is remarkable that it adopted only very

---

[16] See p. 5 and 6 in European Commission (2004b).
[17] See e.g. chapter 3.2, p. 32 of the German Allocation Plan.

few decisions with formal rejections of certain aspects of plans. In fact, while the cap was lowered in fourteen plans and the national debate of which cap to notify to the Commission sparked fierce controversy in many capitals, only in the cases of France and Poland did the Commission adopt a decision rejecting part of the intended total cap. In all other cases Member States have formally amended the notified plan – sometimes a few days in advance of the adoption of the Commission decision – following bilateral contacts with the Commission. Such a formal amendment pre-empting a negative Commission decision was preferred by many governments in order to avoid unfavourable press, as in the end the Commission decision had to be complied with anyway.

The Commission's approach with bilateral contacts may have also been an important factor to avoid the challenge of some of its decisions in Court. A Commission decision may be contested within two months after adoption. Only two Member States have taken the Commission to court.[18] Germany has contested the Commission decision on its plan because the Commission disallowed a number of intended downward *ex-post* adjustments at installation level. The United Kingdom has taken the Commission to court concerning not the initial Commission decision on its plan but the later Commission decision refusing to consider an amendment to its plan, which it notified to the Commission more than four months after the initial Commission decision.

6 Key factors shaping the first allocation round

In the following we elaborate a set of positive political economy lessons that arise out of the first allocation round.

Allocation of allowances has mainly a distributional character

The first allocation of allowances in the context of the introduction of the EU greenhouse gas emission trading scheme created an economic asset in the European economy with an annual value of some €22 to €44 billion.[19] Hence another way of looking at the allocation process is through the lens of distributing an economic asset of that

---

[18] In addition some companies have contested Commission decisions in court, e.g. the German utility Energie Baden-Württemberg (EnBW) has taken the Commission to court for approving the so-called transfer rule in the German plan.

[19] Assuming an allowance price in the first trading period in a range of €10 to €20.

value to covered industry free of charge. Allocation is a distributional exercise of a considerable magnitude. A distributional exercise is inherently controversial and political. This makes it understandable that the Directive agreed at European level contained a loose framework and the distributional conflicts that had to be resolved at some stage were resolved after the agreement on the use of the instrument as such in the implementation phase. The time pressure imposed by the Directive was an important catalyst. Distributional conflicts come in two versions – between companies and between Member States.

Distribution between companies

The economic literature (see e.g. Tullock 1967) describes the phenomenon of rent-seeking that companies will pursue, for example if a valuable economic asset is handed out for free. This rent-seeking behaviour played a major role, in particular with regard to larger companies that could afford to allocate sufficient human resources to influence allocation processes. Smaller companies were largely 'price-takers' in the allocation process. Choices with distributional implications for individual companies related to the base period emissions (as different companies pass through differing cycles of economic activity), the preferential treatment of process emissions (as different companies have differing mixes of energy and process emissions), rules for recognition of early action (as different companies have differing investment cycles) and rules for new entrants and closures (also related to differing investment cycles). Besides these issues where companies covered by the trading scheme have conflicting interests one allocation decision has distributional implications that unites companies covered by the scheme. The setting of the cap is a decision, in a Member State with a binding Kyoto target, of which sectors carry how much of the reduction burden. The looser the cap the more of the burden is shifted to the sectors not covered by the scheme. By the same token the cap-setting in a Member State on track for or even over-achieving its Kyoto target was perceived by companies as a way of gaining access to some of the national surplus.

Distribution between Member States

Distributional considerations at a higher level arose between Member States. This is to some extent surprising, as distributional issues had been at the forefront in deciding on the Burden-Sharing Agreement, an agreement of the fifteen Member States on the national differentiation of the overall Kyoto target to reduce emissions by 8% in the period 2008–2012 compared to the 1990 emission level. While the purpose of the trading scheme is to offer a cost-effective instrument to assist Member States in complying with predetermined targets, to some extent the twenty-five cap-setting processes in the first allocation round had the effect of a new burden sharing exercise for the emissions covered by the trading scheme. While a number of allocation plan design issues were the subject of lively public debate, most Member States decided on the cap based on debate in non-public working groups. In late 2003, before the first draft allocation plans were released for public consultation, there was a reluctance across Member State capitals to be the first to announce a cap figure. The underlying reason for this was the existence of voices in each Member State that opposed being more ambitious and demanding in the cap than other Member States. In order to avoid doing so many Member States were waiting for a few others to announce their figures first and set the standard. The logjam was broken by the United Kingdom when it tabled a draft allocation plan containing a figure for the cap on 19 January 2004 (see United Kingdom 2004).

Another issue in relation to the distributional nature of the cap-setting exercise was the prevalence of economic rather than environmental considerations in the national debate. The twenty-five national cap decisions determined the overall scarcity level and economics of the EU-wide trading scheme. Because of the decentralised nature, only larger Member States could determine to some degree the overall scarcity, not just for the share in overall allowances allocated, but also because many smaller Member States were observing closely the cap-setting process in larger Member States. In sixteen Member States the number of allowances allocated amounts to less than three per cent of the total allowances at EU level (see Table 2.2). For a smaller Member State the overall scarcity in the market can not be influenced by the national decision and the Member States may be tempted to approach the allocation process with the strategy to generate income for covered companies by means of a rather generous allocation. As the section on the Commission assessment of plans elaborates in fact a number of Member States without a binding constraint under the Kyoto Protocol have pursued such a strategy. It is important to keep in mind that the staggering of the release of draft plans, the notification of final plans to the Commission and the assessment of plans by the Commission has meant that several Member States may have assumed moving later rather than earlier to be more advantageous.

Allocation choices, while endless in theory, were in practice rather limited by data and political constraints

Abstracting from data bottlenecks and the need to secure political acceptability one can construct a large number of allocation choices at the sector and installation level. Once the technical work had started the number of feasible options was narrowed down rapidly. Two examples will illustrate this. During the negotiations of the Directive Germany had favoured the harmonised use throughout the EU of 1990 as a single base year for allocation. When it commenced the allocation work it quickly dropped this option even for itself because of the insurmountable challenge of reproducing 1990 emissions data at sector and installation level. The Netherlands favoured during the negotiations allocations based on benchmarking.[20] When it started the technical allocation work it undertook a study (KPMG 2002) which concluded that a total of 120 separate benchmarks would need to be developed for the installations included in the trading scheme. It ended up allocating on the basis of emissions in a base period to which an installation-specific adjustment factor to account for efficiency characteristics was applied.

Free allocation based on recent emissions is widely perceived as unfair, as it results in the allocation of more allowances to less carbon-efficient installations and fewer allowances to better-performing installations. As an expression of the recognition of the downside of such an allocation approach Annex III of the Directive makes reference both to benchmarking (in criteria 3 and 7) and to accommodating early action (criterion 7). However, a strong lesson that emerges from the first allocation round is that the attempt to introduce a wide spread around recent or current emission levels at installation level met strong political

---

[20] Benchmarking is an allocation methodology that works with historic data, but instead of basing the allocation on historic emissions data, it integrates historic input or output data into the allocation formula in order to differentiate the allocation between more and less carbon-efficient installations.

opposition. An example of an allocation formula that met heavy political resistance is so-called 'fuel-blind' benchmarks in the power sector.
One alternative to basing the allocation on emissions in a recent base
period is to use power output in a recent base period multiplied with
an emission factor – a form of benchmarking. If this approach were to
be implemented across all power plants and for example the average
CO_2 per kWh of national power production were used as the emission
factor, a coal-fired power plant would be allocated considerably fewer
allowances than needed, if the plant utilisation did not change much
with respect to the base period, while a gas-fired power plant would
receive considerably more allowances. A variant would be to allocate
on the basis of the emission factor of a gas plant, such that the gas-fired
plant would not receive more allowances than needed, assuming constant plant utilisation, but the coal-fired plant would receive even fewer
allowances than under the first variant. In Member States with a broad
fuel mix in the power sector such fuel-blind allocation formulae did not
receive any serious political attention at all, as they were considered
unfair and heavily opposed by power companies with coal-fired plants
in their generation portfolio.

Allocation aimed for non-achievable perfection

A remarkable lesson from the first allocation round is also the tendency
in many Member States towards a high degree of exactness and perfection of allocation at installation level. The political debate in many
Member States was characterised by industry demands that expected
needs should be covered or that not too much should be diverted from
expected needs. While such an approach can feasibly be implemented at
the sector level, on the basis of projections about output and emissions,
it is not possible to guarantee this at the level of hundreds or thou-
sands of individual installations. And the impossibility of such micro-
planning of expected needs and abatement possibilities by installations
is in fact one of the main attractions of putting in place an allowance
market for the proverbial 'invisible hand' to do the job public authorities cannot accomplish.

An expression of this aiming for perfection is the widespread
intended use of several variants of *ex-post* adjustments of allocations at
installation level. In some Member States a strong desire existed that if
it turned out *ex post* that an installation needed more allowances or had

reduced needs, the government would correct upwards or downwards the number of allowances allocated to such installations. An example: Portugal had witnessed forest fires in 2003. Some Portuguese installations covered by the trading scheme operate installations co-fired with biomass as well as fossil fuels. In the wake of the forest fires it was feared that some installations would not be able to maintain the share of biomass prevailing in the base period years for the allocation because of unavailability of biomass or prices that made a higher share of fossil fuels a rational economic choice. At the time of the elaboration of the Portuguese allocation plan no clarity existed as to what extent the biomass share could be maintained or not. In view of this the allocation plan that Portugal notified to the Commission contained a provision that the government would correct the allocation to installations if it turned out that the share of biomass could not be maintained.

Indirect beneficiaries abstained or advocated preferential allocation rules rather than scarcity

A carbon constraint implemented with an emission trading scheme leads to a change in relative costs and prices such that more carbon-intensive products and production modes are more costly. This means that the economic benefits from a trading scheme for less carbon-intensive products and production modes are directly related to the stringency of the cap in the trading scheme, as a higher allowance price boosts the attractiveness of these products. One would therefore rationally expect that interests representing combined heat and power (CHP), renewable energy, energy efficiency products etc. would lobby actively for more stringent caps to be adopted in the allocation plans. This was however not the case. CHP interests in most Member States advocated preferential allocation rules for CHP installations to reward the carbon benefits that the technology offers. In several cases the allocation approach demanded was to allocate more allowances than needed to CHP installations. This meant de facto that the CHP allocation would relieve market scarcity and depress the allowance price. Interests representing renewable energy and providers of energy efficiency products, which could not be allocated any allowances per se, were largely not interested in the allocation process and did not actively advocate more stringent plans. The only interest group besides environmental organisations which lobbied for more stringent plans were market intermediaries with an interest in an active allowance market.

The prevailing regulatory tendency in some plans to substitute the market with administrative rules

An interesting political economy lesson is the tendency in many plans to devise administrative rules rather than put trust into a carefully designed market mechanism. Emission trading and the allocation plan constituted a major new challenge for authorities and stakeholders. The negotiation of the Directive was focused on the overarching architecture of the scheme and not on details. Hence the Directive does not contain rules where one may alternatively devise an administrative rule or let the market take care of the matter. In the allocation process the prevailing regulatory mindset of the authorities and among stakeholders entered the picture and some of the plans are not simply plans about how many allowances to allocate in total and how to share them out to covered installations, but are supplemented with a set of additional rules that the Directive neither requires nor expressly disallows. Examples for such rules include *ex-post* adjustments in many variations, new entrants and closure rules, and transfer rules. The transfer rule is an interesting example that merits further discussion. As pointed out above the Directive delegated decisions about setting aside a new entrants reserve and closure rules to the national level. After most Member States had decided in principle to put in place a new entrants reserve and to stop allocating allowances upon closure of an installation, in some Member States the question was asked whether there would be a sufficient incentive to close down an old carbon-inefficient installation and build a new carbon-efficient installation if the allocation to the old installation would be lost and the company would receive a lower allocation based on the need of the new carbon-efficient installation. A company may in fact keep on running the old installation at constant capacity or reduce utilisation of the old installation, so as not to lose the allocation, and build a new installation, so as to receive allowances out of the new entrants reserve. In order to overcome the perverse incentive structure several Member States added a so-called transfer rule that would allow a company closing down an installation to demonstrate that it had built within a certain time-frame a new installation to replace production in the old installation and to 'transfer' the allocation entitlement from the closed installation to the new installation. An alternative to the administratively heavy approach with a new entrants reserve, closure rules and transfer rules is to opt for the hands-off approach of not setting aside a new entrants reserve and not interrupting allocation to a closed installation during a trading period. Such an approach would create the very same incentive for a company to close down old and inefficient installations and invest in new efficient installations.

Remarkably high degree of soft harmonisation

As elaborated above a number of decisions were delegated to the national level such that Member States could end up making different choices. Several Member States, industry associations and companies did not feel at ease with this situation and stressed in the run-up to the first allocation round the desire for common choices on some issues. The first allocation round led to a rather high degree of soft harmonisation of choices made by Member States. All Member States set aside allowances in a new entrants reserve, most Member States decided not to allocate allowances to closed installations, most Member States decided not to allow for banking (carrying forward) of unused allowances from the first to the second trading period. Even when it comes to the environmental ambition level and the degree of stringency of national caps the range was not as wide as one would have expected. This is because Member States with a large gap to close to achieve their Kyoto targets undertook substantial commitments to acquire Kyoto units with public funds in order to forgo the need to demand deeper cuts in the sectors covered by the trading scheme.

7 Conclusions

In this chapter we have provided an overview of the first allocation round from a European perspective starting from a description of the legal framework within which it took place and closing with the identification of a set of key factors that shaped the process.

The fact that this was the first allocation round marks a contrast of the EU greenhouse gas emission trading scheme with other cap-and-trade schemes that have been implemented so far. In the American SO_2 allowance trading scheme for example the allocation is of a permanent nature and 30 years' worth of allowances are credited to an installation's registry account on a rolling basis. A second major difference from other schemes, including the American SO_2 scheme, is the fact that in the EU scheme the setting of the cap (as a sum of twenty-five national decisions, rather than at EU level) was an integral part of the allocation process and not superimposed on the process by legislation.

The periodic nature of the allocation in the EU scheme may have simplified the process in the sense that the distributional stakes were lower with sharing out allowances for three years rather than for a longer period or on a permanent basis. The fact that a new round would take place may also have helped to overcome some of the technical issues, as any solution that was adopted could be 'road-tested' and, if considered necessary, changed after a few years. The downside of periodic allocation is that companies may adopt strategic behaviour in order to maximise the number of free allowances to be allocated in future rounds rather than being guided by the invisible hand to the cost-minimum outcome in terms of emissions abatement.

With the prospect of more allocation rounds to follow one may ask the question as to what extent the first allocation process had unique characteristics. We take the view that the first allocation round was from a technical point of view more demanding than further rounds will turn out to be. Some tasks that were outlined in the section on allocation challenges, including identifying covered installations and gathering and processing of relevant data, were one-off hurdles that will no longer exist in later rounds, unless of course the scope of the scheme is extended substantially and another round of such one-off tasks may arise. The experience from the first allocation round in both authorities and among stakeholders should also simplify further rounds. One issue does however stand out as a factor that may be more difficult in future rounds. After the first round the understanding that allocation is a major distributional exercise is universal and the full consciousness of this fact may make it more difficult to secure political acceptability in future rounds.

There are of course options for mitigating distributional conflicts. One could imagine that the first round national caps will be an important marker for national caps in future rounds such that a certain path dependency in cap-setting will develop. In the first round the most important markers were actual and projected emissions. It seems natural that the first issue one would consider with a second-round plan is how the annual cap compares to the one in the first round. Such path dependency, applied not only at the level of the cap but also for sector- and installation-level allocations could help to mitigate strategic behaviour and rent-seeking of companies. Making more use of auctioning as an allocation method could serve the same purpose.

The only certain factor at this stage is that a second allocation round will take place with notification of allocations plans to the Commission by 30 June 2006. Beyond the second allocation round many options are possible. As part of the review of the Directive launched in mid-2006 the legal framework may be changed substantially. So one will have to see to what extent the periodic and decentralised nature of the allocation process will exist beyond 2012. One thing is for sure however: that the allocation of allowances in the EU greenhouse gas emission trading scheme will continue to stimulate a lot of debate and deliver rich material for books like this one.

References

European Commission 2003a. 'The EU emissions trading scheme: how to develop a national allocation plan', non-paper by DG Environment, April.

European Commission 2003b. 'Guidance to assist Member States in the implementation of the criteria listed in Annex III to Directive 2003/87/EC and on the circumstances under which force majeure is demonstrated', COM(2003) 830, January 2004.

European Commission 2004a. 'Emissions trading: Commission clears over 5,000 plants to enter emissions market next January', press release IP/04/862, 7 July.

European Commission 2004b. 'Communication on Commission Decisions of 7 July 2004 concerning national allocation plans for the allocation of greenhouse gas emission allowances of Austria, Denmark, Germany, Ireland, the Netherlands, Slovenia, Sweden, and the United Kingdom in accordance with Directive 2003/87/EC', COM(2004) 500, July 2004.

European Commission 2004c. 'Emissions trading: Commission clears eight more plans paving the way for trade to start as planned', press release IP/04/1250, 20 October.

European Commission 2004d. 'Communication on Commission Decisions of 20 October 2004 concerning national allocation plans for the allocation of greenhouse gas emission allowances of Belgium, Estonia, Finland, France, Latvia, Luxembourg, Portugal, and the Slovak

Republic in accordance with Directive 2003/87/EC', COM(2004) 681, October 2004.

European Commission 2005a. 'Emissions trading: on the eve of kick-off of the scheme Commission cleared five more plans', press release IP/05/9, 6 January.

European Commission 2005b. 'Emissions trading: Commission decides on Polish allocation plan', press release IP/05/269, 8 March.

European Commission 2005c. 'Emissions trading: Commission approves Czech allocation plan', press release IP/05/422, 12 April.

European Commission 2005d. 'Emissions trading: Commission approves Italian allocation plan', press release IP/05/602, 25 May.

European Commission 2005e. 'Emissions trading: Commission approves last allocation plan ending NAP marathon', press release IP/05/762, 20 June.

European Community 2003. 'Directive 2003/87/EC of the European Parliament and of the Council of 13 October 2003 establishing a scheme for greenhouse gas emission allowance trading within the Community and amending Council Directive 96/61/EC', *OJ* L 275, 32–45, 25 October 2003.

Harrison, D. and Radov, D. 2002. 'Evaluation of alternative initial allocation mechanisms in a European Union greenhouse gas emissions allowance trading scheme', report for DG Environment, March 2002.

KPMG 2002. 'Allocation of CO_2 emission allowances in a European emissions trading scheme', report for the Dutch Ministry of Economic Affairs, October 2002.

PriceWaterhouseCoopers and ECN 2003. 'Allowance allocation within the Community-wide emissions allowance trading scheme', report for DG Environment, May 2003.

Tullock, G. 1967. 'The welfare costs of tariffs, monopolies and theft', *Western Economic Journal* 5: 224–32.

United Kingdom 2004. 'EU emissions trading scheme: UK draft national allocation plan for 2005–2007', consultation draft, January 2004.

Zapfel, P. and Vainio, M. 2002. 'Pathways to European greenhouse gas emissions trading: history and misconceptions', *FEEM Working paper* 85, October.

Zapfel, P. 2005. 'Greenhouse gas emissions trading in the European Union: building the world's largest cap-and-trade scheme', in B. Hansjürgens (ed.) *Emissions Trading for Climate Policy: US and European Perspectives*. Cambridge: Cambridge University Press, pp. 162–76.

PART II

Experiences from Member States in allocating allowances

3 | *United Kingdom*

DAVID HARRISON
AND DANIEL RADOV[1]

1 Background and overview

The United Kingdom's Phase 1 National Allocation Plan (NAP) was the first to be published in draft form and it was therefore one of the most influential of the twenty-five Member State plans developed to implement the EU ETS. The UK has made climate change a public policy priority and it had already implemented various policy measures to reduce emissions of greenhouse gases (GHG), including a smaller Emissions Trading Scheme (UK ETS). The UK experience provides insights into the complex issues involved in translating the need to allocate allowances under a cap-and-trade programme into a practical and workable allocation programme.

1.1 Background on UK emissions and climate policies

In 2003, GHG emission sources in the UK were responsible for around 655 Mt CO_2 equivalents (of which 556 Mt, or 85%, was CO_2), making it the second largest emitter of GHGs in Europe. The European Burden-Sharing Agreement (BSA) commits the UK to reducing emissions of all six Kyoto GHGs by 12.5% below its 1990 emissions by the first Kyoto commitment period. In addition to its international commitments, the UK has set for itself a more demanding national target for 2010 to reduce emissions of CO_2 alone by 20% below 1990 levels, and to put

---

[1] The authors advised the UK government in the preparation of the UK's Phase 1 National Allocation Plan over a period of more than two years. The authors are grateful to numerous government officials for their insights on the UK NAP as well as their substantial assistance in the preparation of this chapter; we would particularly like to thank Sayeeda Tauhid, Chris Dodwell, Jill Duggan, and the EU ETS, UK ETS and EPE teams at the Department for Environment, Food and Rural Affairs (Defra). The views expressed in this chapter – and any errors or omissions it may contain – are solely those of the authors, however, and should not be construed in any way to be those of Defra, the UK government, FEEM or NERA Economic Consulting.

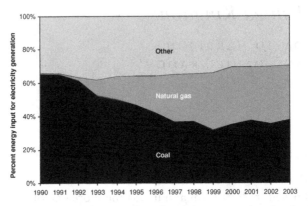

Figure 3.1. Fuel input for electricity generation in the UK, 1990–2003.
Source: Digest of UK Energy Statistics.
Notes: 'Other' includes nuclear, oil, hydro and other fuels.

itself 'on a path towards' a 60% reduction in CO_2 emissions by 2050 (Department of Trade and Industry 2003).

During the 1990s, the UK significantly reduced its GHG emissions. A major factor contributing to this development was substantial fuel-switching in the electric power sector from coal-fired to gas-fired generation. The UK has led the way in Europe in energy market liberalisation, and the shift in fuel use by the electric power sector was facilitated by the liberalisation of the UK's electricity market and by the availability of natural gas from the UK's North Sea oil and gas reserves. Figure 3.1 shows the trends in electricity output from coal, natural gas and other technologies in the UK over the last decade and a half.

Production of oil and gas from the North Sea reserves has now peaked, and is expected to decline significantly in the future. The UK is therefore likely to rely more on imported natural gas. This may mean higher gas prices, in turn affecting the country's 'business-as-usual' emissions, since more coal use and thus greater CO_2 emission could be induced.

In anticipation of the need to meet its obligations under the Kyoto Protocol and BSA and its more ambitious domestic targets, the UK government has put in place a wide range of policies designed to reduce GHG emissions, collectively known as the UK Climate Change Programme (2000), or CCP. The CCP includes policies with relatively limited scope as well as those that apply to very broad collections of GHG-emitting activities, comprising the Climate Change Levy (CCL), the Climate Change Agreements (CCAs), the UK Emissions Trading Scheme (UK ETS), the Renewables Obligation, various energy-saving and efficiency initiatives, efforts to promote combined heat-and-power (CHP, or cogeneration) facilities, and changes to vehicle taxes, among others.

The CCAs and UK ETS are most directly relevant to the subsequent development of the UK NAP. CCAs now exist for nearly fifty energy-intensive industrial sectors, covering around 10,000 facilities. Participation in a CCA gives participants an 80% discount on the CCL in exchange for commitments to reduce energy use and associated GHG emissions. CCAs typically apply to the use of electricity (which does not have direct emissions) as well as the use of other primary fuel sources (which do have direct emissions). Moreover, not all sites within a given industry sector are participants in a CCA (for example, many existing CHP plants are not subject to the CCL, and therefore would not benefit from a discount anyway). The CCAs were recently estimated by the government to have achieved just over 14 Mt CO_2 reductions per year, relative to baselines typically set in 2000.

The UK ETS was designed to cover a set of sites different from the CCAs. So-called 'Direct Participants' in the UK ETS come from about three dozen companies and organisations, some of which have included multiple sites within the scheme. Participants agreed to reduce their emissions of greenhouse gases (including CO_2 and non-CO_2 gases) relative to their emissions during a baseline period (usually 1998–2000), in exchange for incentive payments received from the government. In the third year of the scheme, which is due to run for five years until 2006, Direct Participant emissions were nearly 6 Mt CO_2e below their baseline levels. Also CCA participants are eligible to take part in the UK ETS, either by buying credits from other participants (which can be applied towards meeting their CCA targets) or selling credits they generate when they 'over-achieve' their CCA targets. Although the Direct Participant aspect of the UK ETS finished in 2006, CCA participants may continue to use the UK ETS until 2012. The UK ETS was developed by the government in collaboration with the UK Emissions Trading Group. Much of the institutional and stakeholder knowledge that was accumulated over the course of the development of the UK ETS has subsequently been applied to the UK's implementation

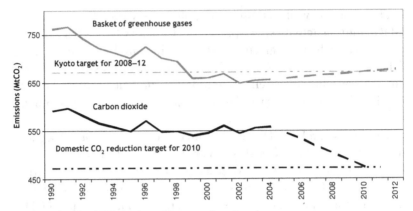

Figure 3.2. The UK's path to Kyoto and national target.
Source: United Nations Framework Convention on Climate Change Green-house Gas Inventory Data and DTI.

of the EU ETS, including registries and approaches to verification of data.

Current projections suggest that the UK will meet its Kyoto target with existing policy measures, but the government has indicated that additional measures may be necessary if it is to meet its more ambitious national targets for CO_2 emissions. Figure 3.2 shows the maximum allowable future emissions for consistency with both the UK's domestic CO_2 target and its international Kyoto commitments.

1.2 Overview of the UK Phase 1 National Allocation Plan

This section provides an overview of the UK Phase 1 NAP, including information on the government entities involved, the key dates, and a summary of the final NAP.

1.2.1 UK government agencies involved in the Phase 1 NAP

The lead government department responsible for the development of the NAP was the Department for Environment, Food and Rural Affairs (Defra), which has primary responsibility for the development of UK environmental policy. Alongside Defra, the Department of Trade and Industry (DTI) also took a major role, providing the UK's energy-use forecasts, serving as a liaison with industry, and later taking responsibility for the New Entrants, Closures and Auctions group. The devolved administrations in Scotland, Wales and Northern Ireland were also closely involved.[2]

The UK's Environment Agency (EA) played a significant role in the development of the NAP. The EA was responsible for issuing permits to sites and therefore ultimately for the determination of which sites would be eligible for coverage under the scheme in England and Wales. The EA also participated in the development of the monitoring, reporting and verification guidelines. Because the UK has 'devolved' certain responsibilities to agencies in Scotland and Northern Ireland, those agencies were responsible for issuing permits to operators located in the relevant regions. The EA assumed overall responsibility for coordinating among the various competent authorities.[3] Representatives from other government departments and authorities were also involved in NAP development.

Coordination of the inputs of the various government departments required the establishment of several inter-departmental working groups. These groups included a high-level group of senior officials that typically met once a month to review progress with the NAP and other aspects of the implementation of the trading scheme. Outside consultants assisted in the preparation of the NAP, reviewed implications for electricity markets, developed detailed growth projections, evaluated consultation response and reviewed the CCA targets.

1.2.2 Key dates in the development of the UK Phase 1 NAP

Table 3.1 highlights key dates during the development of the UK NAP. The development of the UK Phase 1 NAP took place over a period of more than two years, including four major stakeholder consultations on general methodology and specific sector/installation allocations. Additional stakeholder consultations were conducted on benchmarking and the new entrants reserve, government energy-use projections and sector growth rates, and implementation of the Directive. In

---

[2] Extensive information concerning the development of the UK NAP, including intermediate proposals and consultation documents, is available online at www.defra.gov.uk/environment/climatechange/trading/eu/nap/dev.htm.

[3] The Department of Trade and Industry is the regulator or 'competent authority' responsible for regulation of the offshore oil and gas sector, and therefore was responsible for the permitting of these installations.

Table 3.1 *Key dates for the UK NAP*

Event	Date
Stakeholder workshop on data sources and initial allocation options	June 2003
Consultation on allocation options	August 2003
Transposition of ETS Directive into UK law	December 2003
Consultation NAP published	January 2004
UK submits Provisional NAP to Commission	April 2004
UK Provisional NAP published for public consultation following revision of Updated Energy Projections	May 2004
Provisional list of installation-level allocations published	June 2004
Commission approval of UK NAP (subject to the provision of some further information)	July 2004
UK submits proposed amendment to the NAP to the Commission following finalisation of Updated Energy Projections	November 2004
Start of scheme and entry into force of UK regulations implementing Directive	January 2005
UK announces decision to issue allowances prior to resolution of legal dispute with Commission	March 2005
UK publishes approved NAP, installations are issued allowances, registry goes live	May 2005

addition, the scheduled review of the CCAs was moved forward, as was the timetable for the revision of energy projections.

1.2.3 Overview of the final UK Phase 1 NAP

Table 3.2 presents summary information for the UK NAP. The final NAP allocated 245 Mt CO_2 to around 1,100 installations whose 2003 emissions totalled 272 Mt CO_2. Emissions from these installations represented 46% of the UK's overall CO_2 emissions in 2003. The NAP was divided into fifty-two detailed sectors, the largest of which by far was the power sector.[4] The NAP also includes a pool of allowances set aside for new entrants.

---

[4] Sectors were differentiated not only by economic activity but also in terms of whether they were subject to a CCA, so that, for example, there were eight subsectors in the 'Food, drink and tobacco' category.

Table 3.2 *Final UK Phase 1 NAP allocations to EU ETS sectors*

Sector	Historical emissions[a]			Annual total sector allocation	Annual Contribution to new entrant reserve (NER)	Percentage of sector total allocated to NER
	Number of Installations	Average, 1998–2003	2003			
Power stations	123	155.0	174.4	136.9	6.3	4.6%
Refineries	12	17.7	18.0	19.8	0.4	2.0%
Offshore	110	16.6	17.5	19.1	1.5	8.1%
Iron and steel[b]	14	18.3	19.8	23.7	3.7	15.6%
Cement	13	8.8	9.7	11.2	1.6	14.3%
Chemicals	104	9.0	9.4	10.4	0.9	8.8%
Pulp and paper	79	3.7	4.5	5.1	0.1	2.2%
Food, drink and tobacco	138	3.1	3.9	3.9	0.1	3.7%
Non-ferrous	2	2.7	2.8	3.1	0.1	2.1%
Lime	9	2.3	2.2	2.7	0.0	1.4%
Glass[c]	34	1.7	1.9	2.2	0.2	7.9%
Services	208	1.8	2.0	2.1	0.1	2.9%
Other oil and gas	33	1.4	1.9	1.9	0.4	18.3%
Ceramics	112	1.7	1.8	1.8	0.1	4.3%
Engineering and vehicles[d]	55	1.1	1.2	1.3	0.0	2.7%
Other[e]	8	0.3	0.4	0.4	0.0	10.5%
Total	1,054	245.4	271.5	245.4	15.6	6.3%

[a] Excludes sites that closed prior to start of EU ETS.
[b] Iron and steel includes coke ovens and sintering.
[c] Includes mineral wool.
[d] Includes rubber, aerospace, foundries, semiconductors, cathode ray tubes and nuclear fuel processing and production.
[e] Includes other non-metallic minerals, textiles, wood products and wood board, and coal mining.

2 The macro decision concerning the aggregate total

The UK adopted a two-stage structure for allocating allowances to individual facilities, first determining allocations to sectors, and then allocations to installations within each sector, based upon the sector total. Allocation at the sector level thus falls somewhere between the macro-level and micro-level categories that are used in this and other chapters. We therefore discuss aspects of the sector-level allocations in both this section and the next.

2.1 *Initial consultations leading to the January 2004 draft NAP*

Work on the NAP began in spring 2003 with an effort to identify available data sources and to consider alternative allocation methodologies. At this time the government also accelerated the planned updating of its economy-wide energy-use projections, which provided estimates of national emissions based on projected fuel use. These projections had previously been published in 2000 (Department of Trade and Industry 2000).

It was hoped that the results of the updated energy projections (UEP) could be used directly in the development of the NAP. However, with a few exceptions (electric power and refineries) the sectors in the existing DTI model did not correspond well to the sector definitions and thresholds specified in the EU Directive. Indeed, there was no data set available that had been collected with the EU ETS in mind. Unfortunately this meant that there was no simple way of adapting the model – or the existing data – so that it could be used directly for the NAP.

In June 2003 the government held a workshop for interested stakeholders to present the initial findings on data availability and to set out the options for initial allocation that were being considered. Among the options considered were whether projections should form the basis for allocation, and if so, to what extent pre-existing regulations and agreements (for example, the emissions reductions expected from the Renewables Obligation or the CCAs) should be reflected in these projections. At this time, concerns were expressed about the ways that individual sectors would be defined and about what the 'overall cap' or total allocation level would be. In particular, industry representatives were concerned that the cap not go beyond what would be implied by the UK's BSA commitment, while environmental groups argued that this target level had already been achieved and that it should be consistent with a trajectory towards the UK national commitment of 20% reduction below 1990 levels by 2010. UK officials stressed that the decision regarding the overall cap was a separate decision that would be made by the government, and was not dependent upon the specific allocation rules.

The government launched its first formal consultation on the overall NAP methodology in August 2003 with comments due in October. In addition to questions about the type of data to use and the baseline period, the consultation requested views on the incorporation of projections and existing regulations, and treatment of new entrants and closures. Other issues covered by the consultation included the role of 'banking' of allowances, and whether the UK should exercise its right to auction or sell some proportion of the allowances it would make available. The consultation also raised the question of sector treatment – whether the allocation should be a one-stage process without any attention paid to sector classification, or whether allocation should be a two-stage process where allocations were made first to sectors and only then to individual installations.

Responses were received on all these points. A significant majority of respondents, many of whom were CCA participants, preferred a two-stage allocation process. There was significant support for allocations to sectors based on projections rather than historical emissions, but no clear majority. There was little agreement on what baseline period would be the most representative. Where a historical baseline was proposed, the widest range of years (1998–2002) was favoured, but the formula to apply to this range varied. There was broad support for a new entrants set-aside, while forfeiture of allowances upon shutdown was generally not favoured. Most participants indicated that flexible banking provisions would be useful to them. There was strong resistance to auctioning. More generally, many respondents stressed that allocation would strongly influence their competitive position, in particular in an international context.

2.2 January 2004 Phase 1 consultation NAP

2.2.1 Determination of the January 2004 draft Phase 1 UK cap
The UK's Phase 1 Draft or Consultation NAP was released in January 2004 – the first of its kind within Europe. The government proposed an annual 'cap' of 238 Mt CO_2 that was consistent with a reduction in

CO_2 emissions by 2010 of 16.3% below 1990 levels – that is, between its Kyoto target of 12.5% and the more ambitious national target of 20%.

The level of the overall UK cap in the January 2004 Draft NAP was based upon projected business as usual (BAU) emissions, including the expected emissions benefits of policies such as the CCAs and targets for both renewables and CHP. Allocations for sectors other than the power sector were set equal to the level of total projected emissions for 2005–2007. The allocation for the power sector was reduced below BAU to be consistent with the 5.5 Mt CO_2 annual reductions by 2010 that the government had previously stated that it expected to achieve through emissions trading. (By 2007, this translated into an annual reduction in the power sector allocation by 2.75 Mt CO_2.) The reason for reducing the allocation to the power sector but not other sectors was the expectation that the liberalised nature of the UK power market would mean that the power sector would be able to pass on additional costs from the scheme to customers via higher electricity prices without any significant threat of international competition. In contrast, other sectors faced the prospect of international competition that would limit their ability to pass costs on to customers.

The Consultation NAP also included provisions for the creation of a new entrants reserve or set-aside that would be used to supply allowances free of charge to new sites or to existing sites with significant capacity expansions. Each sector would contribute a varying portion of its total allocation to the reserve, but the allowances contributed by a given sector would not be earmarked for that sector.

2.2.2 Uncertainties in the UK Phase 1 cap
The Consultation NAP was based on a two-stage allocation approach, in which sector-level allocations were based on projected business as usual emissions including relevant environmental policies. However, because of the 'misalignment' issues discussed above, most sector emissions projections were calculated in a 'bottom-up' fashion from installation-level historic emissions data. Sector growth rates were then applied to the bottom-up historical emissions totals to arrive at future projections.

One important implication of the proposed two-stage approach was that the macro-level cap would not be determined independently of micro-level data. Because of the limitations of the existing data and uncertainties about the projections, the overall number of allowances distributed by the NAP would not be known until the last installation was identified and its historical emissions verified. Accordingly, the government stressed in the consultation materials that all projections used in the development of the draft Consultation NAP were preliminary and were subject to revision for a number of reasons.

2.3 Consultation based upon the January 2004 NAP

Publication of the first draft installation-level NAP in January 2004 was followed by an eight-week consultation period to seek views on the proposal. More than half of expected EU ETS participants responded to the consultation. Many commented on the overall level of allocation, stressing the perceived adverse effects on UK competitiveness if the government attempted to reduce emissions below the level required by the BSA. Many commented on the categorisation and treatment of individual sectors and urged the use of different sector definitions. Many operators challenged the implementation of the allocation in their sector, in particular the growth-rate projections, the application of energy-efficiency targets or energy-intensity estimates, and the calculation of the new entrants set-aside. A large majority of respondents stated that they objected to auctioning of allowances as a matter of principle.

A broad theme from the consultation process was that many sectors were not happy with the growth projections (for output, energy use or emissions) that had been applied based on the UEP. Many sectors also felt that the level of disaggregation provided by the UEP was insufficient and did not reflect the particular circumstances of their industry, even if they reflected aggregate industrial projections. Sectors participating in CCAs also wished to see their own CCAs reflected in their allocations, and did not want their targets to be combined with other CCAs in a weighted average. In light of the input from stakeholders, the government decided that more disaggregated sector projections were needed. The government commissioned work by independent consultants to produce more detailed sector-level output projections that could be applied to the EU ETS sectors. Consultants were not asked to review projections of energy use or emissions. Their report was completed by the end of summer 2004 after submission of the Provisional NAP to the Commission in April 2004 (as discussed below).

2.4 April 2004 revised provisional Phase 1 NAP

The revised NAP submitted in April 2004 reflected updated growth
projections for the power sector, industrial sectors and the services sector, based on input from stakeholders and modification of the model
itself, but not the results of the consultants' work on projections. The
updated projections indicated that 2010 BAU emissions would be 7 Mt
CO_2 higher than had previously been foreseen. The bottom-up data
used to develop individual sector projections had also undergone substantial additional quality checking to correct errors and anomalies
in the data submitted by operators, and some previously unidentified
operators and installations had been added. The method for calculating
the NER was also modified to be more consistent with overall sector
growth projections, and to apportion expected growth between incumbents and new entrants. The number of allowances to be distributed
annually under the Provisional NAP totalled 245 Mt CO_2, with around
8%, or 19 Mt CO_2 per year, set aside for the new entrants reserve.

The submission to the Commission made clear that the government
still considered the NAP to be a draft subject to revisions – both to the
overall cap and to installation-level allocations – because of the ongoing updating of the UK energy projections and growth rates, review of
emissions factors, treatment of CHP projections, review of CCA targets
and the potential for changes to the underlying bottom-up data submitted by operators. Various other issues related to the installation-level
allocations were also subject to further discussion.

At the beginning of July, the Commission published its decisions on
eight NAPs, including the UK's. The Commission accepted the UK's
provisional NAP, subject to two minor conditions concerning four
installations in Gibraltar that had not been included in the provisional
NAP and clarifications of plans for allocation to new installations.

2.5 November 2004 revised Phase 1 NAP and
Commission response

Work on the NAP continued after submission to the Commission in
April. In particular, as noted above, outside consultants developed
more detailed sector-level output projections for the EU ETS sectors.
After publishing some further decisions regarding the installation-level
allocation during the summer of 2004, the UK published the remaining

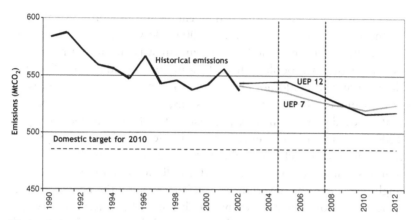

Figure 3.3. Revisions to projected CO_2 emissions and distance from UK target.

Source: United Nations Framework Convention on Climate Change Greenhouse Gas Inventory Data and DTI projections as explained in text.

decisions on issues related to allocation methodology in November 2004. The decisions concerned sector growth rates (based on the work of the outside consultants) that provided more detail than the industry sector growth projections used to predict sector emissions in the prior draft. The final projections (UEP-12) incorporated more up-to-date fuel price and demand estimates, updated some emissions factors, and corrected a double-counting error in the treatment of emissions savings from CHP that had previously led the model to underestimate emissions levels. These updated projections were incorporated into a revised version of UEP and a revised NAP that included an increase in the annual total to 252 Mt CO_2, a 2.9% increase relative to the April 2004 revised provisional Phase 1 NAP.

The new projections indicated that the power sector's annual emissions would be higher than previously thought; in 2005, for example, the updated projection was 17 Mt CO_2 greater, or about a 12% increase. These predicted increases were offset in part by reductions in projected industry emissions. The total allocation was also revised to reflect further verification of bottom-up emissions totals, detailed sector growth rates, as well as revised CCA targets. Figure 3.3 illustrates the revisions to UEP between the version supporting the April Provisional NAP (UEP-7) and the final version (UEP-12).

The UK submitted its proposed amendment to the provisional NAP to the Commission in November 2004. However, the Commission maintained that while the UK government was entitled to revise the details of the methodology for allocating to individual installations, the government could not change the total number of allowances to be distributed. After extended discussions, the UK government decided that it would publish a final NAP consistent with the cap from the April 2004 Provisional NAP, but that it would file a legal challenge against the Commission's refusal to allow the increase in total allocation. To minimise the uncertainty to covered installations that the legal action might engender, the government decided that only the power sector's allocations would be affected by the dispute. The difference between the revised total quantity of allowances and the previously approved provisional total would be reflected by reducing the (revised) power sector allocation, but all other (revised) sector allocations – and therefore all other installations' allocations – would be left unaffected by the Commission's decision and subsequent legal outcome. The eventual outcome of the legal challenge left the total and the installations' allocations unchanged.

2.6 Total allocation under the final UK Phase 1 NAP

The final Phase 1 allocation was approximately 22 Mt CO_2 per year below projected annual BAU emissions for covered sources. This represents an aggregate reduction of 8% relative to BAU for these sources (where BAU already incorporates the reductions expected from the UK's CCP). Based on the government's final updated projections, the final annual allocation to power stations was 22 Mt CO_2 lower than the sector's expected average annual emissions during Phase 1 – a substantially greater reduction than the originally intended 1.8 Mt CO_2 annually (over the three years of Phase 1) that would have been consistent with a 5.5 Mt CO_2 annual reduction by 2010.

3 The micro decisions concerning distribution to installations

3.1 Activities leading to the January 2004 draft Phase 1 NAP

3.1.1 Data development
At the start of the NAP development process in spring 2003, Defra and DTI initiated work to identify sources of information on the

various factors that might be used as the basis for allocating allowances to individual facilities – capacity, inputs, outputs and emissions. The review uncovered ten separate UK data sources, most of them dealing with emissions, but also covering energy inputs and power stations' electricity output. This data review found that emissions data were available for at least one year (2001) for most sites already subject to regulation, but that data for multiple years were limited to only some of the largest sites.[5] Furthermore, earlier data were potentially less accurate, and in fact none of the available data had been verified by independent parties. Moreover, because the EU ETS would apply to sites that had not previously been covered by existing environmental regulations[6] (either EU-wide or national) there was a significant minority of sites – many of them small – for which even a single year of data was not available (and indeed, whose identities were not even known). Electricity output data were available for a number of years for major generators, although the data were complicated by successive changes to electricity regulations that had led to shifts in generator activity. Suitable output and input data for other sectors were not available. In all, somewhere between 1,000 and 1,500 sites were expected to be covered.

Because suitable data were not universally available, and because of the potentially large number of previously unregulated facilities that would be included in the scheme, the government concluded that it would need to request data from all sites that could potentially be covered by the scheme. The relevant departments did not have the legal power to require this information at the time, so the government decided that operators should be asked to submit the information voluntarily in anticipation of a formal legal requirement in the future, to allow the maximum time for review of the newly submitted data. The government indicated that all submitted data would need to be verified by third parties or else operators would risk forfeiting their associated allocations.

During the summer of 2003 operators were asked to submit forms that included address and contact information, identified the covered activity, and showed their fuel use, calculated emissions, and

---

[5] From 2001, UK sites with emissions of CO_2 in excess of 10,000 t CO_2 were required to report their emissions to the relevant environmental authority under the Pollution Inventory.

[6] In particular, the 20 MW_{th} size threshold for combustion installations was lower than the cut-off used for existing regulations.

where relevant, heat and power output for 1998–2002. The forms also included space for comments by operators regarding the site's operations that would be relevant for the allocation. Operators were also invited to submit comments on the summer 2003 consultation concerning the appropriate baseline years and a variety of other issues relevant to installation-level allocations.

3.1.2 Facility-specific allocation

The submission of data by operators for the requested period made it possible to prepare preliminary assessments of a wide range of approaches to installation-level allocation. The government checked the data, considered various approaches to the use of the data and compared the resulting allocations to historical emissions of individual sites. These included both one-stage allocations based exclusively on each site's share of overall UK historical emissions, and two-stage allocations where each sector's share of the total was determined before the installation-level allocations were calculated. Among the two-stage approaches considered were various forms of output-based (or benchmarked) allocations to the power sector, as well as emissions-based allocations to all sectors.

The summer 2003 consultation responses indicated that a significant majority of stakeholders was in favour of a two-stage allocation approach. With the exception of a few sectors where baseline data were available (notably the power sector, the offshore sector and the refineries sector), this approach would require that sector-level allocations be calculated by adding up historical emissions data submitted by individual operators, and then applying the appropriate growth rates and emissions- or energy-intensity adjustment factors to calculate projected sector emissions covered by the EU ETS. The year 2002 was selected as the basis for projecting sector totals, and for certain bottom-up sectors a correct calculation required that the emissions of sites that closed in 2002 also be included. Once the total sector allocation was calculated, each sector's NER contribution was calculated, and the remaining sector allowances were distributed among the incumbent sites.

The government decided that sector allocations should be divided among corresponding installations based on historical emissions. They also chose to make use of the broadest range of data available, in part as a way to credit early action by individual operators. Installation-level allocations would be based on the share of historical emissions over the period 1998–2002, later revised to include 2003 data when this information became available. Instead of using a simple average, operators were allowed to drop the lowest value during the relevant period and then take the average of all remaining years with emissions so that the impact of a 'bad year' on an operator's allocation would be eliminated.

Consequently, the draft Consultation NAP published in January 2004 comprised allocations based on historical emissions shares to 867 installations divided into fourteen different sectors (compared to 1,054 installations and fifty-two sectors in the final NAP). In addition, there were sixteen incumbent sites for which insufficient data were available from which to calculate an allocation (these were sites that began operation in 2003). Finally, there were fifteen named new entrant installations that were expected to require allocations but for which no allocation was proposed because there had been no decision on how new entrant allocations would be determined.

3.2 Facility allocation issues in the final UK Phase 1 NAP

The January 2004 Consultation NAP invited stakeholders to comment on the proposed approach to installation-level allocations. Most individual respondents were unhappy with their allocations, for example because they thought the baseline period was unrepresentative, or that a 'grandfathering' methodology was unfair in their case, or because they felt that special circumstances had not been taken into account. Many sites also argued that they had been classified within the wrong sector, or that a separate sector should be created for one or more sites.[7] These and subsequent consultations led to various modifications to the facility-specific allocations as well as to the development of procedures for allocating allowances to new facilities.

3.2.1 Baseline, commissioning and rationalisation rules
One of the important results of this consultation was the identification of specific cases in which the formula used to determine installation-level allocations put certain sites at what was considered to be an unfair disadvantage. Three rules were developed to deal with these

---

[7] Operators also commented on the appropriateness of growth rates, sector targets and the overall cap, as discussed in Section 2.

exceptional cases: 'baseline changes', 'commissioning' and 'rationalisation'. Each of these applies to slightly different circumstances, but the approach adopted to redress the potential disadvantage was the same in each case. Installations were allowed to eliminate one or more unrepresentative years from the baseline period used to calculate their allocation. The final rules governing each of the three circumstances (collectively known as BCR rules) were proposed in a July addendum to the Provisional NAP submitted to the Commission in April 2004.

Baseline changes. Baseline changes occur at an installation when it undergoes a significant modification to site configuration that results in an increase in emissions. For example, the rule could apply to a site that added significant additional capacity in 2001, which resulted in a doubling of its emissions relative to previous years. If this site received an allocation based on the simple 'drop minimum' rule, its low emissions in the years 1998–2000 would not reflect its current level of operation, and would therefore force the site to incur additional costs to purchase sufficient allowances. Accordingly, such a site was allowed to apply the 'drop minimum' rule to its recent (2001 and later) emissions and to ignore the earlier years. Baseline changes ultimately were judged to be relevant for seventy sites from across a wide range of sectors.

Commissioning rules. Commissioning rules apply to sites that began operation during the baseline period and whose activity levels during the period were lower than would be expected during a 'normal' year of operation as a result of the need to increase utilisation and output gradually. In such circumstances the installations are allowed to ignore the years in which they were commissioning.[8] Commissioning was ultimately judged to be relevant for sixteen power stations and one cement kiln.

Rationalisation rules. Rationalisation refers to the shifting of production from one site (or more) to another to achieve a more efficient level of operations. Under the UK NAP, a site was deemed to be eligible to benefit under the rationalisation rule if the operator could demonstrate

---

[8] The decision was made not to attempt to pro-rate activity during the partial year following a commissioning period, in part because the seasonality of activity would make it difficult to arrive at appropriate scaling factors.

that it had closed one of its own sites and that emissions at remaining sites were at a level sufficiently high to suggest that production had in fact been shifted to the remaining sites. Rationalisation was only applicable to the transfer of production between sites operated by the same firm, and only applied where the product was 'transferable' to the remaining sites. The power sector was not eligible to apply for the use of the rationalisation rules, because both government and a majority of consultation respondents did not believe the concept was relevant for the sector. The rationalisation rule was applied to nine sites in the final NAP, the majority of which were owned by a single operator in the food and drink sector.

3.2.2 Allocation to new entrants and capacity expansion
The rules for new entrants, expansion of existing capacity and closures were developed during the spring and summer of 2004.

Formulas for new entrants. For most sectors, the rules for new entrants were tied to the addition of capacity (whether to a new site or an existing site) of the covered activity explicitly referred to in the Directive. This meant that certain types of capacity expansion (for example 'de-bottlenecking' at chemicals installations) would not qualify for new entrant allocations.[9] The basic formula used for new entrants in most sectors can be summarised as follows:

$$\text{Emissions of } CO_2 \text{ (indicative allocation)} = \text{New capacity} \times \text{Load factor} \times \text{Emissions factor per unit output} \times \text{other factors}$$

The new entrant benchmarks generally were designed to award to new entrants the amount of allowances that they would need to cover their expected emissions over the course of Phase 1. In general, the emissions factors were intended to reflect the best available technology. In some cases the benchmarks were differentiated to reflect the emissions

[9] In November 2004, the government also proposed that some further cases could qualify as new entry. These included development of existing offshore platforms known as 'tie-backs', as well as changes to installations due to compliance with environmental regulations (such as the installation of flue-gas desulphurisation equipment) and increases in levels of 'good quality combined heat and power' from eligible CHP stations would also be able to receive allowances from the new entrant reserve.

characteristics of particular process technologies or inputs, even for installations with the same output. In other cases – notably the electric power sector – the emissions factor used does *not* reflect the actual site process. In the case of power facilities, all new facilities (and facility expansions) received allowances based upon emission factors for a gas-fired power station, regardless of what fuel they used; new or expanded coal units thus would receive allowances equal to only a fraction of their expected emissions.

For the iron and steel and refineries sectors, the consultants concluded that it was too difficult to define a generic or 'fully standardised' approach that could be used for all sites. The reasons are that site expansion is much more likely than entirely new entry and installations in these sectors typically included systems for the recirculation of intermediate fuels, heat, and/or steam that made determination of a generic emissions factor extremely complicated.

Benchmarks for 2003 facilities.   The benchmarks developed for new entrants were also used to calculate allocations to 'incumbent' sites that began operation in 2003, but for which insufficient data were available to apply the 'drop minimum' rule. However, allowances for these sites are not deducted from the new entrants reserve. Instead they are deducted from the residual sector allocations available to other incumbents, before the incumbent emissions are divided among sites for which sufficient historical data were available.

Forfeiture of allowances for closed facilities.   Early on in development of the NAP, the government determined that if it included provisions for a new entrants reserve it would also be necessary to require sites to forfeit allowances upon closure. The Consultation NAP had proposed that sites that closed would be able to retain allowances allocated for the current year, but that they would not receive allowances in subsequent years. In practice, it would be up to the Environment Agency or other competent authority to determine whether or not a site had closed, and therefore whether it could receive allowances as published in the NAP. Any forfeited allowance allocations would be returned to the new entrants reserve.

Subsequently, amendments were added to the NAP (in November 2004) to allow the 'transfer' of allocations belonging to closed activities under certain circumstances. These include rationalisation of production activity (although again this is not applicable to the power sector) and replacement of individual technical units on an existing site (but not at a different site except via the rationalisation rule).

3.2.3 Additional adjustments and requirements

In the final NAP, all installation-level allocations were ratcheted down to reflect one or more additional factors, depending on the sector. The allocations to all sites, including incumbents and new entrants, were reduced by 0.7% to ensure that sufficient allowances are available to provide for the level of new combined heat-and-power (CHP) installations called for by government CHP targets. In addition, incumbent sites whose allocations were determined by the new entrants benchmarks (because of lack of emissions data) had their allocations reduced by an amount equal to the new entrants reserve percentage that had been deducted from other incumbents in their sector. Finally, all benchmarked allocations in the power sector – including both recent incumbents and new entrants – were reduced by the 'adjustment factor' of around 14% that is applied to the power sector to meet the overall allowance cap.

4 Issues of coordination and harmonisation

Two major issues that arose in the context of the UK NAP concern issues of coordination and harmonisation across Member States.

4.1 Determination of the overall cap

The publication of the preliminary Consultation NAP in January 2004 was an ambitious first step in a broader European context, particularly since it went significantly beyond the UK's commitments under the BSA. It was therefore viewed by some commentators as an attempt to influence the development of NAPs in other Member States, so that they too would opt for demanding caps in their NAPs and signal their commitments to reducing GHG emissions substantially in the pre-Kyoto period. The NAPs proposed subsequently by many other Member States did not include caps that were as ambitious (relative to BAU) as the UK January 2004 Consultation NAP. It has been suggested that the subsequent revisions of the UK's NAP and the increase in the overall UK cap reflected an attempt to gain some strategic advantage by first proposing a tight target while always intending to relax it, or more charitably, that the subsequent modifications represented a deliberate backtracking after other Member States refused to follow the UK's more ambitious lead. Neither of these suggestions seems warranted. The revisions to the cap that were proposed over the course of the NAP's development arose as a result of fairly mundane factors involved in implementing the UK's approach, including the finalisation of emission projections, clarification of the targets in the CCAs and the receipt of additional verified emissions data.

Nevertheless, it seems likely that the European Commission did have in mind the potential implications of wider attempts to increase overall allocations when it considered the UK November 2004 revision. However sensible the changes to the UK NAP, the risk that other Member States might attempt to follow suit for other reasons may have been enough for the Commission to oppose the proposed November 2004 changes to the UK's cap.

4.2 New entrants reserve and closure rules

One area where concerns about harmonisation may have played a very significant role is in the development of the rules for new entrants and closure. Providing a new entrants reserve seemed necessary in light of competitive concerns. Many industry representatives in the UK expressed concerns about the potential implications for UK competitiveness if the UK did not provide for a new entrants reserve when other Member States were planning to do so and when other non-EU competitors would not face any mandatory CO_2 constraints at all. New entrant benchmarks were developed to be consistent with best practice. This meant that certain types of technologies or fuels would not receive as high an allocation as they would need to cover all emissions – for example, all power stations would receive an allocation based on natural gas, even if they used a different fuel.

5 Other issues deserving special mention

There are several other issues that arose over the course of the development of the UK NAP that merit some further discussion. We review four of them here.

5.1 Temporary exclusion/opt-out

The Directive includes provisions during the first phase of the scheme that allow Member States to opt out certain installations that otherwise would be covered by the EU ETS, provided the installations meet specific conditions. These conditions require that installations be subject to equivalent emissions reduction requirements, that they face equivalent penalties for non-compliance and that they be subject to equivalent monitoring and reporting requirements.

The UK's CCP includes two significant policies for which the opt-out policies are relevant – namely the Climate Change Agreements and the UK Emissions Trading Scheme.

5.1.1 Climate Change Agreements

The allocations to sectors (and consequently to installations) under the UK NAP explicitly take account of sector CCA targets where these are relevant. This approach was taken in part because it provides a more accurate reflection of BAU emissions for the sites in question, since the CCAs would be in force without the EU ETS. In addition, it was the clearest way to ensure the equivalence of the targets applied to sites covered by both policies and thus establish the basis for facilities to opt out of the Phase 1 NAP.

Translating CCA targets into EU ETS allocations was not always straightforward because the CCAs apply to energy use and some emissions that are not covered by the EU ETS, notably the 'indirect-emitting' use of electricity. At the beginning of the NAP development process a number of options regarding the treatment of CCA participants were considered, including the development of separate targets for different types of energy use by individual sectors. It was subsequently decided that this would be too complex given time constraints. The government also decided that it would not be appropriate to allow sites participating in the EU ETS to drop their CCA targets. As a consequence, sites participating in the EU ETS are still required to meet their CCA targets, in addition to holding emissions allowances.

Sites with CCAs that opted out of the UK Phase 1 NAP had to adopt several modifications of the CCAs designed to ensure compliance with the Directive. For example, to ensure that the monitoring and reporting requirements are comparable to what would be required under the EU ETS, the government proposed in February 2005 that all sites temporarily excluded from the EU ETS be required to submit annual rather than biannual reports as required under the original CCA framework.

5.1.2 UK Emissions Trading Scheme

Facilities participating in the UK ETS also were eligible to apply to be opted out of the EU ETS for the period in which the UK ETS was in operation (between 2002 and 2006). The UK ETS explicitly excluded emissions from the power sector, but there was nevertheless significant overlap among participants – including the cement sector, two offshore oil and gas operators, a food and drink operator and a group of universities. The corresponding sites joined the EU ETS in 2007 as incumbents.

The circumstances of the UK ETS were reviewed to establish that the sites meet the requirements of the Directive – that the sites be subject to equivalent emissions reduction requirements, that they face equivalent penalties for non-compliance and that they be subject to equivalent monitoring and reporting requirements. In general the potential penalty facing participants who did not meet their obligations under the scheme was very high, because it would entail the forfeit of all incentive payments received by a participant's facilities.

5.1.3 Summary of opt-out facilities

In total, the opt-outs from the UK's Phase 1 NAP included 331 CCA participant sites and an additional fifty-nine sites that were Direct Participants in the UK ETS (seven of these were also CCA participants). The UK ETS Direct Participants were eligible for exclusion for 2005 and 2006 (while the UK ETS was still in operation), whereas the CCA participants are eligible for exclusion over the entire Phase 1 period.

In total these sites represented approximately 35% of all sites covered by the scheme. The opt-out facilities as a group accounted for about 24 Mt CO_2 allowances per year (14.1 from CCAs plus 9.8 from the UK ETS), or about 10% of the total annual UK allocation for Phase 1.

5.2 NAP coverage

Many of the data and other complexities surrounding the UK NAP related to small facilities with few CO_2 emissions. A very large number of low-emitting installations account for only a small fraction of total

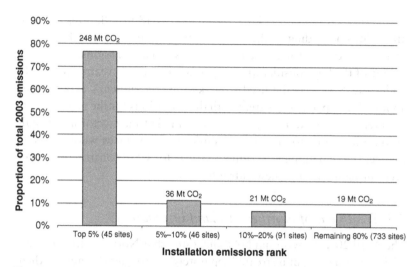

Figure 3.4. Emissions from sites covered by the Phase 1 NAP ranked according to share of UK Phase 1 cap.

Source: NERA calculations based upon 2003 emissions as explained in text.

Notes: Installation emissions rank is based upon ranking all facilities by their emissions and then summing up emissions for the various categories (e.g. 5% of facilities with highest emissions). Note that not all covered sites reported 2003 emissions, so the total is less than the total number of covered sites.

2003 emissions. Figure 3.4 shows that the 5% of facilities with the greatest emissions account for more than 75% of overall emissions while the 80% of facilities with the lowest emissions account for less than 6% of total emissions.

5.3 Treatment of CHP installations

Over the course of the development of the NAP there was considerable discussion about the treatment of CHP facilities. The broad conclusion of the government and of various independent consultants was that CHP would be promoted by the introduction of the EU ETS, because CHP is an efficient technology that could provide electricity (and heat) with lower CO_2 emissions per kilowatt-hour. The government recognised, however, that energy market developments had meant that many existing CHP plants had faced difficult circumstances, independent of anything related to the EU ETS. Moreover, in certain special circumstances, cases were identified in which certain types of CHP might be disadvantaged by the EU ETS. In the end, CHP installations received allocations according to the same formula as all other installations in their respective sectors, based on their historical emissions.

One of the key considerations associated with the treatment of CHP under the NAP was the need to make the government's expectations about CHP capacity consistent with the provisions for the new entrants reserve. As a result, an adjustment was made to all sector and installation allocations to ensure that sufficient allowances were available to meet the remaining government targets for CHP construction, after accounting for known new entrants.

5.4 Treatment of Northern Ireland installations

The April 2004 Provisional NAP suggested that Northern Ireland electricity facilities might be treated separately from others because of their higher growth projections and the lack of interconnection to the rest of UK. Ultimately, however, no distinction was made between installations in Northern Ireland and those in the rest of the UK.

6 Concluding comments

This section provides conclusions regarding the development of the UK NAP and the issues that arose over the course of the process.

6.1 The development of the Phase 1 UK NAP was a major accomplishment

Over the course of the two-year-long process, the UK managed to develop the key elements of a workable allocation programme. The UK's excellent progress in reducing emissions prior to the start of the EU ETS and the First Kyoto Commitment period did not necessarily make the task of developing its NAP any easier. Nor did the early start of the UK in the development of its NAP mean that decisions could be made sooner, because in practice it meant that more options could be considered.

- *Overall methodology.* The UK developed a clear and coherent two-stage methodology for allocation – predicated on the principle that Phase 1 allocations should go to participants rather than be auctioned – based upon allocating first to sectors and then to covered facilities within each sector.

- *Method for sector allocations.* The UK developed a method for determining allocations to individual sectors and then implemented the method taking into account various complications.
- *Overall cap.* The UK used information on BAU emissions for individual sectors to set an overall cap.
- *Detailed database.* The UK developed a detailed facility-level database on emissions for about 1,100 facilities, many of which had virtually no data initially. These data were critical both to the development of the overall cap and to the allocations to individual facilities.
- *Method for facility allocations.* The UK developed allocations to individual facilities within each sector that ultimately were accepted by the vast majority of participants. These allocations included dealing with numerous 'fairness concerns' and other complications that arose for individual facilities.
- *Stakeholder consultations.* The UK provided detailed consultations with stakeholders on draft allocations that clarified the issues to be resolved and provided the basis for an acceptable allocation program.

6.2 Basing the UK NAP sector and facility allocations on clear principles provided coherence but added significant complexity and uncertainty

The UK NAP was based upon two key principles for allocating first to sectors and then to individual facilities.

(1) Allocations to sectors were determined by reference to projected BAU emissions, with non-electricity sectors allocated amounts equal to projected BAU emissions and the aggregate shortage allocated to the electricity sector.

(2) The shares of sector allocations to individual facilities were based upon average historical emissions over as much as a six-year period (dropping one year).

These two general principles provided a coherence and sense of direction that allowed the UK government to focus on obtaining accurate information on historical emissions for individual facilities and on developing reasonable projections for sectors. While these objectives

were relatively straightforward in concept, they proved to be difficult
to implement in practice. Sector projections were complicated by the
misalignment between the sector definitions in the UK model and in
the EU ETS, as well as the very different growth prospects of different subsectors within a given sector. Moreover, as the UK's revised
November 2004 NAP revealed, projecting BAU emissions is itself a
difficult and uncertain enterprise, particularly in light of the importance of fuel prices and other factors that are notoriously difficult to
project. Similarly, determining historical emissions for individual facilities was complicated by many factors, such as unanticipated outages
and significant changes in capacity.

6.3 Conclusions regarding UK NAP allocation method

The UK Phase 1 NAP also provides the basis for some conclusions
regarding the methods that were developed to allocate allowances.

- *The option of initially auctioning up to five per cent of allowances
  was not taken.* The use of an auction to distribute allowances had
  little support among stakeholders since it would decrease the number
  of allowances provided for free and thus increase the costs of the
  program to them. The UK *did* conclude that an auction would be
  used as a way of returning to the market any unclaimed allowances
  set aside for the new entrant reserve.
- *Data limitations substantially constrained the choice of allocation
  methods.* Although the potential range of allocation methods and
  metrics is large, the actual possibilities were circumscribed by data
  limitations. Prior emissions trading programs in the US had used
  benchmark emission rates as the basis for allocation, using facility
  specific information on input or output levels. These options were
  not alternatives for the UK NAP because facility-specific input and
  output data generally were not available.
- *Using sector projections as the basis for sector allocations, while difficult to implement, addressed various requirements for the allocation.*
  The choice to use growth projections presented a number of challenges. Over the course of the NAP development, progressively more
  disaggregated growth projections were developed to account for the
  multiple subsectors, and updated information, including fuel prices,
  was incorporated to ensure allocations meet criteria of fairness. The

multiple sectors, while allowing the UK NAP to reflect the individual circumstances of specific sectors, increased its overall complexity particularly due to the need for detailed site- and sector-specific information. Emissions trading is often cited as a way of avoiding the need for governments or regulators to know details of individual sectors and site abatement opportunities; but some sector-specific and site-specific information was considered necessary to 'get projections right' and ensure that allocations meet criteria of fairness. Although the desire for accurate projections created difficulties, it is not clear that other approaches would have been preferable. Moreover, the difficulties may well be less significant over time as modelling is refined and data issues are resolved.

- *Competitive concerns dictated the decision to have the electricity sector bear the brunt of emission reductions.* The power sector was allocated less than its BAU emissions because of the expectation that it would be able to pass the costs along to its customers. The distinction between the power sector and the other sectors, however, may not be as distinct as not all non-power facilities operate in international markets with prices fixed by international competition. Competitive conditions are likely to differ among sectors, and while some sectors may be more similar to the power sector, others could face serious competitive disadvantages if forced to pay for a significant proportion of allowances.

- *Using emissions as the basis for facility allocations was the only feasible approach, given the difficulties with developing usable benchmarks.* The UK's choice to use emissions as the basis for facility allocations reflected the reality that data to implement a benchmark approach simply were not available for many of the sectors and facilities. This circumstance may change in the future, as appropriate data are collected and benchmarks are developed as the basis for allocations under the new entrants reserve.

- *Concerns of fairness dictated the use of a relatively long period for averaging historical emissions.* A relatively long (six-year) averaging period for historical emissions was chosen as a means of avoiding the use of unrepresentative data for the facility-specific allocations. The longer period also provided a way to recognise early emissions reductions without needing to verify emissions baselines or other complicating factors.

6.4 Conclusions regarding the UK NAP process

The extensive process used by the UK government to develop its Phase 1 NAP also provides some conclusions regarding the importance of process in major public policy decisions.

- *The consultation process was important to educate/motivate stakeholders, to gather necessary data and to obtain political legitimacy for the final UK NAP.* The ultimate political legitimacy of an allocation plan depends on an understanding that the government provides necessary information and listens to stakeholder concerns. The UK NAP succeeded in developing this legitimacy largely because of the extensive consultation process. This process, though costly, was important to develop an accurate database, to educate stakeholders and to obtain meaningful feedback on aspects of the plan.
- *The relatively loose guidance provided by the EC allowed the UK sufficient leeway to develop its NAP.* The Commission provided various ground rules for the NAPs, leaving governments substantial leeway to develop the NAPs. The UK government used this leeway to develop and implement the two-stage allocation approach. Still, the Commission was able to use informal consultations and other means of ensuring that key overall EU-wide objectives for the EU ETS were met.
- *The administrative costs of subsequent NAPs are likely to be substantially smaller than those for the Phase 1 NAP.* The development of a facility-specific database, creation of the information needed to forecast sector-specific emissions, the consultation process, as well as the efforts developed by stakeholders to participate and influence the process all contributed to the considerable administrative costs of the UK Phase 1 NAP. Due to the experience gained both by UK government officials and by stakeholders, administrative costs of subsequent NAPs are likely to be substantially lower. However, the multiple-period approach may increase the administrative burden of the scheme relative to a case in which the allocations are decided only once, since some elements will undoubtedly be revisited and since greater understanding by stakeholders could well mean that some issues become more contentious. Overall, however, the considerable effort to develop the Phase 1 NAP and the framework and data underlying it provide a very solid foundation for future periods.

References

Department of Trade and Industry 2000. 'Energy Paper 68', Energy Strategy Unit, November 2000.

Department of Trade and Industry 2003. White Paper: 'Our energy future – creating a low carbon economy', February 2003.

4 | Germany

FELIX CHRISTIAN MATTHES AND
FRANZJOSEF SCHAFHAUSEN

1 Introductory background and context

1.1 Background and political framework

Among the twenty-five Member States of the European Union, Germany is the country with the highest greenhouse gas emissions in absolute terms. In 2003 the total greenhouse gas emissions of Germany represented a 20.7% share of the total volume of greenhouse gas emissions of the twenty-five Member States of the EU. With regard to per capita emissions, Germany was above the average of EU15 as well as EU25.

For an in-depth understanding of the development of the National Allocation Plan (NAP) in Germany some national circumstances should be considered:

- Germany is relatively close to its target of the European Burden-Sharing Agreement for the Kyoto Protocol which is equal to 21%. The greenhouse gases included in the Kyoto Protocol have been reduced by about 18.5% by the end of 2003.
- In East Germany industry was subject to a fundamental restructuring and modernisation process since 1990, and considerable renovation investments are imminent in the old Federal States in the next few years (Matthes and Ziesing 2003).
- The electricity industry and, above all, coal-powered electricity plants play a prominent role in Germany. Energy and environmental policy regulations on the use of hard coal and lignite constitute a highly sensitive political matter in this country.
- The German Federal States have a great influence on German policy and traditionally own the main administrative competences in the framework of environmental policies.
- Traditionally, German environmental policy has been very markedly characterised by a 'command and control' approach. Significant

powers from politics and the Federal Administration have a very negative stance towards the implementation of the EU Emissions Trading System (EU ETS), or still oppose it even today.

These five factors played an important role in several phases of the German NAP process or have built the background for some crucial provisions in the NAP.

1.2 The political process

The creation of the working group on 'Emissions trading to combat the greenhouse effect' (Emissionshandel zur Bekämpfung des Treibhauseffektes – AGE) on 9 October 2000 within the framework of the National Climate Protection Programme constituted an important milestone in terms of the implementation of the EU ETS. The stakeholders from politics, industry, the Administration, the Federal States, non-governmental organisations (NGOs) and science represented in this working group, which is lead-managed by the Federal Ministry for the Environment (Bundesministerium für Umwelt, Naturschutz und Reaktorsicherheit – BMU), in actual fact organised the process of a permanent hearing, in which the current developments on a European and national level were continually disseminated, analysed, discussed and evaluated and numerous suggestions and recommendations were gathered together. Furthermore, in the run-up to the implementation of the EU ETS, various pilot projects were initiated by or within several Federal States.[1]

Since the beginning of 2003, extensive research and consultancy operations have been brought into being. These projects are devoted to the development of the NAP for Germany (DIW *et al.* 2005), the implementation of the emissions trading system in German law as well as data collection and evaluation for the plants and companies covered by emissions trading.

In mid-2003, an initial systematic overview of the conceivable allocation options for Germany was in hand with a White Paper on NAP design options (DIW *et al.* 2003), in which in particular the interrelation of the different regulations was analysed.

---

[1] The goal of such projects was primarily to prepare affected companies in good time for their participation in EU emissions trading and to test different structures important to its implementation (see Schleich *et al.* 2002, 2003; Ehrhart *et al.* 2003; HMULF 2003; Kruska *et al.* 2003; Schleswig-Holstein Energy Foundation 2003; Bode *et al.* 2004).

Since the autumn of 2003, five different developments have taken place in a largely parallel fashion:

- In a 'high-level negotiating group', the state secretaries of BMU as well as the Federal Ministry for Economics and Labour (Bundesministerium für Wirtschaft und Arbeit – BMWA) and thirteen members of the board of large-scale industrial corporations as well as a member of the Federation of German Industries (Bundesverband der Deutschen Industrie – BDI) had attempted from mid-October 2003 onwards to reach an agreement on the quantity of allowances to be allocated in total as well as on the allocation regulations themselves. After numerous meetings, these negotiations failed to produce a result on 29 January 2004. Following this, the group did not meet up again.
- At the end of 2003, a complete draft for the NAP was drawn up by BMU in cooperation with different research institutes.
- Drafts for the legal groundwork of the emissions trading system were likewise drawn up by BMU. The Greenhouse Gas Emissions Trading Act (Treibhausgas-Emissionshandelsgesetz – TEHG) was adopted as basic law by the German Cabinet on 17 December 2003, so that the parliamentary procedure could be instituted in the German Federal Parliament (Bundestag) and the Federal Council (Bundesrat).
- In the AGE, intensive stakeholder consultations took place. These addressed the different possible designs of the emissions trading system and their prospective or conjectured effects on the individual plant operators and branches of trade.
- As early as the summer of 2003, the emissions data of certain companies were being monitored on a voluntary basis.

A big clash occurred on 29 January 2004 after BMU had submitted the first draft for a NAP (BMU 2004) as a hand-out within the 'high level group'. This draft was submitted in a form that had not been voted upon within the German government. The central points of contention were the predicted total number of allowances intended for the allocation and the type of treatment reserved to the new entrants.

In February and March 2004, long-winded and extremely difficult negotiations ensued under very great time pressure between the ministries, the Federal Chancellery and the associations, as well as a number of companies from German industry who are politically very influential. Announcements of the development of NAPs in other EU

Member States were fundamental in this process; above all, the developments in the United Kingdom and the Netherlands were influential.

Summit negotiations took place on a manager level between the Federal Chancellery, BMU and BMWA; they were extremely complicated and hugely influenced by agents of the business lobby. A 'Ministers' compromise' was reached on the 30 March 2004, enabling the Cabinet to finalise the NAP for Germany on the 31 March 2004 (BReg 2004a). This was then submitted to the European Commission on the very same day.

Since it had already been decided at a relatively early stage that the allocation would be carried out on the basis of a federal act, not only did draft legislation have to be drawn up within a very short time but it also had to be converted into a comprehensive parliamentary procedure distinguished by controversial opinions. On 21 April 2004 the German Cabinet passed the draft of an allocation act (BReg 2004b), which was considered in the German Federal Parliament in April and May 2004. In comparison to the allocation plan submitted to the European Commission, the Cabinet ultimately decided to effect smaller modifications to the allocation rules and criteria, even though significant changes to the allocation rules had been undertaken in the course of parliamentary consultations. Following the failure of mediatory proceedings between the German Federal Parliament and the Federal Council, the German Federal Allocation Act for the 2005–2007 period (Zuteilungsgesetz 2007 – ZuG 2007) was finally passed on 9 July 2004 and came into force on 31 August 2004.

On 7 July 2004, the EC approved the German NAP for the first trading period 2005–2007, which had been submitted to them on 31 March 2004, under a few conditions.

Last but not least, a new institution was created, the German Emissions Trading Authority at the German Federal Environmental Agency (Deutsche Emissionshandelsstelle im Umweltbundesamt – DEHSt), for the implementation of the emissions trading system. This institution has more than 100 employees at present.

2 The macro decision concerning the aggregate total

The initial determination of the aggregate total number of allowances for Germany involved the following problem: data on the CO_2 emissions for the installations covered by the EU ETS in Germany were

Table 4.1 Greenhouse gas emission trends (Mt CO_2e), the Kyoto target and different projections in the context of Germany

	Kyoto Protocol base year	Average 2000–2002		Projection PS III		Projection RWI	
		Actual	Temperature adjusted	2005–2007	2008–2012	2005–2007	2008–2012
Power production	354	316	368	359	353	345	344
Other energy industries	85	52		—	—	—	—
Other industries	197	138	139	136	135	152	154
Including: industrial processes	28	25	25				
Total ETS sectors (energy and industry)	636	505	507	495	488	496	498
Commercial	91	60	64	58	58		
Residential	129	122	132	128	130	181	178
Transportation	159	175	175	180	184	178	173
Total Non-ETS sectors (other sectors)	378	358	371	366	372	359	351
Total CO_2	1,014	863	878	861	860	856	849
Other greenhouse gases	204	127	127	—	116	120	118
Total greenhouse gases	1,218	990	1,005		976	975	967
Burden sharing target					962		962
CO_2 target					846		845

either not available or were only available in an incomplete state and, above all, were only available to the law enforcement authorities of the sixteen German Federal States. During the continuing legislative proceedings and the uncertainties about the interpretation of the EU ETS guidelines, an additional question arose as to which plants from industry and the energy industry would be covered by the EU ETS.

Initial estimations in 2003 assumed that between 4,000 and 6,000 plants in Germany could be subject to emissions trading. The sum of CO_2 emissions from industry that would not be registered by the EU ETS was roughly estimated at 36 Mt CO_2 at this time (Öko-Institut *et al.* 2003).

Against the backdrop of these uncertainties, the total number of allowances for the 2005–2007 period was determined in two stages:

(1) The analyses and discussions about the emission target took place on the basis of national greenhouse gas inventories and corresponding projections. The fact that a share of industry would not be included in the emissions trading system on the one hand, and that a number of plants from other sectors (commerce/trade/services) would fall under the scope of the EU ETS on the other hand, was not taken into account at first.

(2) The actual number of allowances to be allocated was determined by converting the top-down results attained on the basis of inventory data into the bottom-up plant data which had been ascertained in the meantime (on a voluntary basis).

The emissions reduction goal, which Germany accepted within the framework of the European Burden Sharing Agreement (BSA) for the Kyoto Protocol, is a fixed central guideline for determining the cap. A medium emissions level of 846 Mt CO_2 equivalents for the 2008–2012 period results from the 21% reduction target (base year 1990/1995).

Table 4.1 shows the trend of greenhouse gas emissions for Germany as well as two different projections. This overview reveals that the fundamental contributions to the emissions reduction achieved up to 2002 came principally from industry and the energy industry (chiefly attributable to the transformation and modernisation processes in East Germany). By contrast, a massive growth in emissions was recorded in the 1990s, above all in the transportation sector. This can again essentially be explained by the adjustment processes in East Germany. The overview also demonstrates the considerable influence of climate effects on emission levels, above all in the region of private households.

The unusually warm years in the period 2000–2002 caused CO_2 emissions from private households to reach about 10 Mt CO_2 which was below the level to be expected in normal climatic conditions.

The question of emissions projections played a prominent role in the debate on the cap for the EU ETS as well the emissions targets for the sectors not included in emissions trading. Whereas BMU mainly referred to an emission projection ('Policy Scenarios III' study) commissioned by the Federal Environmental Agency (DIW *et al.* 2004), industry as well as parts of the Administration (chiefly BMWA and the Chancellor's Office) drew upon a projection in their arguments which had been drawn up and commissioned by BDI (RWI 2003). Table 4.1 displays the results of both of these projections in a comparative fashion. Central differences are revealed with regard to the future development of emissions from industry on the one hand, and commercial, residential and transportation sectors on the other hand. The business-as-usual (BAU) projection of the industry assumes only a minimal reduction in emissions from industry and the energy industry up to 2012 and anticipates considerable emissions reductions in the sectors not covered by emissions trading, above all in the transportation sector. By contrast, the BAU projection of the Policy Scenarios III Study assumes a pattern of development diametrically opposed to this.

This example shows very clearly how problematic the determination and evaluation of caps from BAU projections can ultimately prove to be.

The analyses and discussions about the total cap for Germany in the first trading period (2005–2007) were initially buttressed by two fundamental premises:

(1) The cap for 2005–2007 should be orientated to approximately regular annual intervals to reach the Kyoto target of 21%. With unfailing consistency, the analyses and discussions kept the targets for the 2008–2012 Kyoto period in view, along with the emissions reduction targets for the 2005–2007 period.

(2) The Kyoto target should be especially geared to domestic action rather than to the emissions credits from the Clean Development Mechanism (CDM) and Joint Implementation (JI) or the use of International Emissions Trading (IET) within the framework of the Kyoto Protocol.

Against the background of these basic assumptions, the future contributions of different sectors to reduction levels constituted a key question in determining the cap for the plants included in emissions trading. The following basic approaches were discussed within the scientific discourse on the preparation of the NAP as well as in political discussions:

- The cap is determined consistently from an economic viewpoint (cost efficiency approach). Under the assumption that the most efficient emission reduction strategy in economic terms arises from equalising the marginal abatement costs for the trading sectors and the non-trading sectors, the cap for the emissions trading system would have to be determined from an economic optimisation calculation. Calculations of this kind were present in a comprehensive study on Germany (DIW *et al.* 2004).
- In two agreements from 2000 and 2001/2004, German industry made a commitment to voluntary emissions reductions. If certain obscurities remained with regard to the sector allocation along with the transfer to the plants included in emissions trading, caps for the emissions trading system could be determined from the voluntary agreements (voluntary agreements approaches).
- The proportional approach was discussed as a further alternative, according to which the fulfilment of the remaining gap to the Kyoto target should be proportionally distributed across all sectors.
- Alongside these systematic approaches, a rather intuitive approach also played a role, whereby the emissions levels of sectors not covered by emissions trading are stabilised without taking a temperature adjustment into account and the remaining emissions reductions are brought about by the plants included in emissions trading (stabilisation approach).

Table 4.2 elucidates the quantitative results of these calculations. The cost efficiency approach leads to an emission reduction for the sectors covered by emissions trading, which for the period 2005–2007 is about 36 Mt CO_2 below the average emissions during the 2000–2002 base period. For the period 2008–2012, the model even produced an emission reduction of 86 Mt CO_2 compared with the base period levels. The prominent role of input data, as well as the specific problems of the optimisation approach of the energy system model which was deployed, is above all apparent in the target value for the transportation sector. Here, the measures to reduce emissions are not cost-efficient from the perspective of the macroeconomic optimisation approach.

However, determining the cap from voluntary agreements led to emission reduction targets that were disproportionately high for the

Table 4.2 *Approaches to ensure Germany's compliance with the emission ceilings (M CO$_2$e) of the Kyoto Protocol*

Historical data	ET sectors	Commercial	Residential	Transportation
Kyoto Protocol base period	636	91	129	159
2000–2002 average	505	60	122	175
2000–2002 adjusted average	507	64	132	175

Discussed allocation approaches	2005–2007				2008–2012			
	ET sectors	Commercial	Residential	Transportation	ET sectors	Commercial	Residential	Transportation
Voluntary agreements Var1	488	61	124	178	479	62	125	180
Voluntary agreements Var2	488	61	124	178	473	63	127	183
Proportional approach	498	60	120	173	495	59	120	172
Non-ET stabilisation approach	493	60	122	175	489	60	122	175
Cost efficiency approach	470	64	130	187	419	72	146	209

NAP approaches	2005–2007				2008–2012			
	ET sectors	Commercial	Residential	Transportation	ET sectors	Commercial	Residential	Transportation
BMU NAP draft of January 2004	488	58	127	178	480	58	127	181
Final NAP March 2004	503	———	356	———	495	———	351	———
Allocation Act July 2004	503	58	298	———	495	58	291	———

sectors included in emissions trading. These targets reached a considerable magnitude with 17 Mt CO_2 for the period 2005–2007 as well 26 to 32 Mt CO_2 compared with the base period levels.

The scope of the necessary emissions reductions for the sectors covered by emissions trading was the lowest with the proportional approach. Here, emissions reductions of 7 Mt CO_2 in the period 2005–2007 and 10 Mt CO_2 in the period 2008–2012 would have been necessary in comparison to the average emissions of the base period.

Somewhat higher reduction guidelines resulted from the stabilisation approach. Here, the cap for the emissions trading segment would have had to amount to 12 or 16 Mt CO_2.

The political discussion quickly arrived at a clear negotiating solution following the presentation of the NAP draft by the BMU on 29 January 2004. Whereas the BMU's NAP draft was conspicuously orientated to the scope of the voluntary agreements, the cap for emissions trading presented in the compromise solution of March 2004 was clearly above the approaches described above, with 503 Mt CO_2 for the period 2005–2007 and 495 Mt CO_2 for the period 2008–2012.

Whilst the German government was unable to reach an agreement in the cabinet resolution on an additional differentiation of the emissions targets for the sectors not covered by the ETS, and only formulated an aggregate target value, a differentiation according to the operations of the different ministries was negotiated in parliamentary proceedings. This enabled an emission reduction target to be established, above all for the measures that fell under the responsibilities of the Federal Ministry for Transport, Building and Housing (Bundesministerium für Verkehr, Bau- und Wohnungswesen – BMVBW).

In the discussions about the cap for the ETS, two political fields of action played a by no means insignificant role:

(1) By virtue of the Nuclear Energy Act (Atomgesetz – AtG) of 2002, a restriction of the operational life of German nuclear power stations was undertaken. Before the end of the period 2005–2007, two nuclear power stations, which have an annual electricity production of around 7.4 billion kWh, are to be shut down. This means that a further fall in electricity production of 94 billion kWh will have to be replaced in the period 2008–2012. The resulting additional CO_2 emissions would substantially influence the allocation negotiations of the power industry.

(2) Electricity production from renewable energies acquires significant economic incentives in Germany as a result of the Renewable Energy Act (Erneuerbare Energien Gesetz – EEG). An expansion of the regenerative electricity production of around 23.5 billion kWh was expected for the period 2005–2007; up to 2010, a further increase of 32.4 billion kWh was anticipated (VDN 2005). The reduction in emissions resulting from incentivised electricity production from renewable energies would thereby benefit in full from the electricity-generating plants which are participating in ETS.

Both changes partly compensate one another and were also correspondingly taken into account in the above-mentioned projections. However, they show very clearly that the target values and the reduction costs for the ETS segment can be significantly influenced by the indirect effects of other energy policy measures.

In summary, the determination of the caps for the first two phases of the EU ETS could be characterised as a mere negotiating solution. Amongst the systematic determinations of the emissions trading cap, only the rather intuitive proportional approach comes closest to the final result. In conclusion, it is striking in this context that the cap for the period 2005–2007 lies above the high emissions level of the BAU projection supported by German industry and the cap for the period 2008–2012 is only slightly below the BAU projection!

In view of this vesting of German industry with emission allowances that are comparatively generous, considerable political efforts to reduce emissions are required in the sectors not covered by ETS if the target of the Kyoto Protocol for Germany is to be reached by use of domestic action. Up to 2012, energy efficiency and extensive measures are also necessary, chiefly with regard to the energy consumption of buildings as well as emissions from the transportation sector.

By taking into consideration the total emissions determined bottom-up of the plants covered by emissions trading which amount to 501 Mt CO_2 for the 2000–2002 base period, a total annual number of 499 million European Union Allowances (EUA) was reached for Germany for the pilot phase of the EU ETS, from the above-mentioned top-down emissions target of 503 Mt CO_2 for the period 2005–2007.

3 The micro decision concerning distribution to installations

3.1 Basic allocation provisions for existing installations

In the discussion on the allocation rules for the individual installations, a multitude of variants were developed, chiefly in 2003, and were considered in detail within the framework of a very wide-ranging discussion.

Allocation on the basis of historical emissions was intensively discussed. Here, it was requested that above all the companies, branches and political powers from East Germany focus on the very early base periods, since considerable emissions reductions had been achieved in East Germany, particularly during the 1990s. By contrast, an allocation based on the most recent base periods was requested by the companies and branches that were not able to refer to emissions reductions in good time, primarily so as not to make the ratio between the free allocation and historic emissions (the so-called compliance factor) too ambitious.

Chiefly in the early phase of the discussion, allocation according to the benchmark principle played a considerable role in Germany. Interestingly, this methodical approach was put forward by the BDI itself. The models discussed here varied in terms of their form. On the one hand, benchmarking models were discussed on the basis of average historical CO_2 emissions according to product unity (average benchmarks). In the course of the discussion, further differentiation according to different criteria led to a multitude of benchmarks for the various classes of installations and products. For the area of electricity production alone, for example, twenty-six different benchmarks for existing installations were requested. Alongside allocation on the basis of benchmarks to be calculated from historical data, allocation on the basis of benchmarks according to the best available technology (BAT) only played a very secondary role. This can be explained not only against the background of the modernisation processes of various types taking place in the different sectors, but also by the different development of East and West Germany and the diverse situation of interests ensuing from this. Benchmarks, however, played a larger role in the context of the early action issue where benchmarks ascertained from historical data at an early point in time were planned to be coupled to the production data dating from a very recent period.

The auctioning of allowances for existing installations was repeatedly brought up in academic fields but was no longer a real option as a result of the government's premature decision before the formulation of a common standpoint regarding rates in December 2002, according to which all existing and new installations would be equipped with allowances ('grandfathering') at no cost up to 2012.

The possible decisions between the different variants of allocation on the basis of historical emissions or on the basis of benchmarks for the existing installations were considerably reduced by a number of pragmatic decisions made in the course of time.

- As a result of the decisions of the German Federal Government and the Federal Environmental Ministry Conference, in March and May 2003 respectively, it was implicitly established that only the data for the years 2000 until 2002, the base period for the first National Allocation Plan, would be collected. Allocation on the basis of a very early base period (e.g. in the 1990s) as well as the determination of benchmarks on the basis of historical emissions from the 1990s were rendered invalid as a result of this decision.
- Given that production data were not collected during the process of voluntary data collection in the autumn of 2003, the implementation of allocation proceedings based on benchmarking was considerably impeded, if not rendered impossible. On the one hand, the determination of benchmarks on the basis of emissions and production data even for the recent period of 2000–2002 was no longer possible. On the other hand, an estimation of the effects of an allocation based on other benchmark concepts (e.g. for BAT) was also no longer feasible for lack of current production data.

Against this background, and in view of the large number of installations affected by emissions trading, allocation on the basis of historical emissions of the reference period 2000–2002 remained the only realistic option for the first trading period 2005–2007.

Alongside the fundamental allocation procedure according to the grandfathering principle, a decision had to be taken as to whether or not the differences for the allocation to individual installations should be made dependent on their sectoral classification. Allocation differentiated by sector – in principle the definition of sector caps – only played a secondary role in the whole discussion process on the allocation methods. In retrospect, two reasons were decisive for this:

(1) The complicated political relationships as well as the considerable problems experienced during data collection led the NAP authors to seek an allocation procedure that was as simple as possible. Discussion of different sector caps would have complicated the process to a much greater extent and would have nourished the redistribution dispute within the German industry which was already raging. Additionally, the question as to whether and to what extent sectors such as for example the electricity industry would be in a position to pass on the opportunity costs of the allowances to consumers was still hotly disputed in 2003 and 2004 in view of the oligopolistic structures of the German electricity market.

(2) On the part of industry, the request of a uniform cap was a fundamental link which was able to hold together the positions of the industry that were, at particular points, very contradictory. The request for different sector caps on the part of the operators would have made a (to some degree co-ordinated) action of industry conclusively impossible and – as already stated – would have further heated up the redistribution dispute. Also there was no interest on the part of the affected industry in requesting a differentiation of the caps, particularly since it became evident in the discussion that a sectoral differentiation between the energy industry and the manufacturing industries was effectively taken into account by the immense significance of the process-related emissions. By contrast, the request for a universal allocation by demand constituted the common denominator – a request whose implementation would ultimately have reduced emissions trading to an absurdity.

The serious conflicts of interest in the process of the NAP preparation, as well as the multitude of practical restrictions (data availability, pressure of time), led to a simple procedure for the basic model of allocation for the existing installations; it proved to be the smallest common denominator. Regarding the allocation of allowances on the basis of historical emissions for the period 2000–2002 without further sectoral differentiation, a very simple base model was created. Here, allocation arose out of the historical emissions for this period multiplied by a so-called compliance factor. This simple model was expanded, however, by a multitude of special rules which had significant consequences for the compliance factor and considerably eroded the uniformity and transparency of the allocation model.

3.2 Special allocation provisions for existing installations

Already during the negotiation of the EU ETS guidelines, the question of *early action* had a particular status from Germany's point of view. The fundamental proponents of a request for special early action provisions were the actors from East Germany; a prominent role was above all played by the electricity provider dominating the East German market, Vattenfall Europe, and its lignite power stations, which have a large capacity and a significant share in the total German CO_2 emissions.

Following very intensive tests, discussions and political disputes about models that were in part extremely complex, a comparatively simple but also undifferentiated early action rule was implemented. The allocation for those emission quantities of the base period 2000–2002, for which the early action facts are taken into account, takes place for a time period of twelve years following the conclusion of the measure with a compliance factor of one. The allocation precisely corresponds to the emissions in the base period of 2000–2002. Accordingly, the following emissions are acknowledged:

- Emissions of plants that were put into operation from 1 January 1994 to 31 December 2002.
- Emissions of plants where the specific emissions achieve an ambitious fixed minimum value of emissions reduction based on the comparison between a reference period (three successive years in the period 1991–2000) and the standard base period by virtue of measures taken after 1 January 1994. This minimum value increases from 7% (with the completion of modernisation measures up to 31 December 1994) in steps, each of which constitutes a percentage point, up to 15% (with the completion of modernisation measures up to 31 December 2002).

As a result, the total volume of emissions covered by the special allocation provision for early action constituted about 22% of the total allocated allowances with approximately 111 Mt CO_2. Three-quarters of the early action allocations went to installations in East Germany (Figure 4.1).

A second central special rule affects the CO_2 *emissions from industrial processes*. From the start, the idea that for those CO_2 emissions not resulting from incineration processes (e.g. in cement, limestone and glass production), an allocation in the region of the corresponding historical emissions should ensue was relatively disputed in the debate on

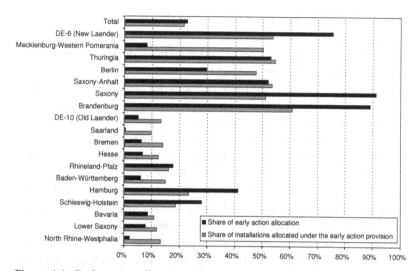

Figure 4.1. Early action allocation: the German East–West divide.

the NAP. The volume of emissions covered by this period was about 21 Mt CO_2. The allocation for these process emissions was planned to be based on historic emissions and a compliance factor of one. In the course of discussions on the NAP, recognition of the CO_2 emissions from blast furnaces and steel mills was requested on the part of the iron and steel industry. A complicated situation developed here. BMU and their consultants took the view that only the CO_2 already contained in the blast furnace gas could be taken into account for the special provision for process emissions. For the energy production plants, in which blast furnace gas is used, measures to increase energy efficiency are certainly possible. Against this background, a special provision for the deployment of blast furnace gas was seen as not justifiable. In the political negotiation process, this position could not, however, be carried through. The iron and steel industry, together with BMWA and the Chancellor's Office, succeeded in finding a solution, according to which the total CO_2 emissions from the deployment of almost all blast furnace gas were defined as process-related CO_2 emissions, whereby an allocation according to historical emissions as well as to a compliance factor of one was adopted. Under this agreement, the volume of allocations falling under the special provision for CO_2 emissions from industrial processes was altogether increased by 40 Mt CO_2. In the follow-up to the dispute with the iron and steel industry, the operators of the German mineral oil refineries were also able to achieve recognition of further emissions as process emissions. Altogether an emissions volume of 71 Mt CO_2 fell under the special provision for process emissions, by which the largest share of allowances for the production trade was allocated with the compliance factor of one, and a sectoral division between the energy industry and the producing trade was effectively created (implicit sectoral differentiation).

In the sum, a volume of 182 Mt CO_2 fell under special regulations, for which allocation only ensued from historical CO_2 emissions, and a compliance factor orientated towards the reduction of CO_2 was applied. The total emissions reduction established in the NAP budget had to be produced by installations which represented only about 64% of the CO_2 emissions covered by emissions trading.

Traditionally, the promotion of *combined heat and power production* (CHP) occupies an important position in the German climate protection policy. Since 1998, a number of economic incentives have been initiated for CHP, by virtue of which the share of electricity production from CHP in German electricity volume is supposed to be expanded. Additionally, special regulations apply within the framework of the ecological tax reform (exemption from the natural gas tax, exemption of small CHP from the electricity tax) as well as special support Acts (CHP Support Act). Since a considerable share of CHP in Germany is deployed for public district heating, the implementation of the EU ETS could produce disadvantages for CHP.

In the allocation procedure, a special allocation for the electricity production from CHP plants was catered for in order to compensate incentives to reduce the electricity production from CHP. The construction of this regulation comprises a special allocation for the historical production of electricity by CHP, which is subjected to an *ex-post* adjustment. If electricity production from CHP is reduced in the course of the period, the return of the special allocation for CHP is requested in the ratio of five times the special allocation for a reduction of 1 million kWh electricity production from CHP. After the special allocation for the electricity production from CHP had borne 35 EUA per million kilowatt-hours in the first draft of the NAP, this special allocation was reduced to 27 EUA as a result of political negotiations, and also in order not to make the compliance factor too ambitious. Annually, 2.02 million EUA were additionally allocated to the CHP plants.

A politically explosive question emerged for the German NAP concerning the *phase-out policy for power generation from nuclear energy.* With regard to determining the remaining operational life of the German nuclear power plants (NPP), the Stade (640 MW) and Obrigheim (340 MW) nuclear power stations are relevant for the 2005–2007 period of the EU ETS. The NPP Stade was shut down on the 14 November 2003 (that is, following the end of the 2000–2002 base period); this was also the case for the NPP Obrigheim on 11 May 2005 (that is, shortly after the start of the pilot phase of the emissions trading system).

The short-term implementation of the corresponding – very limited – electricity production capacities could take the following forms:

(1) A higher capacity utilisation of existing power stations can take place, since only a few larger electricity production plants are to be put into operation for the period 2005–2007.

(2) A share of the lost electricity production can be replaced by newly built electricity production plants (in the area of renewable energies, CHP or other power stations).

(3) It is by no means guaranteed that the volume of electricity of the respective power station operators will really be kept constant and the lost quantities of electricity will not be compensated by imports or income from other power station operators.

All three aspects are and ultimately remain speculative, even when a closer analysis of the affected operators E.ON (NPP Stade) and EnBW (NPP Obrigheim) as well as the burden characteristic of the two power stations provides a certain probability for a variant, according to which at least a portion of the lost capacities of the two power stations is compensated by a higher utilisation capacity of the existing power stations of these companies. This would lead to increased CO_2 emissions from the affected fossil power stations.

Against this background, the operators of the nuclear power stations which have been closed down or are to be closed down requested compensation of altogether 4.6 million EUA for the first trading period (2005–2007) of the EU ETS. In the draft of the NAP from the 29 January 2004, no special allocation for the operators of the closed-down nuclear power plants was stipulated. This fact evolved into one of the central disputes about the German NAP. The solution finally established in the 'Ministers' compromise' of 30 March 2004 represents a mere negotiating solution. For each year in the period 2005–2007, altogether 1.5 million EUA are to be allocated to the two affected nuclear power stations, whereby this special allocation is supposed to be explicitly limited to the 2005–2007 pilot phase. For the years up to 2008, there will be no form of compensation for the phasing out of the nuclear energy in Germany within the framework of the EU ETS. According to the previous notions of the new entrants reserve, the construction of fossil-heated power stations required to replace nuclear power stations is to be provided with emissions certificates at no cost in the second trading period.

3.3 Allocation provisions for new installations and plant closure

There are three general rules for the allocation for new entrants, mostly resulting from the (preliminary) political discussion, but also from the affected companies themselves:

(1) New plants should be allocated free allowances up to 2012, like the older plants.
(2) In order to avoid the so-called 'shut-down premium' (particularly all in the political area), an explicit and very effective shut-down regulation should be created.
(3) The rules for new plants should not hamper the construction of new coal power plants.

The possibility of not providing new plants with free allowances has been discussed widely in the academic sector. Some stakeholders occasionally considered this possibility and some representatives from the industry (e.g. from some presently operating electric utilities) even demanded it in an early phase of the debate. However, this was no longer a realistic option due to the preliminary political decisions. The fact that in the period 2005–2007, in the context of several provisions stemming from environmental policy, a considerable number of new power plants were to be put into operation, which were not going to be built by the four companies that represent the oligopoly on the German electricity market, also played an important role. These power plant projects were very important from a political point of view as well as with regard to their positive impact on the environment and on competition in the German electricity market. These power plant projects were a political priority, received special incentives (e.g. tax breaks) and were not to be endangered by restrictive allocation provisions for newly constructed plants.

According to the BMU's draft NAP of 29 January 2004 (BMU 2004), the operator of a plant was obliged to report its closure, if any. In this case no further emission allowances would be issued for the following year. A plant closure also meant that within one year the CO_2 emission reported by the plant would have to be 10% lower than the average annual emissions in the base period. In this case, the operator would have to prove that it was only a temporary closure (e.g. a long-term repair for reasons of maintenance or interruptions).

A transfer provision was initially drafted for new plants which had been built to replace old plants. If the operators complied with the requirement of the three-month (without any further examination) to the two-year period (if the respective verifications could be provided) from the closure of the old plant and the opening of the new plant, the allocation to the old plants would be transferred to the new plants without restrictions, provided that the new plants produced comparable products. In addition to the transfer rule – which was considered as a standard – there should be a rule that provides additional new plants (as well as plant expansions) a free allocation on the basis of the planned capacity use and an emission benchmark. The emission benchmark should be orientated towards BAT procedure. Above all, it should not be differentiated by fuel or technologies. As far as electricity production is concerned, a benchmark orientated towards a modern natural gas-fired combined cycle gas turbine (CCGT) power plant (365 g CO_2/kWh) was scheduled. The allocation calculated in this way should be subject to an *ex-post* adjustment in the course of the following years. Thus, a special procedure was scheduled for CHP plants. The allocation should be determined with the help of the benchmark for electricity generation as well as the benchmark for heat generation.

Concerning the terms of the transfer rule and the provisions for additional new plants, the procedure had been specified until the end of the second trading period. The transfer provision should, for plants put into operation within the period 2005–2007, also be applied to the trading period of 2008–2012. The benchmark version for additional new plants should be valid until the end of 2012, without applying a compliance factor.

The benchmarks for additional new entrants and the 'guarantee periods' for the transfer provision, together with the regulation for additional new entrants, became the central factors of the dispute concerning the NAP. Above all, most representatives of the industry considered the fuel-specific new entrant benchmarks as a *conditia sine qua non*. In the political process the allocation benchmarks for additional new plants were often seen as a prohibition to build power plants other than the natural-gas-fired CCGT ones. Key actors also discredited the transfer rule, which was initially unlimited, as an investment support for natural-gas-fired power plants that was no longer justifiable.

The NAP also shows significant changes in the version of the cabinet resolution (BReg 2004c). Here, the transfer rule is limited to a period of four years. Thereafter, the allocation is planned for another fourteen years on the basis of the annual average CO_2 emissions in the reference period valid for the respective trading period, without considering a compliance factor.

The allocation provisions for additional new plants (and plant expansions) were changed in two essential ways, compared to the draft BMU plan of 29 January 2004 (BMU 2004). First of all, a fuel differentiation of the emission benchmarks was introduced. Then a maximum value of 750 g CO_2/kWh was fixed as an electricity benchmark. If the affected power plants showed an emission value lower than 750 g CO_2/kWh due to the characteristics of the fuel, the allocation should not be higher than the real emissions. However, a benchmark of at least 365 g CO_2/kWh should be applied. This means that for the power plants that function with hard coal or with less emission-intensive fuels, a fuel differentiation would be introduced.

For the additional new entrants, the allocation based on production data and the new entrants benchmark should be applied for a period of fourteen years, without considering a compliance factor. The allocation should still be subject to an *ex-post* adjustment to the lower as well as to the higher levels.

The plant rule involving the compulsory closure notification and the shutdown supposition should the emission level be lower than 10% of the annual average emissions within the base period has been replaced in the political process by extensive *ex-post* adjustments. The issue of emission allowances is reduced, if the annual emissions of a plant due to production cuts are lower than the threshold value of 60% of the annual average emissions within the base period. From this point, the operators must return the surplus allowances to the authorities.

With the introduction of a free allocation for new plants, it was necessary to create a set-aside for new entrants. An analysis of the foreseeable new construction projects in the energy sector and the manufacturing industries led the first NAP of 29 January 2004 to establish a reserve estimation of approximately 5 million EUA per year. As a result of the negotiation process, the size of the new entrants reserve was reduced to a value of 3 million EUA, above all to create a compliance factor that was seen as politically acceptable. Should the allowances in the new entrants reserve be insufficient, the final NAP established that an authority independent of the Federal Government would buy the missing amount of allowances on the market and transfer them to the Federal Government thereby enabling a free allocation for all new plants. In the following second trading period, the respective authority should receive the respective amount of allowances from the current reserve and should be able to sell it for its own refinancing.

Altogether, the final regulation in the field of new plants/plant closures shows a clear erosion of the incentives of the EU ETS. In particular, considering the extensive application of *ex-post* adjustments, given that the benchmark system is increasingly differentiated according to technologies and fuels, the EU ETS only exercises a limited influence on the implementation of the price signals of CO_2 emissions for plant operation and new investments in central sectors in Germany. The provision for the transfer of the allowances of closed plants to new plants is an important factor of the incentive effects of the ETS in this context. At the same time, the associated advantages for incumbents led to fierce disputes as well as court actions against the transfer provision.

3.4 Flexibility provisions and allocation adjustment

In the context of multifaceted special demands of some sectors and companies, a series of special allocations and flexibilisation options was integrated into the regulations of the first German NAP.

First of all, hardship rules were created in order to find answers to specific cases (e.g. in the glass industry) for a series of special factual situations. If the special and very specific conditions are met, the respective plants can benefit from an allocation on the basis of forecasted instead of historical emissions. However, the overall sum of the allowances, including those to be allocated as a result of these specific hardship rules was limited to 3 million EUA for the overall period of 2005–2007. If this sum is exceeded, the special allocations in the context of the special rules concerning cases of hardship would have to be correspondingly reduced.

Table 4.3 Results of the German allocation process compared to planned data

	Average allocation (mln EUA per year)		Remarks
	Allocated	Planned in NAP	
Overall allocation	499		Base period emissions 501 Mt CO_2
New entrants reserve	3		Reduced in NAP negotiations
Reserve for hardship clause	1		
Allocation to existing installations	495		
CHP	2.0	1.5	Reduced in NAP negotiations
Nuclear phase-out	1.5	1.5	Introduced in NAP negotiations
Special allocation	3.5	3.0	
Allocation to existing installations w/o special allocation	492	492	
Process emissions	72	69	Increased in NAP negotiations
Early action	111	114	
Allocation with compliance factor 1	183	183	
Allocation with compliance factor <1	309	309	
Options provision[a]	77		Introduced after NAP approval
Cap overrun	14		Mainly resulting from options provision
Memo items			
Cap versus historic emissions[b]	0.996		As laid down in Allocation Act
Uniform compliance factor (CF)[c]	0.971		As applied in allocation procedure
Cap adjustment factor (CAF)[d]	0.954		
Maximum effective compliance factor[e]	0.926		

[a] Allocation based on new entrants provision instead of historic emissions.
[b] Cap of 499 mln EUA vs. base period emissions of 501 Mt CO_2.
[c] Allocation with CF <1 vs. non-privileged emissions (501 − 183 Mt CO_2).
[d] Cap overrun vs. non-privileged emissions.
[e] Resulting from CF × CAF.

A general rule for cases of hardship was added to the special rule. According to this rule, an allocation based on predicted emissions was possible in the case of unreasonable economic charges. The necessity of this general rule concerning cases of hardship arose above all for reasons of constitutional law.

The flexibilisation principle introduced just before the decision of the German Federal Parliament in the context of the latest parliamentary negotiations carried serious consequences for the allocation procedure. Following this principle, the operators of existing plants can choose whether the allocation is assigned according to the general rules on the basis of historical emissions or alternatively according to the rule for additional new entrants. If the capacities of certain plants were only minimally used in the base period of 2000–2002, or if the rules for new plants were advantageous for other reasons (e.g. thanks to the application of a double benchmark for CHP plants), the operators had a tremendous incentive for maximising the number of allocated allowances following the allocation rules for additional new entrants. However, an additional *ex-post* adjustment is foreseen to adjust the allocation in case the actual production is less than the planned data submitted in the allocation application procedure.

In the context of the diverse special rules in the German NAP, an adjustment of the allocations to installations was necessary, as a means to limit the overall number of allowances allocated to the plants to the fixed cap.

The German NAP therefore uses the so-called 'proportionate allocation cuts' procedure. If the total sum of the allowances allocated to individual installations (except cases of hardship) is higher than 495 (499 million EUA minus new entrants reserve minus allowances for hardship clauses), the allocations for all the plants are correspondingly reduced, if the allocation does not have the compliance factor one.

Table 4.3 gives an overview of the role of special provisions and the resulting allocation. The allocation for plants which cannot receive further special allocations has changed as follows:

- On the basis of the defined cap, there is an allocation which is 0.4% lower than the historical emissions of the period 2000–2002 (contribution to reduction by climate protection policy).
- The special provision for process emissions reduces the allocation by a further 0.39 percentage points.

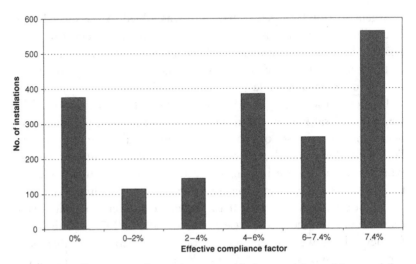

Figure 4.2. Effective compliance factor considering special provisions and the cap adjustment according to §4(4) ZuG 2007.

- The special provision for early action reduces the allocation by a further 0.64 percentage points.
- The special provision for CHP emissions reduces the allocation by a further 0.3 percentage points.
- The special provision for nuclear phase-out compensation reduces the allocation by a further 0.3 percentage points.
- The new entrants reserve as well as the reserve for specific hardship clauses reduces the allocation by a further 0.8 percentage points.
- As a result, the intentionally planned uniform compliance factor was 0.971.[2]
- However, the adjustment of the allocation to a total of 495 EUA annually leads to a further reduction of the allocation by the factor of 0.954.

Thus, there is a plant which cannot be entitled to further special provisions or special allocations, and an allocation which amounts to 92.60% of the historical emissions in the base period 2000–2002. The multitude of special provisions, special allocations and flexibility options causes above all immense distributional effects compared

---

[2] The compliance factor in the original BMU proposal from January 2004 amounted to 0.927. The difference from the uniform compliance factor of 0.971 is mainly a result of an inflation of the cap, the reduction of the new entrants set-aside and the special allocation to CHP.

with a very limited overall CO_2 reduction effect. Nevertheless, many installations gained from the variety of special provisions (Figure 4.2).

4 Issues of coordination and harmonisation

4.1 *European coordination and harmonisation*

Germany is one of the states in which the work for the establishment of a National Allocation Plan was started a long time ago (the planning was started and the first steps for its implementation were taken just after the Common Agreement of the Council in December 2002). From the middle of October 2003 onwards, negotiations were taking place between the Federal Government and high-ranking representatives from industry. At the beginning of January 2004, the first draft of the NAP had to a large extent been finalised internally.

With this chronology in mind, the Guidance Paper of the European Commission of 7 January 2004 could not have a significant impact on the essential structures of the German NAP. As a result, the European Commission was first able to play a manifestly more important part in the approval procedure, in which the German *ex-post* adjustments were particularly criticised. Ultimately, however, these matters of authorisation did not have an immediate impact on the allocation procedure of autumn 2004. The Federal Government, however, did in fact take legal steps against the requirements of the Commission concerning the *ex-post* adjustments.

As a result, the formal coordination within the European Union beyond the EU ETS Directive did not essentially influence the structure and the concrete rules of the German NAP. At the same time, the discussions in the other Member States played a role that should not be underestimated. Above all, information on the continuing discussions concerning the NAPs of the United Kingdom and the Netherlands was substantially important at a certain point. However, in spite of the very intensive debates at the various levels, the central points of the discussion in other states, like the allocation and the consideration of competition-related questions in the United Kingdom or the allocation on the basis of benchmarks in the Netherlands, did not find their way into the German NAP. Nevertheless, within the political debate focused on the individual rules of the NAP, different stakeholders always stressed examples from the discussions taking place in the other Member States. Policy by rumours constituted a facet of the political dispute concerning the German NAP which should not be underestimated and was, in fact, undeniably hazardous.

4.2 Coordination of macro and micro decisions

Furthermore, the interaction of the macro decision on the aggregate total and the distribution of allowances to the installations (micro plan) were significant for the development of the German NAP. In the scientific preparation of the NAP, it had always been assumed that it would be possible to take the decisions on the two fields independently of each other. Along this line, it was repeatedly pointed out that changes in the special provisions and special allocations had to lead to a smaller allocation for the other installations (interdependency as a zero-sum game). Accordingly, in the initial negotiations between the Federal Government and the high-ranking representatives of industry between October 2003 and January 2004, there was an attempt to reach an agreement, at first with regard to the cap. The allocation rules would be formulated later.

It became clear, however, that such an approach could only be partially implemented within the political process. Ultimately, the effective compliance factor, i.e. the relation between the allocated allowances and the historical emissions of the base period, played the central role in all of the negotiations. In order to meet the maximum fulfilment factor of 0.97, which constituted the political target, special allocation rules were modified on the one hand (e.g. a reduction of the special allocation for CHP), and the new entrants reserve was reduced on the other hand. Furthermore, the aggregated emission cap for the pilot phase 2005–2007 was increased. Nevertheless, contrary to the stated political objective, the underestimation of the effects of the flexibilisation option finally led, to rise up to a 0.926 of the effective compliance factor for the affected plants.

4.3 Harmonisation of the NAP with national energy and climate policies

The embedding of emissions trading in the policy mix of the energy and environmental policy played an important role in the drawing up of the NAP at various points.

This is directly apparent when taking into account the phase-out of nuclear energy in Germany. The associated effects were considered both when establishing the cap and also when determining the allocation regulations.

The effects of the massive expansion of renewable energy on the development of emissions was in fact taken into consideration in the first draft to the NAP; however, they then became increasingly less significant during the further development of the NAP.

By contrast, the support given by German energy policy to electricity generation from coal played a central role. The rules regarding free allocation for new plants in particular turned out to be very strongly aligned with the question of whether strong incentives are created for replacing coal-fired power stations with plants fired by natural gas.

Also in view of the allocation rules for new plants, the structure of the German electricity market had a strong influence upon decision-making. The strong oligopolistic structures in the field of electricity production and the reinforcement of oligopolistic trends since the opening up of the German electricity market formed a significant background to attempts to create additional entrance barriers for new electricity producers with the allocation rules.

Also the allocation provisions for the CHP are to be primarily understood against the background of national CHP policy. The political concern to promote CHP implicitly created great pressure to pay particular attention to the CHP problem within the context of the EU ETS.

In addition, considerable changes within national climate protection policy came about as a result of the introduction of the EU ETS. The obligations of industry up to 2010, which were entered into within the context of the voluntary agreements for climate protection policy, are to be found again in the emission goals of the German NAP, only to a clearly lesser extent.

5 Issues deserving special mention

The allocation of allowances to the individual installations in the first trading period 2005–2007 was characterised in the main by two fundamental restrictions:

(1) The necessary data for the individual installations were not available or only existed in a very incomplete state.

(2) The allocation regulations had to be drawn up, discussed and passed under great pressure of time.

The actual availability of data for the allocation process was extremely unclear for a long time. For this reason an iterative and very pragmatic approach had to be ultimately pursued; as a result, a considerable degree of uncertainty could not be avoided.

The basis for this approach to data collection was created by the decision of the German government on 28 March 2003 as well as by the resolution of the 60th Federal Environmental Ministry Conference on 15 and 16 May 2003. Since the cabinet resolution established that the data for the 2000–2002 base period should be collected, other alternatives for the delimitation of the reference periods were effectively ruled out at this stage.

The collection of data, that was initially carried out on a voluntary basis, followed a three-layered approach to create a database:

(1) The emission declarations were evaluated in the summer of 2003 by the responsible state authorities in accordance with §27 BImSchG of the Federal Immission Control Act (Bundes-Immissionsschutz-Gesetz) of 2000.

(2) A direct data investigation was carried out by the plant operators. This query encompassed the total time period of 2000–2002 and was effected with a huge deployment of external consultancy experts. It was, to a great extent, completed in December 2003 (when still encumbered with considerable missing information). Following the corresponding evaluations, a (provisional) final result of the data collection for the years 2000–2002 was reached on 11 February 2004.

(3) The final and legally binding third phase of the data collection was carried out in the summer of 2004 within the framework of the application process for free allocation.

In the first half of 2004, the scientific and political discussions had to draw upon data from the second phase of data collection which materialised on a provisional and voluntary basis and did not bear further quality assurance (certifications etc.).

Ultimately, the preparation of the allocation rules for the existing installations had to be so effected that the rules led to calculable results on the basis of the minimum available data.

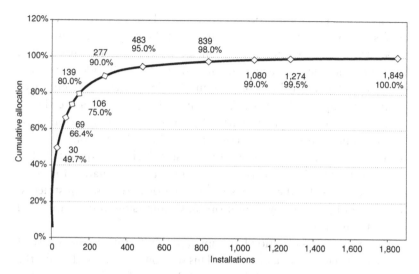

Figure 4.3. Installations covered by the EU ETS in Germany and the structure of the allowances allocation.

However, the large number of installations should be seen against their contribution to the total emissions. As Figure 4.3 indicates, 139 installations received 80% of the total allowances allocated to individual installations.

6 Concluding comments

The introduction of the EU ETS in Germany was characterised by a series of difficult general conditions (time pressure, large number of installations, data, inexperience with market-based instruments etc.). In Germany, a country which is characterised by a strong corporatist policy style and by a diversity of diverging interests, the political system's ability to manage complex political processes of this kind, on the basis of very uncertain and incomplete information, is remarkably limited.

Against this background, considerable problems arise in establishing the allocation rules for the individual installations. Whilst very clear and straightforward coalitions developed with regard to the definition of the cap, a very complex situation resulted with regard to the individual regulations of the NAP. Since the NAP had to be implemented by means of a negotiation process between the Federal Government and representatives of industry, associated with fierce disputes within the Federal Government and, in addition, a complex legislative procedure, the representatives of particular interests had a multitude of starting points to influence the body of rules and regulation. Thus, the notion of creating a simple and rather undifferentiated regulation for the allocation was at least partially thwarted by a variety of special regulations and flexibility options. Altogether fifty-eight different combinations for the various allocation provisions arose, which consequently led to a largely incalculable allocation result and to a hazy situation for many plant operators. Furthermore, some regulations (*ex-post* adjustments) have been objected to by the European Commission in their approval of the German NAP.

The information situation has improved significantly in the course of the NAP development in Germany. This will, on the one hand, make the consequent development of the future NAP considerably easier. On the other hand, the information situation on the part of the representatives of particular interests has also improved considerably.

The theoretical approach that an aggregate emission ceiling should be determined first, followed by the allocation of allowances to the plants, has only been sketchily implemented in reality. Finally, the ratio between the allocation to individual installations and their historical or expected emissions formed the decisive indicator, towards which the decision-making processes were orientated. Thus, the increase of the emissions ceilings in reality perfectly correlated to the parameters with which additional allocation situations were compensated in their allocation effect. Even if the degree of freedom for such adaptations in the Kyoto period were to be reduced, an early fixing of the cap remains a central challenge for future allocation processes.

The reality of a multi-period emissions trading scheme with free allocation for new installations in connection with the (mainly policy-driven) attempt to avoid shut-down premiums for plant closure demonstrates that the allocation rules could have a considerable influence upon the efficiency of the emission trading system. As a result of differentiated new plant allocations, as well as a series of *ex-post* adjustments, an erosion in the CO_2 price signal for operational or investment decisions by the economic entities within the scope of the EU ETS could arise. The development of the allocation provisions for new entrants is one of the key decisions for future NAPs.

In the assessment of the German NAP, the diverse uncertainties for the operators are again and again put forward as a central point of criticism and both clear and straightforward specifications are required, as well as long-term reliability. At the same time, diverse new special provisions are being demanded by many actors. The simplicity, transparency and calculability of the allocation rules, both for the operator and for the political decision-makers, should take on a special role with regard to future NAPs.

The approval of the NAP by the European Commission formed a central and indispensable corrective to the design of the NAP at a national level. Since the decisions of the Commission for all NAPs of the Member States will be available henceforth, the transparency of the criteria and their operationalisation by the Commission will become an extremely important point.

Notwithstanding all the difficulties and imperfections, the introduction of the EU ETS in Germany remains a success story. In a very short period of time, the system for a very large number of plants was implemented, and an allocation system established that offers sufficient prerequisites for improvement and further development. The utilisation of the specified potentials, options and opportunities form the central challenges for the next steps towards stabilisation and establishment of the EU ETS in Germany and within the European Union.

References

BMU (Bundesministerium für Umwelt, Naturschutz und Reaktorsicherheit) 2004. 'Nationaler Allokationsplan (NAP) für die Bundesrepublik 2005-2007', Entwurf vom 29. Januar 2004, 17.30 Uhr. Berlin.

Bode, S., Butzengeiger, S., Lehmkuhl, D., Bode, A., Rumberg, M., and Zisler, S. 2004. 'Der Hamburger CO_2-Wettbewerb', *Umwelt-WirtschaftsForum* 12(4): 59–63.

BReg (Bundesregierung) 2004a. 'Entwurf eines Gesetzes über den Handel mit Berechtigungen zur Emission von Treibhausgasen (Treibhausgas-Emissionshandelsgesetz – TEHG)', Gesetzentwurf der Bundesregierung. Bundesrats-Drucksache 14/04.

BReg (Bundesregierung) 2004b. 'Entwurf eines Gesetzes über den Handel mit Berechtigungen zur Emission von Treibhausgasen (Treibhausgas-Emissionshandelsgesetz – TEHG)', Gegenäußerung der Bundesregierung. Bundestags-Drucksache 15/2540.

BReg (Bundesregierung) 2004c. 'Nationaler Allokationsplan für die Bundesrepublik Deutschland 2005-2007', Berlin, 31 March 2003.

DIW (Deutsches Institut für Wirtschaftsforschung), Öko-Institut and ISI (Fraunhofer-Institut für Systemtechnik und Innovationsforschung) 2003. White Paper: 'National allocation plan (NAP): overall concept, criteria, guidelines and fundamental organisational options'. Berlin, Karlsruhe, 7 July 2003.

DIW (Deutsches Institut für Wirtschaftsforschung), STE (Forschungszentrum Jülich, Programmgruppe Systemforschung und Technologische Entwicklung), ISI (Fraunhofer-Institut für System- und Innovationsforschung) and Öko-Institut 2004. 'Politikszenarien für den Klimaschutz. Langfristszenarien und Handlungsempfehlungen ab 2012 (Politikszenarien III), Final Report for Umweltbundesamt. Berlin, Jülich, Karlsruhe, July 2004.

DIW (Deutsches Institut für Wirtschaftsforschung), Öko-Institut and ISI (Fraunhofer-Institut für System- und Innovationsforschung) 2005. 'Entwicklung eines nationalen Allokationsplans im Rahmen des EU-Emissionshandels', Draft Final Report for Umweltbundesamt. Berlin, Karlsruhe, May 2005.

Ehrhart, K.-M., Schleich, J. and Seifert, S. 2003. 'Strategic aspects of CO_2 emissions trading: theoretical concepts and empirical findings', *Energy and Environment* 14(5): 579–97.

HMULF (Hessisches Ministerium für Umwelt, Landwirtschaft und Forsten) 2003. 'Hessen-Tender Initiative für den Ankauf von CO_2-Emissionsminderungen', Ergebnisbericht Pilot- und Demonstrationsprojekt zur Erprobung von Instrumentarien eines Emissionshandelssystems. Wiesbaden.

Kruska, M., Hahn, M., Klein, M. and Barzantny, K. 2003. '10 Forderungen an den nationalen Allokationsplan', *Energiewirtschaftliche Tagesfragen* 53(7): 478–82.

Matthes, F. Chr. and Ziesing, H.-J. 2003. 'Energiepolitik und Energiewirtschaft vor großen Herausforderungen', *Wochenbericht des DIW Berlin*, 48: 763-9.

Öko-Institut, DIW (Deutsches Institut für Wirtschaftsforschung), and Ecofys 2003. 'Impact of the European emissions trading system on German industry', Final Report for WWF. Berlin, Cologne, September 2003.

RWI (Rheinisch-Westfälisches Institut für Wirtschaftsforschung) 2003. 'Klimagasemissionen in Deutschland in den Jahren 2005/2007 und 2008/12', Forschungsvorhaben im Auftrag des Bundesverbandes der Deutschen Industrie (BDI). RWI Materialien Heft 2, Essen.

Schleich, J., Betz, R., Wartmann, S.C., Ehrhart, K.-M., Hoppe, C. and Seifert, S. 2002. 'Simulation eines Emissionshandels für Treibhausgase in der baden-württembergischen Unternehmenspraxis (SET UP)', Final Report for Ministerium für Umwelt und Verkehr, Baden-Württemberg, Fraunhofer ISI, Universität Karlsruhe. Karlsruhe: Takon GmBH.

Schleich, J., Ehrhart, K.-M., Hoppe, C. and Seiffert, S. 2003. 'Üben für den Ernstfall: Der Emissionsrechtehandel als Planspiel', *Energiewirtschaftliche Tagesfragen* 53(1/2): 104–8.

Schleswig-Holstein Energy Foundation (Energiestiftung Schleswig-Holstein) 2003. 'Emissionshandel Nord – Nutzen für Wirtschaft und Umwelt', Final Report. Kiel. VDN 2005.

VDN (Verband der Netzbetreiber beim VDEW) 2005. EEG-Mittelfrist prognose 2000–2010, Übersicht der wichtigsten Daten. Berlin.

5 | Denmark

SIGURD LAUGE PEDERSEN[1]

1 Introductory background and context

This chapter presents lessons learnt in connection with the national allocation plan for Denmark (Ministry of Environment 2004). Section 1.1 outlines some particular features of the Danish energy sector and the historical CO_2 emissions. Section 1.2 describes the previous experience with emissions trading in Denmark, and Section 1.3 describes the process and milestones in the production of the Danish National Allocation Plan (NAP).

1.1 Country-specific features related to emissions trading

Denmark has no hydro and no nuclear power. Historically, this has led to a relatively large share of coal in the power sector and a high per capita CO_2 emission. Power and heat production are heavily integrated through the use of combined heat and power (CHP). The extensive use of district heating with CHP combined with high power plant efficiencies[2] has led to an energy efficiency in the power and district heating sector that is among the highest in the world. The manufacturing industry and commercial sector have a comparatively low energy intensity. The domestic sector has a stable and low energy demand.

Figure 5.1 shows the total primary energy supply (TPES) by fuel types over the last twenty years. Since the early 1980s, use of natural gas and renewables has increased in the electricity and heating sectors. As a result, use of coal decreased – in particular after 1995. Oil has also decreased in the heating and electricity sector, but the decrease has been offset by an increase of its use in the transport sector.

---

[1] The ideas and statements in this chapter have been discussed with Danish colleagues and stakeholders. However, the findings presented are the sole responsibility of the author and do not represent the views of any organisation or firm.

[2] All major steam power stations are supercritical.

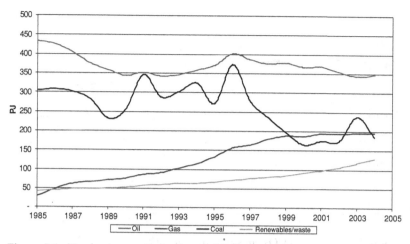

Figure 5.1. Total primary energy demand in Denmark.

The TPES has remained largely constant over the last thirty years as a result of a number of policies: various promoting schemes for CHP (both central and small-scale), mandatory use of specific fuels and production forms in certain geographic locations, high energy taxes in the domestic and transport sectors, energy-saving subsidies in industry and domestic sector, building codes, renewable energy subsidies etc.

Since the first oil crisis, energy independence has been a central topic in Danish energy planning. Today, there is a large offshore oil and gas sector, and Denmark is currently a net exporter of oil and gas.

If one compares the actual 1990 CO_2 emissions with actual present emissions (2003–2004), it is easy to get the impression that the Danish emissions are increasing. However, emissions corrected for variations and trends in electricity exchange, which are influenced by random variations in Norwegian and Swedish rainfall[3], and fluctuating heat demands in Denmark due to outdoor temperature variations (less important), in fact are 'on track' to Kyoto compliance and have been so ever since greenhouse gases (GHG) became a political issue in the 1990s. This should be clear from Figure 5.2.

---

[3] Since Scandinavia is electrically interconnected, variations in rainfall in Norway and Sweden lead to large variations in Danish electricity imports and exports. Danish energy statistics therefore operate with corrected as well as actual emissions. The corrected emissions are calculated using zero electricity exchange as a basis.

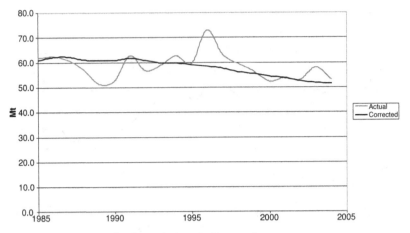

Figure 5.2. Historical CO_2 emissions in Denmark.

The Danish target under the Burden Sharing Agreement (BSA) is
−21%. As 1990 was a year with heavy rainfall and therefore high
electricity imports, the Danish CO_2 emissions were artificially low in
that year. Denmark has therefore maintained the position that a contin-
gency such as significant electricity imports in a single year should not
imply that Denmark's reduction contribution of 21% to the EU burden
sharing should be calculated on the basis of incidentally low emissions
in 1990. Negotiations are pending regarding a possible compensation
to Denmark for this effect, in connection with the final determination
of the Kyoto targets in tonnes of CO_2 for the EU15 as well as each
Member State. The final targets depend on the final 1990 inventories
which will be fixed in 2006. The difference between using corrected
and uncorrected base year emissions is close to 5 Mt of 2008–2012
GHG emissions.

Further country-specific details are given in the Danish NAP
(Ministry of Environment 2004).

1.2 Previous experience with emissions cap-and-trade in Denmark

In the four-year period 2001–2004, a domestic CO_2 cap-and-trade
scheme was operating in Denmark. In many ways, this scheme resem-
bles the European ETS now in operation. The main differences are that
the Danish domestic scheme only applied to electricity producers over

100,000 tonnes of CO_2 per year, that it had a modest non-compliance penalty (for reasons of competitiveness) and that the regulated entities were companies rather than installations. The domestic scheme is described in further detail in Lauge Pedersen (2000, 2002).

Thus, CO_2 emissions cap-and-trading was already a well-known instrument, at least for electricity producers, and had also been discussed for a number of years outside the electricity sector. This fact is likely to have contributed to the smooth implementation of the emissions trading directive in Denmark (see the next section).

Emission caps on SO_2 and NO_x are also applied to the power sector and have been for a number of years. The caps apply to power plants over 25 MWe and are set at company level (not at installation level). The caps are not tradable.

1.3 The implementing process

The primary responsibilities for implementing and administering the EU ETS Directive in Denmark are divided between two ministries:

(1) The Ministry of Transport and Energy[4] is responsible for the implementation act itself, issuing of permits[5], sector and installation level allocation, monitoring plans, auctioning etc.

(2) The Ministry of Environment[6] is responsible for the overall climate policy and the coordination of the preparation of the NAP and is in charge of the registry (including the Danish Kyoto account).

The Ministry of Finance takes a general interest in climate policy and emissions trading due to the socio-economic impact and state budget implications. The Ministry of Taxation is indirectly involved due to tax adjustments for some emissions trading installations.

Denmark, together with a small group of other Member States, managed to implement the Directive within the deadlines laid down in the Directive. The fast implementation was possible because:

---

[4] More specifically the Danish Energy Authority (DEA). In February 2005, the DEA was transferred from the Ministry of Economics and Business Affairs to the new Ministry of Transport and Energy.

[5] Involving the regional and local authorities in the issuing of permits was considered. However the tight time schedule and other considerations tipped the decision in favour of a more centralised approach to permits.

[6] More specifically the Environmental Protection Agency (EPA).

- Cooperation between the involved ministries was smooth, both formally and informally, and there were no serious disputes over competence between ministries.
- Previous experience with CO_2 caps and emissions trading enabled the DEA and EPA to arrive at a working draft legislation relatively fast without too many detours and investigations.
- The 2002 Government Climate Strategy pinpointed emissions trading as the most important greenhouse gas abatement measure. Thus, the debate on whether or not emissions trading was a suitable instrument had already been taken.
- The NAP and legislation were produced in basically one process by more or less the same people.
- The various formal and informal EU guidelines (allocation guidelines, registries regulation, monitoring and reporting guidelines etc.) could be used as references.

Box 5.1 presents the most important milestones in the implementation of the ETS Directive in Denmark.

Box 5.1 Milestones in the implementation of the EU ETS Directive in Denmark

- Emissions trading directive adopted July 2003.
- First draft of implementation act and NAP principles August 2003.
- Inter-ministerial consultation autumn 2003.
- Public consulation February–March 2004.
- Final NAP sent to the European Commission 31 March 2004.
- Implementation act including NAP provisions adopted by Parliament 9 June 2004.
- EC approval of NAP 7 July 2004.
- Registry online January 2005.
- Linking directive implementation act adopted by Parliament 1 June 2005.

---

2 The macro decision concerning the aggregate total

The total number of allowances was fixed as the result of the combination of a top-down and a bottom-up exercise. The top-down exercise was based on (a path towards) the Kyoto target and climate strategy considerations in general. The bottom-up analysis was based on a business-as-usual (BAU) forecast of emissions combined with estimates of the competitive position of various sub-sectors and company types. The allocation aimed at seeking a balance between environmental integrity and industrial competitiveness.

In general, if potential CO_2 reductions in Denmark are prioritised according to the least-cost principle, most reductions would generally be expected to occur in the emissions trading sectors, since most of the inexpensive potential is exhausted in the non-trading sectors due to significant economic and administrative pressure on energy savings in general and GHG reductions in particular in the past years. Historically, emissions from the trading sectors have been subjected to considerably less pressure and therefore offer a greater and less expensive reduction potential. International purchase of allowances and CO_2 credits was also rated among the most inexpensive GHG reduction measures.

2.1 Kyoto target

The Danish GHG reduction obligation under the EU Burden Sharing Agreement[7] is 21% compared to 1990 emissions. The total quantity of allowances should, according to Annex III of the emissions trading directive, be 'consistent with a path towards achieving or overachieving' the target. However, the translation of this criterion into the final number of allowances in the Danish of 2005–2007 NAP was not straightforward.

First, the unresolved 'base year issue' mentioned earlier raises uncertainty as to what the Danish target really is: the actual GHG emissions in 1990 were 69.7 Mt, whereas the import-corrected emissions were 76.0 Mt (Ministry of Environment 2004). The target is ~55 Mt without base-year correction and ~60 Mt with base-year correction. Hence the target itself was not seen as well defined. More seriously, this problem may still be unresolved well into the process of negotiating the second NAP.

Second, the fact that a 'path' is not necessarily a straight line left an (undefined) interval of possible allocations – and speculations as

---

[7] Decision 2002/358/EC.

to what the other Member States might do. There was some discussion between various ministries, stakeholders and non-governmental organisations (NGOs) on how to find a balance between industrial competitiveness and environmental integrity with respect to the total allocation. Naturally, the final result represents a compromise.

2.2 Climate strategy implications of the allocation process

In February 2003, the government issued a climate strategy (Danish Climate Strategy 2003). The climate strategy included a projection of Danish GHG emissions to 2017. According to this projection, there was a gap of some 20–25 Mt of CO_2 equivalents in the period 2008–2012 between the BAU emissions and the Kyoto target[8]. The BAU forecast from the climate strategy was used as a starting point for the first Danish NAP. This projection was made at a time when the ETS Directive was only a proposal. Hence, the projection can be seen as unbiased – at least from an emissions trading point of view.

The climate strategy listed a number of GHG reduction measures and the associated costs. The overall conclusion was that use of flexible mechanisms, including emissions trading and government purchase of credits, constituted the cheapest way to reduce emissions. It was also concluded that the largest, cost-effective potential for domestic CO_2 reductions existed within the sectors that are now included in the EU ETS Directive. The non-trading sectors typically pay high energy taxes. Therefore, the remaining reduction potential is expected to be smaller than in the trading sectors.

At the time when the climate strategy was discussed, consultants estimated allowance prices of €7–14 for the first and second period, i.e. up until 2012 (Copenhagen Economics and National Environmental Protection Agency 2002). In mid-2005, allowance prices of around €25 have been seen. With these prices, the economic potential for domestic CO_2 reductions should be higher than expected in the climate strategy. As a consequence of the built-in logic of emissions trading, the market should be able to adjust to these higher allowance prices, thus producing more domestic CO_2 reductions and fewer allowance purchases abroad.

---

[8] The 5 Mt uncertainty in the gap is linked to the 5 Mt uncertainty in the target.

2.3 Sector and bottom-up considerations

Four sub-sectors were considered: power supply, district heating, industry and offshore. These four sectors have different regulatory regimes and different competitive positions.

2.3.1 Power sector

The power sector was considered to have the largest reduction potential. Furthermore, analysis showed that the electricity was likely to earn a net profit from emissions trading. The logic behind this is that even though the extra costs of burning coal and gas (expressed by the allowance price) would result in a reduced production volume for the Danish power producers[9], this would be more than offset by increasing electricity selling prices[10] and the value of free allowances to the electricity sector. With respect to the total number of allowances to the electricity sector under the EU scheme, the number of free allowances allocated under the 'old' domestic emissions trading scheme was considered an indication of the upper limit. The district heating contribution to CO_2 emissions was not included in the domestic scheme. In order to compare with the EU ETS, the 20 million allowances that the power sector received in 2004 under the domestic scheme are equivalent to ~23 million allowances under the EU scheme. The actual allocation to the power sector in 2005–2007 is 21.7 million allowances per year.

2.3.2 District Heating Sector

In the district heating sector, the reduction potential was believed to be relatively low as considerable effort had already been put into the sector, and since it was under a number of administrative regulations (of fuel and/or technology) and furthermore is regulated by a CO_2

---

[9] Danish electricity production is more carbon-intensive that electricity production from competitors abroad.

[10] The first estimate from the Climate Strategy 2003 (Danish Climate Strategy, 2003) was that an allowance price of DKK100 (€13–14) would lead to an increase in electricity prices on Nordpool of DKK40 (€5–6) per MWh. A more recent estimate from the Energy Strategy (2005) showed a price increase of around DKK70(€9) per MWh. These estimates were made on the assumption that power producers would increase the bid prices corresponding to the allowance prices.

tax. Passing on extra costs of allowances to consumers was considered politically undesirable. The district heating sector ended up receiving the same number of allowances as their historical emissions.

2.3.3 Industry Sector
In the industry sector, is was estimated that the potential to pass on costs of CO_2 allowances to consumers through commodity selling prices would be lower than in the electricity sector. Moreover, the industry usually *buys* electricity (as opposed to the power sector that *sells* electricity), which becomes more expensive as a result of emissions trading. This problem was solved to some extent by reducing the carbon tax[11]. The total number of allowances to other industry was set as identical to the historical emissions.

2.3.4 Offshore oil and gas sector
The offshore oil and gas sector was treated in the same way as the manufacturing industry in the NAP. The fact that the energy consumption of an oilfield increases over the lifetime of the field was estimated to be counterbalanced by the possibilities of reducing flaring emission.

2.4 *State purchase of credits*

State purchase of credits (Joint Implementation (JI) and the Clean Development Mechanism (CDM)) could not directly influence the 2005–2007 NAP, since these credits would only be of use after 2008. However, in order to ascertain that the Danish NAP was on track for Kyoto compliance, the Danish government stated that it would be prepared to close any gap in 2008–2012 with government purchases of JI and CDM credits. Indeed, state purchase is already well on the way. In 2003–2005, the state budget has a reservation of €71 million, and for the period 2006–2009, the state budget has a preliminary reservation of €107 million. Since the first NAP was published, the involvement in state purchase of credits has continued. One example is the creation of the Danish Carbon Fund[12] as a joint venture between a few major companies and the Ministry of the Environment and the Ministry of Foreign Affairs.

---

[11] The emissions trading installations were also given the option to be relieved from their voluntary energy-saving agreements on heavy processes.

[12] http://carbonfinance.org/router.cfm?Page=html/danishcarbonfund.htm

Decisions on annual funding for government purchases are taken on a three-year rolling basis, hence it is too early to say whether similar levels of government purchases will be continued into the period 2008–2012. From the government's climate strategy as well as from indications in the first NAP, it seems likely that the sectors covered by the ETS should expect a further tightening of the allocated amount of allowances in the period 2008–2012, but the final decision on how to secure the compliance to the 2008–2012 target, by a combination of further measures outside the ETS sectors, by tightening allowances for the ETS sectors and by government purchases of project credits will only be known when the second NAP has been elaborated and approved.

2.5 Overall allocation

The bottom-up analysis (BAU forecast without emissions trading) led to projected annual 2005–2007 emissions of 39.3 Mt CO_2 for the emissions trading sectors. As a comparison, the historical emissions in 2002 were 30.9 Mt, and in 2003 they were 36.6 Mt. On the basis of the considerations outlined above, it was decided that the following allocations for the first Danish NAP, covering the period 2005–2007, would be consistent with a path to Kyoto compliance:
- 33.5 million total allowances per year, of which
- 1.7 million auctioned allowances per year;
- 1.0 million allowances per year set aside for new entrants;
- 30.8 million remaining allowances per year free of charge to existing installations.

This allocation is about 15% lower than the BAU forecast, and therefore 15% lower than the 'demand' for allowances in the emissions trading sectors. Some flexibility was introduced by allocating 40% of the allowances in 2005 and 30% in 2006 and 2007. Effectively this means that existing installations will have access to 70% of the period's total allowances before having to surrender allowances for 2005.

The scarcity of allowances expressed by a 15% reduction compared to BAU was received with mixed emotions by various stakeholders. It was not surprising that some 'green' organisations expressed the view that there were too many free allowances in the NAP. Neither was it surprising that some stakeholder organisations felt that their sector in particular was given too few allowances.

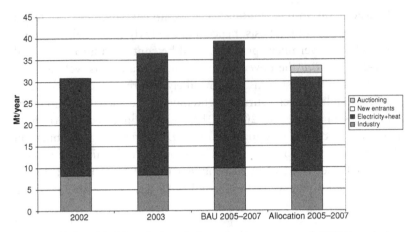

Figure 5.3. Danish historical emissions, business-as-usual (BAU) emission forecast and 2005–2007 allocations in Mt CO_2 from installations covered by the Emissions Trading Directive.

Figure 5.3 shows the average annual allocation in 2005–2007 compared to the historical emissions and the BAU forecast from the climate strategy.

3 The micro decision concerning distribution to installations

In determining the allocation to individual installations, several problems had to be solved: selection of a fair and representative base period, collection of data of sufficient quality and production of a (number of) good and simple allocation formula(e).

3.1 The base period

In choosing the base period, a number of different – and partly conflicting – factors had to be considered:
- Data should be 'new' – otherwise there would be too many existing installations without historical data.
- Data should cover a long period in order to take into account early action and the potential for CO_2 reductions (see Emissions Trading Directive, Annex III, points 3 and 7).
- The choice of base period should not contain an incentive to increase emissions just to get more allowances.

After considering these factors, the five years 1998–2002 were chosen as the base period. More specifically, the historical average over the period 1998–2002 was used. However, in order to handle relatively new installations with only a short period of operation, 2002 is used as the base period, if 2002 emissions were higher than the average emissions in the period 1998–2002.

This choice of base period still left a few 'semi-new' installations, i.e. existing installations that had not been in operation for at least one full year in the base period. Allocations to these installations (primarily combustion installations) were calculated as the average allocation per unit of installed capacity to the same type of existing installations for which historical data exist. It could be argued that the allocation to semi-new installations is not 'fair' in the sense that semi-new installations would 'need' more or fewer[13] allowances than the average of existing installations. However, no political problems seem to have surfaced – perhaps because of the simplicity and transparency of the method.

There is an ongoing debate related to the choice of base period for the second (and third) phase of the EU emissions trading directive, i.e. 2008–2012 and 2013–2017: some stakeholders have expressed fear of rolling base years, since rolling base years would tend to punish operators that reduced their emissions in the first period. The problem is most severe if 'grandfathering' is used. It could be mitigated to some extent by using output-based benchmarking, but this may not be possible in all cases. From an environmental point of view, the use of a fixed base period is the best choice, if it can thereby be avoided that operators could claim that their actions in the first period should influence the allocation in the next period, thus creating skewed incentives. A really good long-term solution to this problem has not been found. Therefore, it must be dealt with again in connection with the second NAP.

3.2 Data

The data used in allocation to installations came from various sources. Most data already existed in-house in the DEA:

---

[13] More, because a semi-new installation would tend to run longer than average installations because they use new(er) technology; fewer, because they would tend to use newer, hence more efficient, technology.

- For electricity and heat producers, a detailed database at installation level[14] had already existed since 1994 and even to some extent before that. The data had been collected on a legal basis arising from the Electricity Supply Act and the Heat Supply Act.
- For the oil and gas sector, data had been collected through the licensing and monitoring of offshore oil and gas installations carried out by the DEA.
- For most of the remaining installations, data had been collected since 1993 in connection with voluntary agreements with energy-intensive industries and subsidies to industrial energy-saving projects.

Thus most of the data work consisted of checking already existing data, rather than collecting whole new sets of data. The relevant companies and installations were typically contacted twice (using a questionnaire for potentially affected installations) and asked to check their data. No legal objections to the use and verification of data have been raised.

The number of installations covered by the ETS Directive was 357 when the NAP was submitted to the European Commission in March 2004. The number of installations has since increased to ~380. This increase has a negligible effect on the total coverage of the directive, which is about 62% of Danish CO_2 emissions. The increase was caused by

- new entrants
- combustion installations that were originally below 20 MW_{th} but expanded later
- combustion installations that were originally *believed* to be below 20 MW_{th} but later – after study of the data in more detail – turned out to be over 20 MW_{th}.

The existence of the last category shows that the NAP process should be able to cope with minor data errors and adjustments.

3.3 Allocation at installation level

3.3.1 Benchmarking vs. grandfathering

During the allocation process, it was argued that some kind of *benchmarking* allocation (allocation per unit of output) would be 'fair' – in the sense that it would automatically take into account early action and

---

[14] In fact, data are collected at plant unit level, which in some cases is more detailed than needed for CO_2 allocation.

the potential for CO_2 reductions. Hence, operators who had improved their environmental standards historically would be rewarded – or rather not punished – for doing so. Benchmarking is appealing for all sectors in principle, but to implement it in practice, a unique, homogenous output has to be identified.

- In the electricity sector, a MWh of electricity serves as a unique, homogenous unit of output. However, in order not to give unnecessary rewards to renewable electricity, that already was under various CO_2-motivated subsidy schemes, it was decided to benchmark on the basis of the number of *fossil* MWh of electricity produced. In this context it must be remembered that the mere existence of an emissions cap-and-trade scheme gives indirect subsidies to renewables through increasing electricity selling prices[15].
- In the heating sector, a MWh of heat might serve the same purpose as a MWh of electricity in the power sector. However, the Danish heating sector has been – and to a large extent still is – heavily regulated with respect to fuel choice and even choice of technology. The district heating installations therefore may have very little scope for fuel and/or technology change. Benchmarking was therefore seen as 'unfair'.
- In the oil and gas sector, a barrel of crude oil or a cubic metre of gas might serve as the allocation benchmarking unit. However, each individual oil or gas field has special circumstances relating to porosity, age, fuel processing etc., which gives the field an individual CO_2 emissions profile. Moreover, many fields produce both gas and oil, which further complicates benchmarking. After some consideration, it was decided that benchmarking of the oil and gas sector in 2005–2007 would be premature.
- In the remaining industry sectors, the number of output products was quite large, hence benchmarking would have needed a large number of 'categories'. In some cases, one category would cover only

[15] This is only true if the renewable electricity selling price is directly related to the electricity market price, but not if the electricity selling price is fixed by some legal provisions. It should be noted that allocating allowances to all electricity producers, including those using renewable energy, seems to be an option that is not allowed in the Directive. However, granting free allowances to new fossil-fuel producers (new entrants), but not to renewable producers, constitutes a de facto subsidisation of the former, which means that some continued support for renewables, apart from the indirect support achieved by the increase in market rates for electricity, seems justified.

one company or one installation – in which case benchmarking does not appear meaningful or worthwhile. Process emissions caused a further complication in some cases. Thus, benchmarking allocation to existing industrial installations appeared too complicated for the first phase (2005–2007).

It was finally concluded that, with the exception of the power sector, benchmarking allocation to existing installations would not be feasible for the first period (2005–2007). Thus, free allowances to existing installations ended up as a mix of benchmarking and grandfathering.

The resulting allocation to installations can be formulated as follows: (a) an electricity-producing installation receives 0.56 allowances per MWh of historical, fossil electricity production; (b) any other installation receives one allowance per tonne of historical CO_2 emission.

Benchmarking in the electricity sector lead to some redistribution of allowances among *installations* compared to the historical situation – but resulted only to a minor extent in redistribution between *companies*. A few small electricity producers with only gas-fired CHP plants received more allowances than their historical emissions.

It must be emphasised that the electricity allocation described above is not really benchmarking in the strict sense, as the number of 0.56 allowances per MWh does not represent a specific technological standard. Rather, the allocation represents something between the emissions per MWh from coal and gas. Apart from the distributional effects, this allocation led to the power and district heating sector getting 26% fewer allowances than they would need according to the BAU forecast or 14% fewer than actual emissions in 2002–2003.

3.3.2 Simplicity

An important factor that influenced the allocation process in Denmark was the desire to create simple, transparent allocation rules. The fewer special provisions, the fewer cases of operators claiming special circumstances and extra allocations. The two formulae above are in fact very simple and transparent, and the use of the formulae in practice has resulted in only two formal complaints over allocation[16]. It is worth noting that simplicity has been kept throughout in the sense that allocation to installations:

[16] These were filed by a power plant operator with an old CHP plant in the Copenhagen area. The complaints are currently being investigated by the Energy Complaints Board.

- has no special provisions for process emissions,
- uses the same compliance factor for all installations,
- has no special provisions for early action and potential for CO_2 reductions, other than the long base period and electricity benchmarks (some stakeholders have argued that these factors are not sufficiently taken into account);
- has no opt-in provisions and no opt-out provisions, and
- has no special provisions for CO_2 capture[17].

3.3.3 The complexities that were not avoided

Though the allocation formulae are simple, administrative complication was not entirely avoided.

- A number of biomass combustion installations >20 MW_{th} with little or no fossil fuel consumption were 'captured' by the directive provisions. These installations receive very few allowances – or no allowances at all – as they only use oil for start-up (if at all). Nevertheless, these installations are covered by the EU ETS Directive and therefore have to carry out reporting etc. To relieve these installations from part of the administration costs, they were exempted from the administration fee to the DEA.
- A similar situation applies to a large number of peak or reserve boilers in the district heating sector with potentially large emissions but historically low emissions. Administrative costs for installations receiving 100 allowances or less could easily be higher than the values of the allowances themselves.
- A number of combustion installations with integrated municipal waste/natural gas combined cycle exist in the Danish energy system. As municipal waste is not covered by the Directive, but natural gas is, these installations had to be 'split' into two parts. This was done pro rata, using fuel consumption as the key. Even though this appears unique, there was complex discussions with some operators on how to apply the pro-rata principle.
- In the district heating sector, there is an unresolved competition issue between installations covered by the Directive and (small) installations outside the scope of the Directive. If these installations supply heat to the same district heating network, there may be an economic

---

[17] Some companies showed interest in CO_2 capture (limestone, desulphurisation in the power sector).

incentive to run the small installation outside the Directive rather than the installation under the ETS. The opt-in provision might have been used to solve this problem[18]. However, this was rejected in order to avoid the administrative burden of a large extra number of small installations.

3.3.4 Allocation to district heating and combined heat and power

Due to the large district heating sector and the fact that most district heating is supplied from CHP plants, a special discussion has taken place in the major cities in Denmark on who should receive the 'heat allowances': the CHP producers or the district heating suppliers[19]. As CHP has supplied most heat in the base period, the CHP operators receive most of the allowances, whereas the district heating suppliers only receive few allowances due to few operating hours of peak boilers in the base period. If, at some point in the future, CHP becomes uneconomic due to low electricity prices or for other reasons, the district heating suppliers will have too few allowances, and the CHP operators will have too many. On the reasoning that the one who receives allocations should also be the one who needs them, this led the district heating suppliers to come up with the idea that they should receive all 'heat allowances' and use them as part of the payment for heat from CHP plants[20]. This suggestion was considered for some time but rejected in the end because it would appear to result in serious over-allocation to heat producers with little or no historical production. Instead, a certain number of the allowances to CHP plants are defined as 'heat allowances'. It was then put into the Heat Supply Act[21] that a CHP-based heat supplier cannot increase the heat selling price motivated by extra costs of CO_2 allowances before all the 'heat allowances' have been used for heat production. In this way, some protection of

---

[18] This solution was chosen by Finland and Sweden.

[19] A district heating supplier in the major cities will typically be a district heating grid company that buys heat from the power stations (centralised CHP-plants) and supplies it to consumers. The district heating supplier also operates a number of peak boilers.

[20] That the district heating sector wants more direct control of allowances can be seen from the fact that a pooling arrangement under Article 28 of the Directive has been made in one of the major cities.

[21] The Heat Supply Act regulates supply of district heat from installations over 1 MW. It also regulates heat prices, which must be fixed on a non profit basis.

heat consumers in the major cities was provided. This regulation is enforced by the Danish Energy Regulatory Authority (DERA)[22].

4 Issues of coordination and harmonisation

The EU ETS Directive suffers from lack of precision in a number of areas. This is mostly a result of the fact that the harmonisation effort under the ETS Directive negotiations could only be taken so far as Member States would agree at the time.

The Commission has put a lot of work into guidelines – both formal and informal. But guidelines cannot go beyond the provisions of the Directive and hence cannot solve harmonisation issues that were left unsolved in the Directive. The discussions in Working Group 3 (WG3) under the Commission has served to some extent as a catalyst in 'soft harmonisation'. One example of successful soft harmonisation was the question of whether or not banking of allowances between the first and second phase – i.e. from 2007 to 2008 – should be allowed. In other areas, soft harmonisation has not quite succeeded, but the discussions in WG3, workshops etc. have led to a degree of common understanding. Examples are the discussions of allocations to new entrants and on the scope of the Directive. These two subjects are discussed in further detail below.

4.1 Allocation to new entrants

The easy way to allocate allowances to new entrants[23] is to allocate no allowances. This would mean that new entrants meet the full environmental costs of their investment decisions. Though many experts would agree to this in principle, most NAPs ended up with a large number of rather complicated – and not harmonised – rules for allocation to new entrants.

---

[22] See www.energitilsynet.dk/english/

[23] In the Danish implementation act, a new entrant is defined as an installation commissioned after 31 March 2004. For combustion installations, capacity extensions of more than 10 MW or (for CHP) more than 20% are treated as new entrants. For non-combustion installations capacity extensions of more than 10% are also treated as new entrants. There is no distinction between known and unknown new entrants.

There are a number of reasons for this. A potentially sound reason lies in the fact that some European companies compete with companies outside the EU. A less sound reason is that if one Member State supplies free allowances to new entrants, all other Member States will be compelled to do the same, to avoid distortion of incentives across Member States to establish new power plants and new industrial facilities. After playing hide-and-seek for some time, all Member States chose to have a new entrants reserve.

In Denmark, the new entrants reserve (NER) was set at 1 million EUA per year or close to 3% of the total annual allocation. The size of the NER was estimated from the number of expected new entrants in the first period[24]. Currently, around 25% of the NER has been used. The reserve is administered on a first-come–first-served basis. Unused allowances after the first period are cancelled[25]. No new allowances are given to an operator to a closed installation, and freed allowances from closures are transferred to the NER.

When allocating free allocations to individual new installations, Denmark uses benchmarking allocation in all cases. The benchmarks are based on the capacity of the new installation, i.e. MW installed for combustion installations and tonnes per hour of production capacity for (most) non-combustion installations.

The benchmarks are independent of actual fuel choice and actual operation pattern. Thus a *strict ex-ante* allocation principle is applied. In case of, for example, a new power plant, the operator receives 1710 allowances per MW of installed capacity per year of operation. An allocation of 1710 allowances is what the operator will need, if the plant has an electrical efficiency of 60%, runs on natural gas and has 5,000 annual equivalent full-load operating hours. If, in actual operation, the plant runs 6,000 or 4,000 hours, it still receives 1710 allowances per MW. If the plant uses coal and/or has a lower efficiency than 60%, the operator still receives 1710 allowances per MW. Benchmarks to industrial installations are constructed according to the same *ex-ante* principle and based on already existing 'key numbers' applied under the voluntary agreement/CO_2 tax exemption regime. The benchmarks are listed in Annex 2 of the Danish law on the implementation of the

---

[24] No new power plants were expected in the first period. This was a reasonably certain expectation due to the long lead times of power plants.

[25] This is a consequence of the fact that Denmark has chosen to auction the full 5%, thus excess allowances from the NER cannot be auctioned.

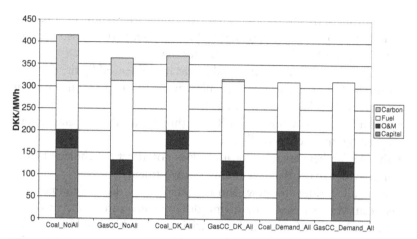

Figure 5.4. Long-range marginal costs (LRMC) of gas combined cycle and coal steam turbine with and without CO_2 costs. (Allowance price: € 20. Gas/coal price ratio: 2. Annual operating hours: 5500.)

ETS Directive[26]. These differentiate between thirty different heavy processes[27]. Thus, new entrants in Denmark will not receive exactly the number of allowances they need, but usually less than that. How much less depends on how they perform compared to the benchmark.

Free allocations to new entrants might influence investment incentives. This can happen in two ways: (a) by speeding up investments compared to what would be optimal without free allowances, and (b) by distorting the investment choice between technologies. Though the total CO_2 caps remain unchanged, the total socio-economic cost involved in complying with the caps could increase.

An example of skewed investment incentives is shown in Figure 5.4. The figure shows the long-range marginal costs (LRMC) of two power plants: a gas-fired combined cycle plant and a coal-fired supercritical steam turbine plant. The LRMC should – together with the electricity price – in principle constitute the basis for the investment decision. The LRMC are shown in three variations: on the left, the full CO_2 costs are included (corresponding to a situation where no free allowances are given to new entrants). In the middle, the free allocations are as in

[26] Law on CO_2 allowances of 9 June 2004, www.ens.dk/sw17278.asp.
[27] New entrants do not receive free allowances for light processes and space heating; the value of the carbon tax reduction was considered sufficient.

the Danish NAP. On the right, the two plants are assumed to receive exactly the allowances they 'need'. In this case, carbon costs are not part of the LRMC. The figure indicates that:

- the choice between gas and coal might be influenced by the decision on whether or not to give free allowances to the plant (if the plant receives the allowances it 'needs'),
- free allowances will reduce LRMC, thus (other things being equal) speeding up investments[28], and
- the Danish benchmark allocation does not affect the choice of technology but may affect the timing of investments.

Free allocations to new entrants may also distort competition between Member States by giving different investment incentives in different Member States. In the electricity sector, this could have a serious effect on security of supply in some Member States. As an example, a comparison of the first NAP versions showed that a new power plant would get more free allowances in Germany and Finland than in Denmark – and fewer in Sweden than in Denmark. This is a point of great concern in the Danish power sector. The Association of Danish Energy Companies has estimated that a new gas-fired combined cycle plant in Denmark receives only around 80% of the free allowances it would receive in Germany and Finland. A Danish coal-fired power plant receives only half of the allowances it would receive in Germany and Finland. For both types of plant, a Swedish power plant receives even fewer allowances.

Thus, free allowances to new entrants may, if they are left unharmonised, lead to 'wrong' technologies being promoted in the 'wrong' places.

4.2 Scope of the Directive

It came as a surprise that it was not entirely clear from the Directive which installations were covered by emissions trading. Denmark has – like a little more than half the Member States – chosen to use the Commission's definition of 'combustion installation', whereas the remaining Member States implemented a more narrow definition. This has led to

---

[28] As a result, long-term electricity prices will tend to be lower than if new entrants were to buy all their allowances. This has the effect that renewable suppliers will get a slightly too low incentive through the cost of carbon signal in the electricity prices.

a situation in the first phase where a combustion installation may be covered by the Directive in one Member State but not in another.

This could potentially lead to distortion of competition within the EU. The Commission has maintained that there is only one definition of scope. Nevertheless, this does not appear to be so in real life. Hence more harmonisation of scope is desirable.

In the political discussions prior to the first Danish NAP, this problem was raised by some stakeholders and politicians. It was stated in the Danish emissions trading legislation that if the Commission should change its definition of the Directive's scope, the Danish government would be prepared to change the legislation in order to secure that the scope of the Directive was the same in Denmark as in other Member States.

5 Issues deserving special mention

In this section, a number of other subjects related to the first NAP are discussed. The subjects treated are special Danish NAP-related circumstances or discussions related to NAPs and emissions trading in general.

5.1 Auctioning or sales of allowances

If emissions trading were applied worldwide, by far the simplest way to allocate allowances would be for Member States to sell or auction all allowances and let the market decide who should get how many allowances.

In the real world, the Directive ended up allowing only up to 5% auctioning in the first period and up to 10% in the second period. Denmark decided on maximum auctioning (5% of total allowances, amounting to 5,025 million allowances) in 2005–2007. No decision has been made with respect to auctioning in the second period.

The main reason for choosing the auctioning option in Denmark was that it seems to be the correct allocation method in principle, that it is consistent with the 'polluter pays principle' and that it might be worthwhile to gain some experience. It goes without saying that the auction revenues will not be entirely unwelcome to the Treasury. Interestingly, the major Danish power producers have suggested 100% auctioning across the EU. This suggestion presupposes that their competitors are faced with the same conditions, i.e. full auctioning.

It has not been decided how and when to auction the 5 million EUA. In fact, some allowances might be sold on the market in other ways than by auctioning, e.g. on one or more of the emerging CO_2 market places. As a few other Member States have auction set-asides, and many Member States plan to auction unused allowances from their NERs, some international coordination or discussion is desirable – and has in fact started.

5.2 Taxing issues

Emissions trading and carbon tax can be seen as mutually exclusive, in the sense that having both will constitute double regulation. The use of free allocations could on the other hand be seen as a tax exemption on (part of) the emissions, leaving the double regulation less problematic. In connection with the Danish NAP, double regulation still exists to some extent, though a carbon tax reduction has been introduced in the industrial sectors.

Another taxation issue is related to the Danish Constitution, which states that no one but the Parliament can institute a tax. The non-compliance penalty in the emissions trading directive of €40 per tonne of excess emissions was interpreted as a tax in Denmark (and not a penalty in the sense of the Danish penal code). As a consequence, it was decided that the detailed allocation rules had to be included in a legal document and be approved by parliament. Thus, the Danish emissions trading act contains not only the administrative regulations implementing the articles of the EU Directive but also all the allocation rules. This has proved a great advantage in the later administration of emissions trading in Denmark, by reducing the (potential) number of disputes.

5.3 The environmental incentives of emissions trading

Once the number of total allowances has been decided, the distribution of free allowances to *existing* installations is just an exercise in income-distribution policy. The free allowances to existing installations represent a financial subsidy that contains no environmental incentive whatsoever. The way free allowances are allocated to existing installations may thus reward companies for a *historical* effort in CO_2

reductions – but it is of no consequence for *future* CO_2 reductions[29]. Nevertheless, it is tempting to view for example grandfathering as environmentally 'reactionary' and allocation methods like benchmarking as 'progressive'. This sometimes confuses the debate, where it is really the (scarcity in) total allowances that provides the environmental incentive.

The above reasoning applies when free allocations to existing installations are regarded as a stationary exercise. However, if (companies' expectations of) free allocations in the next period(s) are included, the issue becomes more complex. If, for instance, rolling base years are used (or companies expect that to happen), or if allocation principles are shifted from one period to the next (e.g. from grandfathering to benchmarking), this may induce companies to (environmentally) suboptimal behaviour in one period in order to get more free allowances in the next period. This is an important concern among regulators as well as industries.

Another important issue that influences the environmental incentives of emissions trading is related to *closures*. If operators lose 'their' allowances to closed installations[30], they have an incentive to keep old, inefficient installations running – or at least not close them completely. This may be environmentally and/or economically inefficient. If, on the other hand, companies were allowed to keep allowances to closed installations, they might get an incentive to move part of their production to countries outside the EU. 'Earmarking' of allowances from closed installations to new installations could solve this problem. This was discussed in connection with the first Danish NAP, but not implemented[31].

Thus, even though the main environmental effect in emissions trading is a result of the total caps, it is important to ensure, that individual allocations do not result in environmentally perverse incentives for companies, especially when emissions trading is considered over a longer period.

---

[29] The allocation to new entrants is different, as free allocation may influence investment decisions (see the previous section).

[30] This is what happens in Denmark and a number of other member states in 2005–7.

[31] For three reasons: (1) it was slightly complicated legally, (2) it was unclear if it was consistent with the Directive, and (3) it was not considered as a very urgent problem at the time.

5.4 Long-term framework for emissions trading?

A desire to have a stable, long-term regulatory environment is often expressed by companies, industrial organisations and some NGOs. The first NAP exercise has not really fulfilled this desire, as it by its very nature is concerned with only the three-year period 2005–2007. Companies are therefore faced with considerable uncertainty regarding the number of allowances that will be available in the second period (2008–2012) – and at what cost extra allowances can be purchased. The uncertainty is even higher when the period after 2012 is concerned.

It appears that there is not much to do in order to reduce uncertainties at least in the near future. Uncertainty regarding future allocations can however affect the investment decisions by companies. A recent report from RISØ National Laboratory (Morthorst *et al.* 2005) indicates that uncertainties in the electricity market, including uncertainties in future carbon prices and allocations, may postpone investments in the power sector and reduce the security of Denmark's electricity supply.

6 Concluding comments

To summarise, the main lessons from the first Danish NAP appear to be the following:

- Emissions (cap and) trading is a simple environmental instrument in principle. But there is a serious risk of complexity due to lobbying and desire to make 'fair' allocations in all cases. Simplicity can however be achieved to some extent – and has clear merits in the ensuing administration.
- Data quality was good in general and no serious data problems were encountered.
- Benchmarking to existing installations proved too difficult, with the exception of electricity production.
- Applying different aspects of the Directive is not seen by Danish stakeholders as acceptable in the long term due to distortion of competition between Member States.
- Free allocations to new entrants entails a serious risk of competition distortions between Member States. Moreover, free allocations to new entrants can lead to loss of environmental integrity as it may affect choice of technologies. If free allocations to new entrants are maintained, *ex-ante* benchmarking should be preferred.

- Lack of formal harmonisation of allocation and scope across EU was in some areas mitigated by 'soft harmonisation' via WG3, workshops etc.
- There is a commonly expressed demand for a long-term NAP framework to secure a stable investment climate. But this is very difficult to deliver in practice, due to the short-term nature of the 2005–2007 exercise.

References

Copenhagen Economics and National Environmental Protection Agency 2002. 'Prices and risks in international markets for flexible mechanisms', November 2002.

Danish Climate Strategy 2003. 'Oplæg til klimastrategi for Danmark', Regeringen Februar 2003. www.fm.dk/1024/visPublikationesForside. asp?artikelID=5354&mode=hele

Energy Strategy 2005. www.ens.dk/sw11628.asp (currently only in Danish).

Lauge Pedersen, S. 2000. 'The Danish CO_2 emissions trading system', *Review of European Community and International Environmental Law* 9(3): 223–31.

Lauge Pedersen, S. 2002. 'Danish CO_2 cap and trade scheme. Function and experience', SERC Workshop, Tokyo, 3 December 2002.

Ministry of Environment 2004. 'Danish national allocation plan', March 2004. www.mst.dk

Morthorst, P. E., Grenaa Jensen, S. and Meibom, P. 2005. 'Investering og prisdannelse på et liberaliseret elmarked', Risø-R-1519(DA). www.tisoe.dk/vispub/SYS/syspdf/vis-r-1519.pdf

6 | Sweden

LARS ZETTERBERG

1 Introductory background and context

1.1 The development of Swedish energy use and climate policy

1.1.1 Important change in energy use since the 1970s

The energy supply in Sweden has changed dramatically since the 1970s. The major change has been the decrease in oil from 77% of the total energy supply in 1970 to 33% in 1997. This change was made possible mainly thanks to the development of hydro power and the nuclear programme. Also, the use of bio-fuels and peat has increased from 9% of the total supply in 1970 to 15% in 1998. Energy consumption has been reduced in the industry and housing sectors and increased within transportation between 1970 and 1998. It is mainly the use of oil that has been reduced in the industry and housing, and increased in transport. The use of electricity has increased considerably within industry and housing (SOU 2000a).

1.1.2 High use of energy

Sweden has a relatively high energy use per capita in comparison with other OECD countries. In 1997 the average energy use per capita was 17,000 kWh as compared to 7,500 kWh in the OECD. This can mainly be attributed to the fact that Sweden has natural resources such as forest, iron ore and hydropower, which in turn has resulted in a large share of energy-intensive industry. The geographical position and low population density has contributed to Sweden having large heating needs and long transportation distances. Sweden is a country with a high share of hydro and nuclear in its power production. In 1997, only 4.5% of the electricity production was based on fossil fuels (SOU 2000a).

1.1.3 Carbon dioxide emissions

Even if energy use is high in Sweden, CO_2 emissions per capita are low, thanks to a large share of hydro, nuclear and bio-fuels. CO_2 emissions in 1997 were 6.6 t/capita in Sweden. This can be compared to about 8.5 t/capita in EU15 and about 12 t/capita in the OECD (STEM 1999). Since the early 1970s CO_2 emissions have been reduced by almost 50%, mainly due to the conversion from oil to electricity and bio-fuels used for heating (Regeringens Proposition 2003).

In the trading sector, CO_2 emissions were 20.2 Mt on average over the years 1998–2001. This is about 29% of the total greenhouse gas (GHG) emissions in Sweden, if all six Kyoto gases are included. This is a considerably higher share than in 1990, when the share from the trading sector was only 20% of the total emissions. There are several reasons why the share of the trading sector is increasing. First, emissions from individual house heating have 'moved' to district heating, thus being transferred into the trading sector. Second, heavy industry has experienced growth in production throughout the 1990s after the recessional period of the early 1990s. Heavy industry represents a large share of the total industry in Sweden. Finally, increased emissions in the refinery sector can partly be explained by the development of low-sulphur gasoline processes, but will be compensated by lower CO_2 emissions in the transport sector (SOU 2003).

1.1.4 Carbon dioxide taxes in Sweden

In 1991, Sweden introduced a tax on CO_2 emissions corresponding to SEK0.25 per kg CO_2 (about €26 per ton CO_2). This tax has been continuously increased and is today SEK0.91 per kg, which corresponds to about €95 per ton CO_2. As a general rule, this tax is paid on all fuels except bio-fuels and peat. There are however several important exceptions to this rule. Emissions of CO_2 from industrial processes, such as coke ovens, blast furnaces, lime kilns, cement production, refineries and the use of carbon electrodes, are exempt from tax. Industrial use of fuels for heating and transportation pay 25% of the tax, i.e. SEK0.21 per kg CO_2. Before 1 January 2004 fossil fuels used for electricity production were also totally exempt from CO_2 tax. So, in fact, the large industrial CO_2 emitters are either exempt from tax or pay a fraction of the full tax. The main sectors left that are paying full CO_2 tax are transportation, housing (private and public) and heat production (Regeringens Proposition 2003; SOU 2003).

There has been a discussion on whether the CO_2 tax should be kept or not when emission trading starts. If the tax is omitted, the loss of tax revenues from the trading sector has been estimated at between SEK600 and SEK1200 million net (SOU 2003), corresponding to about €70–130 million. These tax revenues come from non-process CO_2 emissions, largely the use of oil and gas for heating within the pulp and paper, iron and steel, cement and energy sectors. This does not include the large process emissions from the trading sector, mainly from coke and steel production, cement production and refineries. The total tax revenues can be compared to the total value of allocated allowances, 22.9 Mt, which assuming a price of €20 per ton totals €460 million. As we now have entered the trading period Sweden has kept the CO_2 tax on the trading sector, so some sources both need to pay the old CO_2 tax and in addition they will need allowances. But it should be noted that there are currently discussions on reform of the CO_2 tax rules for the trading sector.

1.1.5 Swedish energy policy

Swedish energy policy states that the objectives are to safeguard the supply of electricity and other energy carriers in a short- and long-term perspective at internationally competitive prices. This energy policy should create conditions for efficient and sustainable energy use and a cost-efficient Swedish energy supply with low impacts on health, environment and climate as well as facilitating the transition to an ecologically sustainable society. The energy policy should consider the Swedish environmental and climate targets (Regeringens Proposition 2003). Nuclear energy corresponds to about 46% of Swedish electricity production in 2003 (Energiläget 2004). The energy policy says that nuclear energy is to be replaced through increased efficiency in energy use, conversion to renewables and to environmentally acceptable electricity production technologies (see also Section 2 on how Swedish nuclear policy influenced the allocation).

1.1.6 Swedish climate policy

According to the EU Burden Sharing Agreement (BSA) of the Kyoto protocol, Sweden's emissions for 2008–2012 may increase by 4% compared to the 1990 level. This is in contrast to the EU Kyoto commitment of a reduction of 8%. The reason why Sweden could negotiate an increase in emissions can largely be explained by the fact that Sweden after the oil crises in 1973 and 1979 undertook measures to reduce its dependency on fossil fuels, mainly through a nuclear programme. Sweden is therefore in a position where further reductions would be more costly than in other states.

However, the Swedish parliament decided in 2002 that the Swedish emissions for the period 2008–2012 should be at least 4% lower than 1990. This target should be reached without using flexible mechanisms or uptake in carbon sinks. However, if the emission trends turn out to be less favourable or if the reduction measures are not as effective as anticipated, the government can suggest additional measures and/or reassess the target. The government should consider here the consequences for Swedish industry and the international competition (Regeringens Proposition 2003). At the very least, the government is faced with the potential consequences of participating in an international emissions trading system through which allowances may be imported into Sweden to cover emissions in excess of the allowances allocated to the trading sector. This ought not to be a problem since corresponding emission reductions are realised elsewhere in the EU, but it poses an obvious threat for a target that is limited to Sweden. There is currently a debate in Sweden whether the national target should be kept or revised.

The climate policy from 2002 also includes a long-term target. Sweden will try to direct international climate work towards the objective of stabilising the atmospheric concentrations at 550 ppm. In the year 2050 the Swedish emissions will be lower than 4.5 t per capita per year and thereafter decrease further. In order to accomplish this target, international cooperation and measures in other countries will be necessary (Regeringens Proposition 2003/04:31).

1.2 Description of the National Allocation Plan process

1.2.1 Phases, timetables and actors
March 1999: The FlexMex Commission was appointed by the Swedish government to investigate ways to implement the Kyoto flexible mechanisms. The Commission presented a proposal in April 2000. In March 2000 the EU presented its 'Green Book' on emission trading, proposing that the ETS should include CO_2 from six industrial sectors: electricity and heat production, iron and steel, refineries, pulp and paper, chemical and cement, glass and ceramics. Kjell Jansson, heading the FlexMex

investgation, recommended that Sweden should develop an emission trading system in collaboration with other EU Member States. But in contrast to the EU 'Green Book', Sweden proposed that the emission trading system should be widened to include all sectors that presently paid CO_2 tax in Sweden, namely emissions from the transportation and housing sector. Moreover, process emissions should not be included initially at the start in 2005 due to international competition, but could be included from 2008. From 2008 other greenhouse gases (GHGs) should be included. Kjell Jansson proposes that the allowances should be auctioned, following the 'polluter pays' principle and also giving new installations and incumbent installations equal conditions (SOU 2000b).

Spring 2002: A new parliamentary commission, FlexMex2, was appointed by the government to deliver a proposal for a Swedish emission trading system. The Commission, also chaired by Kjell Jansson, consisted of members of parliament representing all parties, but also representatives from industry. The purpose was to achieve a high level of acceptance for the coming propositions and for the parliamentary bill so it would pass easily through parliament. The scope of the Commission changed as the EU ETS evolved. Since Sweden had to follow the EU legislation the work of the commission was directed towards implementing the EU ETS for Sweden (SOU 2003).

May 2003: The FlexMex2 Commission presented a report to the government on principles for allocation within the EU ETS. These principles had been developed through analyses of different allocation methods at sector level, using currently available emission data for about 450 installations (corresponding to about 90% of the emission volume in the Swedish National Allocation Plan (NAP)). In this proposal, allocation was based on historic emissions. For process emissions (metallurgy, cement, catalytic cracker), the projected increases were added to the allocation. An allocation scheme based on these principles at sector level was presented (SOU 2003).

October 2003: The EU Commission and the EU parliament presented the Directive on the EU ETS.

4 December 2003: After the FlexMex2 Commission had presented its proposal on principles to the government, the Ministry of Industry and Trade was requested to develop a government bill on allocations. This bill largely followed the same principles as earlier, but after political negotiations the cap was reduced from 25 Mt to

19–22 Mt. This bill did not present a final NAP (Regeringens Proposition 2003).

10 March 2004: The bill was passed by the Swedish parliament.

22 April 2004: Sweden's NAP was delivered to the European Commission.

7 July 2004: The Swedish NAP was approved by the European Commission.

1 August 2004 and 1 January 2005: The Swedish laws regulating the allocation and emission trading entered into force (SOU 2005).

The current organisation. The Swedish Environmental Protection Agency (EPA) (Naturvårdsverket) is responsible for allocation, sanctions and certain tasks concerning verification and control.

A council represented by the EPA, the Swedish Energy Authority (STEM) and the Swedish Business Development Agency (NUTEK) prepares the applications from installations for permits and allowances.

The regional authorities (Länsstyrelser) issue permits to installations. The Länsstyrelser were chosen because they are normally the authority responsible for issuing environmental permits to industrial installations. STEM is responsible for developing and running the national register.

1.2.2 Political dimensions, lobbying and turf battles

Government changes the cap and overrules the parliamentary FlexMex2 Commission. The proposal from the FlexMex2 Commission of May 2003 had the necessary political support, through the Commission delegates, to pass through parliament. This was one of the purposes of creating a parliamentary commission in the first place. After the FlexMex2 Commission presented its proposal, the issue is moved over to the Ministry of Industry and Trade who in December 2003 presented a bill to parliament. In this bill, the allocation principles were somewhat changed and more notably so also was the total cap which is reduced to 19–22 Mt from the FlexMex2 proposal of 25 Mt. This was the result of consultations at a higher political level between the governing Social Democrats and the Green Party (Miljöpartiet). In Sweden, in 2003, the Social Democratic government ran a minority government, and normally seeks parliamentary support from the Left Party and the Green Party. These parties are referred to as collaboration partners to the government. Through this political agreement, the Greens were able to put pressure on the Social Democrats to change the cap that had originally been recommended by the FlexMex2 Commission. Hence, the proposal from FlexMex2 from June 2003 was de facto overruled and as a consequence, the mandate of the FlexMex2 was thereafter undermined.

Interaction with industry.   There is a tradition in Sweden of trying to include industry in the process of investigating environmental issues and developing abatement strategies for pollutants. When government bills are prepared, representatives from industry, environmental organisations, research organisations and the general public are normally invited to deliver their views and propose alternative solutions. In the case of the process of developing a Swedish NAP, the participation of industry was even stronger. Industry had three permanent representatives in the FlexMex2 Commission, coming from different industry sectors: the energy sector (represented by Svensk Energy), the refineries sector (represented by Svenska Petroleuminstitutet) and from the Confederation of Swedish Enterprise (Svensk Näringsliv). Information was also delivered continuously from different sector and lobby organisations in Sweden and assessed by the FlexMex2 Commission. In February 2003 industry was invited to participate in a hearing on principles for the allocation of allowances. There were a number of bilateral meetings between the FlexMex2 Commission and representatives of industry. After the FlexMex2 proposal was presented in May 2003, industry, environmental organisations, research organisations and government authorities were asked to submit written reports presenting their views. After May 2003, responsibility for allocations was moved to the Ministry of Industry and Trade. The Ministry of Industry and Trade later considered these reports when developing the bill on allocations. The interaction with industry continued throughout this process, and up until the NAP was presented industry was very active in informing the government in their views on the allocation.

Lobbying from industry: the battles.   The lobbying came mainly from sector organisations, rather than from individual companies. Lobbying was normally done by presenting technical facts to the FlexMex2 Commisison or the government at bilateral meetings or hearings, or by sending reports. The tone at these meetings was in general friendly. My impression is that there was basically an understanding within Swedish industry of the need to reduce CO_2 emissions and a positive view on emission trading. The main concerns within industry were connected to the issues of fairness, competition and uncertainty. Within the government and the involved authorities there was also an understanding of the concerns put forward by industry. However, some specific lobby campaigns are worth mentioning.

Steel industry: The steel industry claimed that the 'CO_2 emissions from blast furnaces was close to the physical limit.' However, this was far from true. In blast furnaces in Sweden the input of metallurgical carbon and consequent CO_2 emissions are about 470–480 kg carbon per ton pig iron, which is about three times higher than the chemical limit according to calculations by the the Swedish Environmental Research Institute (IVL). Moreover, they lobbied strongly for the introduction of international benchmarks and for the Swedish cap to be increased accordingly. Besides submitting technical documents the steel industry was the sector that was most visible in media, threatening that jobs would disappear. These tactics most probably had an effect, since they were given a generous allocation based on projected production increases.

Energy: At first, the concern of the energy sector was on the selection of base years. The base years used in the Swedish NAP represented years with warm winters, high hydropower capacity and consequently low emissions. The energy sector also argued that many emission reductions had been realised in the early 1990s (early actions) for which they had not been given credit in the allocation plan. When details of the NAP became available, the sector was very concerned about the downscaling of the allocation to 80% of historic emissions. Moreover the sector was upset about the considerable asymmetries between Member States in allocation rules for new entrants and the preservation of the CO_2 tax for installations participating in the emission trading system. The arguments were well formulated, but the energy sector lost the battle.

Refineries: Refineries lobbied that the emissions from the catalytic crackers are needed for the production of low-sulphur gasoline and should therefore be considered as process emissions. This in turn meant that they should be subject to allocation based on projected emissions. They submitted technical documents, maintained a low profile and received the allocation they suggested.

Cement: The cement industry lobbied for process emissions and won this fairly easy case. However, they also lobbied that the use of shredded rubber tyres as a fuel in cement kilns should be considered process emissions, but this argument was turned down.

Pulp and paper: This sector did little lobbying. Allocation was not considered as important as the expected increase in the price of electricity. There was a certain concern for the allocation to industrial electricity and heat production. They won this battle since industrial power units received 100% of historic emissions, in contrast to the other energy units receiving 80%. The sector was content with the result.

2 The macro decision concerning the aggregate total

2.1 Considerations in determining the cap level and interdependence with allocation to sources

Deciding the level of the Swedish cap was an iterative process in three steps:

(1) Bottom-up, based on old data of CO_2 emissions at installation level. In the spring of 2003 the FlexMex2 Commission investigated different allocation schemes. In (most of) these schemes, allocations were based on historic emissions with adjustments for projected increases in process related emissions. These principles where applied to a data set giving CO_2 emissions for 450 installations over twelve years, 1990–2001. In the final NAP, these 450 installations accounted for about 90% of the total emissions of all Swedish installations in the EU ETS. These data were calculated from existing energy statistics that had been previously delivered by the installations. A complementary data set was developed showing projected increases in process-related emissions for a limited number of installations. The testing included analyses of outcome for different base year periods. In May 2003 the FlexMex2 Commission presented a proposal for allocations based on these analyses (see Table 6.1 below). These data had not been collected for ETS purposes and contained significant uncertainties. In preparing the final NAP, new data would have to be collected from the participating companies. In the FlexMex2 proposal allocations were based on historic emissions in the period 1998–2001 with special provisions for projected increases in process-related emissions. A new

Table 6.1 *Development of the Swedish cap*

Allocation (Mt CO_2) due to	FlexMex2 May 2003	Government Bill Dec 2003	NAP Apr 2004
Emissions 1998–01	18.3	17–18	20.2
Process emissions increase	2.3	2–4	1.8
New entrants	2.0		1.8
Other	0.6		
Uncertainty reserve	2.0		
Reduction in energy sector			−0.9
Total	25.2	19–22	22.9

entrants reserve was set to be 2.0 Mt, and 0.6 Mt was added corresponding to CO_2 leakages from the steel industry. These leakages
had not been included earlier in the Swedish report to the United
Nations Framework Convention on Climate Change (UNFCCC).
The uncertainty in the data set used was estimated to be about 20%,
possibly as large as ±4 Mt. Due to this uncertainty an 'uncertainty
reserve' of 2.0 Mt was added to the cap, in case the new emissions
inventory yielded higher total emissions (which it later did). The
resulting cap was 25.2 Mt.

(2) Top-down in a political process with somewhat new principles.
After the FlexMex2 Commission had presented its proposal, the
issue moved over to the Ministry of Industry and Trade who in
December 2003 presented a bill to parliament. In this bill, the principles for allocations were changed, and consequently the cap was
changed from 25.2 Mt to 19–22 Mt, after consultations between
the governing Social Democrats and the Green Party. Allocations
were still based on average historic emissions in the period 1998–
2001 (17-18 Mt) with special provisions for projected increases
of process-related emissions. However, the increases in process-
related emissions plus the new entrant provisions were to be within
2-4 Mt was instead of 4.3 Mt earlier. The uncertainty reserve of
2 Mt is omitted as well as the leakage provisions for the steel industry, 0.6 Mt. The resulting cap was 19-22 Mt.

(3) Bottom-up based on new data from the participating installations
and reductions in the energy sector, and adjusted after political

consultations. In preparation of the NAP, new data were retrieved from the participating installations in connection with the applications for permits. This data inventory increased the average historic emissions for 1998–2001 from 18.3 to 20.2 Mt. This was largely due to the fact that the number of identified installations increased, but also due to discrepancies between these data and the older energy statistics on which the earlier allocation schemes were based. Projected increases in process emissions were 1.8 Mt. If the new entrants reserve were to remain at 2 Mt, this would add up to a total cap of 24 Mt. This created problems since it meant that the politically agreed cap of maximum 22 Mt would be overshot by 2 Mt. This situation resulted in new political consultations, which is one of the reasons why the submission of the Swedish NAP was delayed. The result of these consultations was to decrease the allocation from energy installations to 80% of historic emissions, which reduced the cap by 0.9 Mt, and in addition to limit the new entrants reserve to 1.8 Mt. This rendered a total cap of 22.9 Mt in the final NAP that was submitted in April 2004.

The development of the Swedish cap is shown in Table 6.1 above.

2.1.1 Did Swedish nuclear policy influence the allocation?

Sweden's energy policy says that nuclear energy is to be replaced through increased efficiency in energy use, and by conversion to renewables and to environmentally acceptable electricity production technologies. Of Sweden's original twelve nuclear reactors (at four sites), in 2003, one reactor had been taken out of operation, and there are currently negotiations between the government and the operators to close a second reactor. So what importance did the phase-out of nuclear power have for the Swedish cap? There was no open debate on the question of how the planned nuclear phase-out might influence the allocation. There was no debate to re-evaluate the nuclear policy due to the carbon restriction that emission trading introduces.

2.2 The cap in relation to the Swedish Kyoto target

In Figure 6.1, the first column (from the left) shows how the average emissions of the period 1998–2001 were split between trading (TS) and non-trading (NTS) sectors. Note that the trading sector only includes CO_2 emissions whereas the non-trading sector includes all six

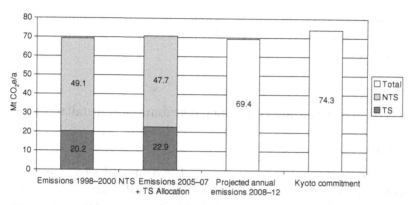

Figure 6.1. The national greenhouse gas emission budget of Sweden. NTS, non-trading sector; TS, trading sector.

greenhouse gases (measured in CO_2 equivalents per year: Mt CO_2e/a). The second column shows the total emissions of GHGs as projected for 2005–2007. A projection for the non-trading sector is used whereas the allocation is used for the trading sector. The third column shows total GHG emissions projected for the period 2008–2012 (Kyoto commitment period). The fourth column shows the amount of emissions allowed in accordance with the national commitment according to the EU BSA.

If we compare the first column with the fourth, we can see that Sweden can increase its current emissions (69.3 Mt) by 5.0 Mt (equivalent of 7% increase) before Sweden reaches its Kyoto commitment of 74.3 Mt. The second column shows that the sum of allocated allowances to the trading sector and projected emissions in the non-trading sector adds up to 70.6 Mt. Even if this includes an allocation to the trading sector which is 2.7 Mt higher than the emissions for 1998–2001, the sum is still well below (3.7 Mt or 4%) Sweden's Kyoto commitment (Zetterberg *et al.* 2004).

If compared to the Kyoto commitment, Sweden is well on track, even given the relatively generous allocation. However, the Swedish parliament has taken on a more rigorous national emissions target that is 96% of the 1990 emissions, or 68.5 Mt in absolute values. According to this target Sweden's allocation in the trading sector plus projected emissions in the non-trading sector, 70.6 Mt, overshoots this more stringent target by 2.1 Mt or 3%. Moreover, the parliament's decision states that this reduction should not include the use of 'flexible mechanisms', which can be very difficult to achieve if emission trading is allowed.

3 The micro decision concerning distribution to installations

3.1 Data availability

First data set, FlexMex2, May 2003: The data used were delivered to the FlexMex2 Commission by the Swedish EPA, who in turn had ordered the data from the so-called SMED project. SMED is a consortium that consists of the Swedish Environmental Research Institute (IVL), Statistics Sweden and the Swedish Meteorological and Hydrological Institute (SMHI). The project retrieved and compiled data from those installations that were known to be included in the emission trading system, at the time 450 installations. The data were taken from currently available Swedish energy statistics; Sweden had collected data on energy use at installation level from 1990 or earlier through yearly inquiries to all companies. Another data set was retrieved from legal environmental reports from the same companies in order to investigate the uncertainty in the data. This uncertainty was estimated to be ±20% or larger! This insight was quite surprising to the Swedish authorities and resulted in a separate initiative to make an in-depth overview of the data collection routines for UNFCCC reporting. A complementary data set was developed showing projected increases in process-related emissions for a limited number of installations, mainly from the steel, cement and refinery industries. As this data set became available to the FlexMex2 Commission in early 2003, it became an important data source when investigating the consequences of alternative allocation methodologies. The whole data set was not made available to the public, only the figures at an aggregated sector level.

Second data set, NAP, February 2004: After the proposed allocation principles had been presented by the government in the parliamentary bill in December 2003, a new data set needed to be compiled. The first data set consisted of data that originally had been collected for other purposes. This new data set was to be collected specifically for the purpose of developing the NAP and the companies needed to be aware of this. This was a big challenge, since time was extremely short.

The data were collected by sending enquiry forms to all installations, at that time known to number about 500.

Third data set, summer 2004: When the legal framework was in place, companies applied for permits to join the EU ETS and to receive allowances. In connection with this application companies were asked to confirm the data submitted earlier that year in order to give the data a legal foundation. After the application process the number of installations with permits was now close to 700.

The emission data used for the EU ETS will be harmonised with national UNFCCC reporting. This presents some difficulties since sector boundaries and calculation methods differ.

3.2 Criteria for internal allocation of the cap

3.2.1 Basic principle

The basic principle is that allocated allowances = average emissions of the period 1998–2001, with the amendments listed below.

3.2.2 Amendments

(1) Installations with process-related emissions (i.e. metallurgy, cement etc.) receive an addition of allowances corresponding to the expected increase in process-related emissions from 1998–2001 to 2005–2007. Emissions that are defined as process-related are from the following processes:

- Cement and lime
- Manufacturing of glass, glass wool and ceramic products
- Blast furnaces, iron and steel production, coke ovens (including carbon injection and carbon electrodes, but not oil and gas for heating and cutting purposes)
- CO_2 emissions from limestone and dolomite in relation to the manufacture of pellets for steel production
- Emissions from the combustion of residual gases from the steel industry, i.e. coke oven gas, LD converter gas and blast furnace gas
- Combustion of coke for the regeneration of catalytic crackers in refineries.

These process-related emissions account for about 35% of the total emissions in the period 1998–2001.

(2) Electricity and heat-producing installations receive 80% of emissions 1998–2001 with the exception of industrial combined heat and power (CHP) which receives 100% of emissions 1998–2001. Industrial CHP is predominently sourced from the pulp and paper industry, using mainly forest residues for fuels. The reason for the 80% reduction was to limit the total cap. The reason for directing this reduction to the energy sector is that this sector was considered to have the lowest reduction costs and best opportunity to pass on costs to its clients.

(3) Installations with exceptional events during the period 1998–2001 leading to at least 10% lower emissions this particular year than the 'normal' years in the period 1998–2001 receive allowances based on the 'normal' years. An exceptional event could for instance be that the operation had been closed due to rebuilding or operation failure.

(4) New installations that are known as of today receive allowances based on benchmarks and expected production. A total of 1.0 Mt CO_2 was allocated to these known new entrants in the NAP.

(5) New installations that are unknown as of today will receive allowances based on benchmarks and expected production up to a maximum of 0.8 Mt CO_2 on a first-come–first-served basis.

These adjustments were decided on through a political process considering the costs for reductions, ability to pass on costs, needs in the non-trading sector and the obligation to meet the Kyoto and national target (NAP 2004).

3.3 Issues in determining base years

When investigating what base year period to choose, the following effects were identified:

(1) Early base year:
- Actors that have carried out early actions will benefit, compared to the situation if a late base year is chosen
- There are more problems with new entrants and production changes that have occurred since the base year, compared to the situation if a late base year is chosen
- Problems in obtaining reliable data are greater the earlier the base year.

(2) Late base year:
- Fewer problems arise concerning new entrants and changes in production that have occurred after the base year
- Risks for stranded costs are reduced
- Actors that have carried out early actions will benefit less.

In the Swedish NAP the period 1998–2001 is used as the basis for the allocation. In the FlexMex2 report of May 2003, it was recommended that a base year period of 1998–2001 should be used, where 2001 was the year for which the latest data were available. This was said to best achieve the balance between rewarding early actions while satisfying the needs for installations with a significant increase in production. However, it can be questioned whether this base year period rewards early action at all.

4 Issues of coordination and harmonisation

4.1 Influence and the role of EU guidance and review

The government was clear in indicating to the European Commission that Sweden did not see the NAP guidance as a legal binding document. Nevertheless, the guidance was mainly followed when the NAP was written. The suggested structure of the NAP, given in an appendix to the guidance, was used. But the NAP guidance was made available too late in relation to the Swedish NAP process. When the EU NAP guidance was published, the government bill had already been presented to the parliament in December 2003. Sweden gave a number of inputs to the draft version of the guidance, as did other Member States (T. Borgström, pers. comm.).

4.2 Influence of signals and rumours from other Member States

It was difficult to consider other Member States' NAPs since they were to be submitted at the same time as the Swedish NAP. There was, however, some exchange of information between Member States in Working Group 3 (WG3) and in other informal fora. This did contribute to the result that Sweden decided to have a new entrants reserve, that Sweden did not auction, and that Sweden did not allow for banking between the first and second trading periods (T. Borgström, pers. comm.).

4.3 Influence of other EU or Member States policies

Three EU directives were identified by the government as having influenced the allocation:

(1) Directive 1999/32/EC with amendments in Directive 1993/12/EEC on reduction of sulphur content in certain fuels. Partly as a consequence of this Directive a new installation will be constructed in Lysekil consisting of a hydro cracker and a hydrogen manufacturing unit. The emissions from these installations will be about 0.8 Mt per year during 2006–2007. These installations are given special treatment in the allocation plan and will be allocated allowances corresponding to their projected emissions for 2006–2007. This is motivated by the fact that emissions from the transport sector is expected to decrease by 1.0 Mt as a direct consequence of using the low-sulphur fuel. Most of these reductions will take place abroad, while about 0.26 Mt of the reductions will be realised in Sweden.

(2) Directive 2003/96/EG on restructuring of the common framework for taxation of energy products and electricity. Sweden has a long history of taxing energy and CO_2 emissions. The EU Directive is not likely to change the Swedish emissions, since the minimum taxation rates are already satisfied.

(3) Directive 2001/77/EG on promotion of electricity from renewable energy sources. Sweden's strategy is to introduce a system for certificates for electricity from renewables. This system is expected to reduce the future CO_2 emissions (NAP 2004).

5 Issues deserving special mention

5.1 Auction

Sweden does not use the option to auction allowances.

Sweden's view on auctioning is in general positive. The first FlexMex report (April 2000) proposed that all allowances should be auctioned. In the EU negotiations on the Directive Sweden advocated the use of auctioning as a basis for allocation. In the government bill of December 2003 Sweden states a number of advantages of auctioning. Auctioning is consistent with the polluter-pays principle. With free allocation, a government may over- or under-allocate. With auctioning, the installation will buy as many allowances as needed. In free allocation there is an incentive for the operators to exaggerate the need of allowances.

When Sweden's NAP was submitted it was therefore surprising that Sweden did not use the auctioning option, in spite of the directive allowing up to 5% of the allowances to be auctioned. There are two reasons for this. First, the FlexMex2 Commission did not investigate the possibility of auctioning allowances, since according to the EC and the EU parliament all allowances should be issued freely. This was different from the final Directive under which up to 5% of the allowances may be auctioned. Second, at the time when the government presented the bill (December 2003) there were no indications that any Member State would use the option of auction. At this point, if Sweden had unilaterally used the auctioning option this would have put Swedish industry at a disadvantage relative to other Member States.

The government therefore suggested that Sweden should not use the option of auctioning. In the parliamentary bill, the government stated that the focus of the first trading period should be to establish a well-functioning, trustworthy system. The government further stated that it was of high importance to analyse the possibility of using auctioning in the period 2008–2012. In the final FlexMex2 report of 2005, Sweden suggested that for the period 2008–2012 all Member States should auction 10% of the allowances.

5.2 New installations

A total of 1.8 Mt CO_2 are reserved for new entrants. Of this, 1 Mt is reserved for installations that already have permits according to Swedish law, while 0.8 Mt is reserved on a first-come–first-served basis for installations that do not have permits according to Swedish law or are unknown to the authorities.

The allocation to new entrants will be based on benchmarks when these are available. When benchmarks are not available, best available technology (BAT) will be the basis for allocation of allowances.

The allocation of allowances to new energy installations will be restricted to CHP. Neither new hot-water plants nor new condensing power plants will receive free emission allowances, which is a rule unique to Sweden. Benchmarks have been developed for this purpose. CHP is allocated allowances for both heat and electricity production. For electricity the benchmark is 265 t CO_2/GWh and for heat production 83 t CO_2/GWh. These benchmarks have been based on data of actual emissions and production from Swedish CHP, condensing plants, heat boilers and industrial CHP, including both fossil and

bio-fuels. Since bio-fuels are included in the calculation of the benchmark, and bio-fuels are a large part of Swedish electricity and heat production, the resulting benchmarks are very low. The allocation to new
CHP will be calculated as the projected production times the benchmark times a scale factor of 0.8. For a gas-fired CHP, this allocation
will satisfy approximately 60% of the total emissions, which is considerably lower than the other Member States' new entrants reserve. The
purpose of the restriction to allocate only to CHP is probably to create incentives that new hot-water plants should be based on bio-fuels.
Concerning condensing power plants, there are no plans for building
such new installations in Sweden.

There is a problem in having different allocation rules for incumbents
and new entrants. For example, new entrants in the energy sector will
receive a smaller allocation than the corresponding plants that were
built before 2002 (O. Hansén, pers. comm.).

5.3 Opt-in

Sweden has used the possibility of opting in energy installations below
20 MW in district heating nets where total net capacity is above
20 MW. This means that about sixty installations have been opted in to
the emission trading system. According to the FlexMex2 Commission
a district heating net can be regarded as one installation even if there
are several boilers connected to it. Therefore, the government is of the
opinion that the trading sector in Sweden should include boilers with a
capacity below 20 MW where the total capacity of the heating net is at
least 20 MW. The rationale for this may be that boilers belonging to the
same net should be subject to the same allocation rules. Large installations (above 20 MW) should not have a competitive disadvantage
compared to small installations (below 20 MW).

5.4 Opt-out

No opt-out has been notified. The government does not rule out that
an opt-out application may be submitted to the EC at later stage, but
the NAP has not identified such an exception.

5.5 Ex-post adjustments

No *ex-post* adjustments will be made. The perhaps most important
issue here is the *treatment of closures*. Sweden and the Netherlands

are the only Member States that let the owner of a closed installation keep the allowances that have been allocated previously. In Sweden, if a plant is closed, the emission allowances continue to be the property of the plant owner as long as the owner keeps its permit to emit CO_2. If allowances were to be confiscated when a plant is closed, the allocation could be considered as a production subsidy, since it is conditioned by the fact that production continues. This would in turn create the perverse incentives to keep installations running that are inefficient from a CO_2 emission point of view. The fact that the plant has been closed can also be seen as a form of reduction of emissions. Moreover, what should be considered a closure? How large does the reduction need to be to be considered a closure? This is clearly a difficult question of definition.

5.6 Early action

Sweden has not rewarded emission reduction actions taken before 1998 in the NAP. On the other hand installations that have implemented reduction measures in the period 1998–2001 will be rewarded for these 'semi-early actions'.

5.7 Clean technology

The allocation of allowances to new energy installations shall be restricted to CHP. Neither new hot-water plants nor new condensing power plants will receive free emission allowances, which is uniquely a Swedish rule.

The allocation to new entrants will be based on benchmarks when these are available. When benchmarks are not available, BAT will be the basis for allocation of allowances (NAP 2004).

6 Other comments on the allocation process

6.1 Alternative methods of allocation

At an early stage the FlexMex2 Commission invested a lot of effort in investigating alternative allocation methodologies. In order to assess different options a Swedish list of criteria was created and used as guidance along with the EU guidance document. Different options were discussed with experts, industry and politicians. There was clearly a high ambition to develop the 'best possible' allocation principles. It was therefore something of an anticlimax when the final proposal was presented, proposing an allocation methodology that was quite simple with hardly any considerations of CO_2 efficiency.

6.2 An unambitions cap?

It was also apparent that the first cap of 25.2 Mt (May 2003) was not very ambitious. It was based on historic emissions, with a number of add-ons, like projected increases in process emissions, provisions for data uncertainty and provisions for new entrants. This 'generous' cap was later reduced to the final 22.9 Mt after political negotiations. The reason for this may have been that the FlexMex2 Commission was afraid that a lower cap would put Swedish industry at a disadvantage compared to that of other Member States. In some aspects this turned out to be true. When the other NAPs were presented, Sweden's allocation to the energy sector and especially to new entrants was much stricter than in other Member States.

6.3 The steel sector

It was surprising how generous the allocation was to SSAB, the Swedish steel company which dominates this sector. The allocation corresponded to their projected emissions and was based largely on unverified data supplied by SSAB.

6.4 The energy sector

In contrast, allocation to the energy sector (electricity and heat production) was more restrictive. First, the selection of base years was a disadvantage to the sector since emissions, due to climatic reasons, were low in these years. Second, allocation had been set to 80% of base year emissions. Third, allocation to new entrants in the energy sector is considerably lower than for all other Member States in the Baltic region. New condensing power plants and hot-water boilers will receive no allocation at all.

6.5 The trading sector

In total, the allocation to the trading sector is 13% above current emissions. However, Sweden can still increase emissions in the non-trading sector before reaching its Kyoto ceiling.

6.6 The non-trading sector

There has been very little focus on consequences on the non-trading sector. FlexMex2 commissioned computer simulations to be done and they clearly showed potentially large consequences on the non-trading sector (SOU 2003). But in public debate little attention was given to this issue. This is surprising, since the allocation to the trading sector will have a direct effect on how much pressure needs to be put on the non-trading sector in order to reach compliance with Kyoto targets.

6.7 Incentives for abatement

In Sweden, there has been a lot of attention on the allocation process rather than on the effects of the emission trading system itself. Some companies argue that if CO_2-efficient companies are not rewarded in the allocation there will be no incentives for abatement. I do not agree with them. It is true that allocation can be a very important revenue. For new entrants, the value of the allocation can be comparable to the annual cost for the investment (Åhman and Holmgren 2006). But the incentives for abatement are created from the cost for carbon emissions and not from the individual allocation. One can ask whether this is what companies really believe or whether they were just making noise in order to increase their allocation.

6.8 Benchmarking

There is currently a considerable support for using benchmarking in Sweden, both among governments and industry (see for instance SOU 2003; SKGS 2004). The main arguments are that benchmarking rewards CO_2 efficiency, it is seen as a driver towards more CO_2-efficient processes and it is perceived to be fair. However, experiences to date have shown that benchmarking is often associated with problems concerning data retrieval and the definition of product groups. As the EU ETS moves into its second and consecutive periods, benchmarking may provide the means of updating the allocation of allowances without introducing incentives for increased emissions.

6.9 Other sectors

The steel, refinery and energy sectors were very active throughout the whole process preceding the Swedish NAP. It was therefore surprising how little was heard from the Swedish pulp and paper industry and from the cement industry in this context. Were they unprepared or did they actively choose to maintain a low profile? The pulp and paper industry says that allocation was not their main concern. They were much more concerned about the effects of the EU ETS on electricity prices.

6.10 Increased electricity prices

The increased electricity prices seem to have come as a surprise to many and this issue is currently the subject of a massive debate.

6.11 Credibility of the ETS system

In the opinion of the public in Sweden, there are a number of factors that have undermined the credibility of the system. First, if the trading sector can increase their emissions according to business as usual and Sweden/Europe still can reach its Kyoto target, then the Kyoto target is probably not stringent enough. Second, it is becoming clear to the public that the price of electricity and heat will increase due to the cost of carbon emissions. In the end, the consumer will pay for this difference. So even if energy producers have received allowances freely, they will be able to increase their revenues and increase their profits. Moreover, over 95% of Swedish electricity production is carbon-free, largely based on hydro and nuclear generation. District heating is also to a large extent CO_2-free. These companies will also increase the price of energy to the same level as the fossil-fuel-based production and thereby considerably increase their profits, referred to as windfall profits. It is a challenge to explain to the public that an increase in the price of electricity and heat is a natural consequence of introducing a price on carbon. This will in turn lead to the introduction of CO_2-free energy production and the phase-out of fossil-fuel production, which is one of the objectives of the EU ETS.

6.12 Taxation

Historically, CO_2 tax and energy taxes have been very important in reducing CO_2 emissions in Sweden. However, the tax pressure imposed by the state has been unevenly distributed over Swedish society. While process-based industry has been exempt from CO_2 tax, a high tax has been introduced in the transportation sector, in industrial heat production, in the housing sector and in private consumption. This imbalance in pressure is also reflected in Sweden's allocation plan where process-based industry is given a generous allocation, while energy production is given a much tougher allocation.

6.13 Peat

Peat is used in Sweden as a fuel in electricity and heat production and corresponds to about 1.3 Mt CO_2 per year. Currently, peat is exempt from CO_2 tax. Moreover, if peat is used as a fuel for electricity production, the operator will receive green certificates, which can be sold. With the introduction of the EU ETS, operators using peat will need to acquire allowances, which will increase the operators' costs for using peat considerably. Calculations have shown that if allowances prices rise above €20-25 per ton CO_2, operators will shift from peat to alternative fuels, such as coal (cheaper), or natural gas or bio-fuels (lower emissions). As a result it is possible that the use of peat for energy supply will be significantly reduced.

6.14 Harmonisation

The allocation process has demonstrated that some issues may need tighter rules/harmonisation, in particular the transparency of cap-setting and projections calculations, harmonised rules for allocation to new entrants, for the treatment of closures and for the use of auction and requirements on data quality and verification.

References

Åhman, M. and Holmgren, K. 2006. 'Harmonising principles for allocating emission allowances to new entrants in the Nordic electricity market', IVL report B1679, commissioned by the Nordic Council of Ministers.

Energiläget 2004. Report from the Swedish Energy Agency, ET 17:2004. www.stem.s

NAP 2004. 'Sveriges Nationella Fördelningsplan (The Swedish NAP)', April 2004, Promemoria 2004-04-22. Näringsdepartementet. Regeringskansliet, 10333 Stockholm.

Regeringens Proposition 2003. 'Riktlinjer för genomförande av EU:S direktiv om ett system för handel med utsläppsrätter för växthusgaser', Regeringens Proposition 2003/04:31 (government bill of 4 December 2003), Näringsdepartementet. Regeringskansliet, 10333 Stockholm.

SKGS 2004. 'En Europeisk benchmarkingmodell för tilldelning av utsläppsrätter', Promemoria from SKGS, 10321 Stockholm.

SOU 2000a. 'Förslag till Svensk klimatstategi (Proposed Swedish climate strategy)', Statens offentliga utredningar, 2000:23, Miljödepartementet. Regeringskansliet, 10333 Stockholm.

SOU 2000b. 'Handla för att uppnå klimatmål (Report from first FlexMex Commission, April 2000)', Statens offentliga utredningar, 2000:45, Näringsdepartementet. Regeringskansliet, 10333 Stockholm.

SOU 2003. 'Handla för att uppnå klimatmål (First report from FlexMex2 Commission, May 2003)', Statens offentliga utredningar, 2003:60, Näringsdepartementet. Regeringskansliet, 10333 Stockholm.

SOU 2005. 'Handla för att uppnå klimatmål- från införande till utförande (Fourth and final report from FlexMex2 Commission, January 2005)', Statens offentliga utredningar, 2005:10, Näringsdepartementet. Regeringskansliet, 10333 Stockholm.

STEM 1999. 'Energiläget 1999', Statens energimyndighet.

Zetterberg, L., Nilsson, K., Åhman, M., Kumlin, A.-S. and Birgersdotter, L. 2004. 'Analysis of national allocation plans for the EU ETS', IVL report B1591.

7 | *Ireland*

CONOR BARRY

1 Introduction

1.1 Background

1.1.1 Economic growth

Ireland's economic growth over the past decade has been well documented; however, it is worth restating the key facts here to provide background to the challenges faced by Irish policy-makers in general and the context in which allocation decisions were being made. As 1990 is the Kyoto base year and 2001 and 2002 were the latest years for which data were available during the decision-making process on the Irish National Allocation Plan (NAP) these are the years that will be most often cited throughout this chapter.

- Ireland's gross domestic product[1] more that doubled between 1990 and 2002
- The index of industrial production in 2002 was almost four times that of 1990
- Electrical output increased by 67%
- Unemployment has fallen by a factor of four since 1993 (16% to 4%).

1.1.2 Emissions profile

Driven by the high rate of economic growth Ireland's reported emissions in 2001 were 31% above the 1990 level, and while the EU15 committed itself to a reduction of 8% below the 1990 level under the Kyoto Protocol, the individual target for Ireland, as agreed under the EU's Burden Sharing Agreement (BSA), was a limitation in emissions growth to 13% above 1990. Stakeholders, external to the Irish administrative

---

[1] GNP has grown more slowly due to the impact of foreign direct investment. However as the Kyoto Protocol applies to emissions within the national territory GDP is the most accurate indicator against which to mark performance.

system, hold two widely divergent views of this target. The first, held by officials from other EU Member States, non-governmental organisations (NGOs) and some independent commentators, is that the target is a major concession to Ireland in addition to already significant transfers under the structural and cohesion funds. The second, held in the main by Irish industry, is that the target was badly negotiated and does not sufficiently account for Ireland's economic expansion. Proponents of this second view focus primarily on the more lenient targets of Greece, Spain and Portugal, to the exclusion of the more severe challenges faced by countries such as Denmark. The advent of emissions trading serves to exacerbate these views as the difference between the communal -8%, the notional $+27\%$ and the actual $+13\%$ can now be monetised. Pre-allocation estimates of the market price of allowances of between €10 and €20 show potential differences in the target to be of substantial value before the end of the first commitment period.

Ireland's greenhouse gas (GHG) emission profile is unique among European countries given the large contribution to total emissions from the agriculture sector and thereby the relatively low proportion from activities covered by the emission trading scheme in particular and of CO_2 in total emissions in general.

Agricultural emissions account for 35% of Kyoto base year emissions, and despite the ongoing industrial growth and increased electricity consumption, this sector still accounted for 29% of 2001 emissions. The activities covered by the ETS Directive account for approximately 33% of total emissions annually.

1.1.3 Political background

The current coalition government, made up of the centrist Fianna Fail (thirteen cabinet positions including prime minister) and their economically liberal partners (two cabinet positions including deputy prime minister) the Progressive Democrats (PDs), was first elected in 1997 and was re-elected in 2002. This government has won both elections predominantly on economic issues, including its focus on low taxation and pro-business policies.

Following its re-election in 2002, the government faced an almost immediate fall in popularity due to deterioration in public finances caused primarily by the downturn in the global economy and the high rate of inflation, including wage inflation; there were also concerns over the competitiveness of Irish industry, in particular the ongoing ability of Ireland to attract foreign direct investment to the same extent as previously. This political dynamic was present throughout the period from the 2002 general election to the 2004 local and European elections.

1.1.4 Policymaking background

Social partnership, credited by many as being the genesis of Ireland's economic growth, has been a key constituent of public policy in Ireland since 1987. The principle of social partnership is agreement by trade unions to industrial peace and wage moderation in exchange for reductions in income taxation. However, through numerous rounds of 'partnership agreements' the agenda has spread into most aspects of public policy. The most visible involvement of the social partnership model in emissions trading was the establishment of an Emission Trading Advisory Group (ETAG), through which the views of Irish industry on every aspect of the ETS Directive were sought. However following the adoption of the Directive, as the policy system moved from negotiation to implementation, the channels for direct input of industrial stakeholders became more formalised. This was a result of the 'entity-specific' nature of the input being received, i.e. it became less equitable and practical to rely on the views of a representative sample of views with respect to decisions regarding individual companies.

An additional ingredient to the policy-making mix during the development of Ireland's NAP was Ireland's Presidency of the European Council during the first six months of 2004.

1.2 Regulatory structure for emissions trading

Ongoing investigations by tribunals of enquiry into various aspects of political corruption in the 1980s lead to an early decision that the distribution of CO_2 allowances would be conducted by a body independent of the political process with government retaining responsibility for determining the total quantity of allowances to be allocated and the power to give general policy direction regarding their distribution. The establishment of a new agency for emissions trading was not possible due to restrictions on public service recruitment, and though various energy and industry agencies were considered the Environmental Protection Agency (EPA) was empowered with the task of distributing the total quantity. This would provide synergy with the task of permitting the participants and as the EPA already undertook regulatory decisions through a Board of Executive Directors would not require organisational restructuring.

2 Total quantity of allowances

2.1 Background

2.1.1 National Climate Change Strategy

Ireland's National Climate Change Strategy (NCCS) was first published in October 2000, in advance of the seventh conference of the Parties (COP7) to the United Nations Framework Convention on Climate Change (UNFCCC) at Marrakech. While this strategy contained references to the Protocol's flexible mechanisms and a commitment to develop a carbon tax by 2002, it focused on identifying actual measures to reduce emissions in each sector without identifying the policies to be used to ensure that such measures were taken.

The NCCS projected a gap of 13 Mt between business-as-usual (BAU) emissions and Ireland's burden-sharing target, and identified 15.4 Mt of reductions. These reductions included switching the coal-fired Moneypoint power station to gas (3.4 Mt), increased use of renewable energy (1 Mt), reducing fuel tourism (0.9 Mt), process substitution in cement production (0.5 Mt), reduction in fertiliser use (0.9 Mt), additional forestry sequestration (1 Mt), improved waste management (0.85 Mt) and reduction in the national bovine herd and altered feeding regimes (1.2 Mt). These, and all of the other reduction measures identified in the NCCS, were to be delivered by the relevant government department with coordination by a cross-departmental Climate Change Team.

Uncertainty regarding the cross-sectoral instruments (emission trading and carbon taxation) and the entry into force of the Protocol undermined efforts to progress policies sufficient to deliver the identified measures.

2.1.2 Carbon tax

The NCCS commitment to introduce carbon taxation 'on a phased and incremental basis' from 2002 was restated in the Programme for Government after the 2002 general election. Pre-budget discussions regarding changes to taxation are conducted through a Tax Strategy Group (TSG) of senior officials from relevant government departments.

In the late 1990s, due to the increased prevalence of ecological tax reform in OECD countries, a 'Green Tax' sub-group of the TSG was established to prepare proposals on environmental taxation, including carbon taxes. The first detailed proposal regarding carbon taxation was presented to the TSG in late 2001 in advance of December's Budget. Given that 2002 was an election year the announcement of a new tax in the Budget was highly unlikely. However, this proposal did set out a framework for carbon taxation including the need to introduce it on a pre-announced phased basis, its interaction with other policy instruments (including voluntary agreements and emissions trading), and the need to address revenue recycling and fuel poverty. No rate of taxation was suggested in this proposal.

Further proposals were brought to the TSG in advance of the Budget in December 2002. This proposal suggested a rate of €7.50 per tonne rising over three to four years to €20 per tonne (the assumed price of CO_2 allowances), with a complete exemption for those engaged in emissions trading. Despite concerns at the time over the impact on inflation, due to the potential-longer term impacts of the Irish Exchequer of not reducing emissions, in his December 2002 Budget speech the Minister for Finance announced that 'the Government has asked the relevant Departments to advance the plans for a general carbon energy tax, with a view to introducing this from the end of 2004'. Due to this announcement, preparations for Ireland's Presidency of the European Council, and the ongoing negotiation and implementation of the ETS Directive little further work was devoted to carbon taxation in 2003. It was anticipated that a final carbon tax proposal would be developed in tandem with energy tax changes required by the Taxation of Energy Products Directive in advance of Budget 2005 in December 2004, once the pressures of the Presidency were over.

2.1.3 Flexible mechanisms

As the NCCS was published before the Marrakech Accords and ETS Directive were agreed it contained no detailed provisions with respect to the use of flexible mechanisms. The lack of a defined policy in this area was a matter of concern to business interests who considered it essential for the Government to engage in international emission trading to remove the need for domestic action, pointing at the activities of the Dutch ERUPT and CERUPT programmes in particular. However, the Department of the Environment did not particularly favour such a method of compliance, nor would it have been likely that the Department of Finance would have agreed to the release of funds to comply with a Protocol that had not yet entered into force.

2.2 Process

The decision to appoint the EPA with responsibility for implementation of the ETS, including the distribution of allowances, was made by the Cabinet of Ministers in July 2003. A government commitment to cut the number of public-sector employees by 5,000 made this task particularly difficult and in the end the additional staff and funds required by the EPA had to be met by reductions in the key government departments involved in the Climate Change Team: Environment, Heritage and Local Government; Enterprise, Trade and Employment; Finance; Communications, Marine and Natural Resources; Agriculture and Food; and Transport. The latter two departments had not previously been engaged with the ETS, as their areas of responsibility had no direct involvement. The decision regarding the total quantity to be allocated would however have implications for the burden to be borne by other emissions sources including the transport and agriculture sectors.

It was also agreed in July 2003 to appoint consultants to advise the government as to the appropriate total quantity of allowances to be made available. A steering committee for this consultancy was established involving all key departments above. The first task of this committee was to draft terms of reference for the consultants, a task that would determine to large extent the outcome of the study. The importance placed on the competitiveness effects of the ETS on Irish industry was the most difficult issue to resolve with a dispute over whether or not minimising the macroeconomic cost of reductions or the microeconomic effects of allocation would be given priority. Given that the distribution of allowances would be a separate body of work to be undertaken by the EPA focus was kept on the macroeconomic costs.

Throughout the period of the decision-making process on the total quantity (2003/early 2004) a number of related policy decisions were also being brought before government, including *inter alia* medium-term energy policy issues such as the development of an electricity interconnector to Britain and the retrofitting of Moneypoint with SO_2 scrubbers and selective catalytic reduction for NO_x emissions as part of Ireland's compliance strategy with the National Emission Ceiling Directive, the finalisation and implementation of the Energy Product Tax Directive, and the mid-term review of the Common Agricultural Policy. The impact of Ireland's Kyoto commitments on each of these, and other, policy areas served to heighten the profile (within administrative circles) of the decision on the total quantity. The impact of this increased profile was to provide a ready-made pathway to cabinet, through a sub-committee on Housing, Infrastructure and Public Private Partnership through which most energy policy issues were being progressed. A high-level working group, chaired by the Department of the Taoiseach (Office of the Prime Minister), reported to this sub-committee and offered pre-established forum inter-departmental negotiations on the total quantity. A representative from the Department of the Taoiseach was also appointed to the consultancy steering committee.

The selection of consultants who would satisfy the analytical requirements of the varied government departments proved particularly difficult, and eventually a consortium of ICF Consulting, Byrne O'Cleirigh (BOC) and the Economic and Social Research Institute (ESRI) was appointed. As stated above the terms of reference for the consultants focused on the macroeconomic elements of the decision. This was to be achieved through two distinct tasks:

- the preparation of 'with current measures' projections on a sectoral basis to 2012
- an assessment of the cost of available emission reduction measures in each sector.

On the basis of this analysis the terms of reference required a recommended allocation for the first commitment period and the pilot phase of the ETS. The terms of reference also required recommendations in relation to a quantity of Kyoto credits to be purchased by government and in relation to the treatment of new entrants in the ETS.

2.3 Emission projections

The preparation of economy-wide sector-level emissions projections to 2012 was hampered by the data constraints, time constraints (the task needed to be completed in three months) and the need for agreement among departments on basic input assumptions. The overall gap between Ireland's BSA target and projected emissions over the first commitment period was estimated as 9.2 Mt per year, an overshoot of 15%. This BAU projection was the sum of a number of sector-level projections harmonised by the use of economic growth forecasts from the ESRI's Medium-Term Review. Each of these sector-level projections was based on a number of simplifying assumptions.

(1) **Energy**

Future emissions from electricity generation were derived from ICF Consulting's *Integrated Power Model (IPM)*, which is a full model of the European electricity market including data on all generating units and transmission systems in the EU. The projections were the result of both the IPM and the imposition of the following exogenous variables:

- Base case electricity demand is imposed, to ensure consistency with projections for all other sectors. This demand is modelled by the ESRI as part of their Medium-Term Review of the Irish economy and predicts annual demand increases of 3.4% until 2009 and 2.3% thereafter, derived from ESRI projected economic growth rates.
- The three peat-fired power stations, supported by a Public Service Obligation Levy for security of supply purposes, were imposed as 'must-run' generating units.
- The achievement of Ireland's target under the RES-E Directive (2001/77/EC) of 13.2% of gross national electricity consumption being supplied by renewable sources by 2010.
- The construction of a 500 MW DC interconnector between Ireland and the UK in 2009.

The final projections regarding the importation of electricity from Northern Ireland were the subject of much debate and the outturn will be of substantial interest in preparing subsequent allocation plans.

(2) **Industry**

Projected emissions from manufacturing industry were estimated from a combination of a bottom-up survey by ICF/BOC and top-down modelling of energy demand by ESRI.

Verification of these data and the estimation of the split between ETS participants and non-participants was conducted by ICF/BOC with reference to work carried out by the Statistical Support Unit of Sustainable Energy Ireland. This work is based on historical energy consumption profiles obtained from data underpinning the Central

Statistics Office's Census of Industrial Production. The annual rate of increase in base case emissions is 3%.

(3) **Agriculture**

Full decoupling of agricultural subsidies from production took place in Ireland in 2005. Teagasc, the national agricultural research and advisory body, estimated the effects of this policy on production and consequently on emissions. This estimation was conducted using models developed by the FAPRI–Ireland Partnership at Teagasc. Both macro and farm-level models were used in assessing the outlook for Irish agriculture, with input from the Global Agricultural Outlook provided by FAPRI at the University of Missouri. The effect of decoupling will be to reduce emissions in 2010 to 1.8 Mt CO_2e below 1990 levels, and 1.3 Mt CO_2e below the *no decoupling scenario* modelled by Teagasc.

(4) **Transport**

Growth in emissions from the transport sector reflects an increase in the stock of cars from 796,000 in 1990 to 1.38 million in 2001. However the growth in fuel consumption in Ireland also reflects the lower retail price of transport fuels here than in the United Kingdom, and to a lesser extent continental Europe. Cross-border fuel trade has increased, and is continuing to increase Ireland's reportable emissions from the sector. The projections of demand for transport fuels were based on the ESRI Medium-Term Review. Demand for fuel is then derived as a function of the stock of cars and of the fuel price differential relative to the United Kingdom.

(5) **Waste**

Projected emissions from the waste sector were based on levels of regional waste generation and on the implementation of Regional Waste Management Plans. The implementation of these plans has resulted in increased recycling rates and this trend is projected to continue. However due to the time lag in decomposition (CH_4 emissions from landfills peak three to seven years after materials are deposited) the benefits of this recycling will not accrue immediately. The uncertainty regarding the diversion of waste from landfills to planned municipal waste incinerators also complicated projections in this sector.

(6) **Forestry**

The production of removal units (RMUs) in Ireland under Articles 3.3 and 3.4 of the Kyoto Protocol have been modelled by the Irish

Table 7.1 *Ireland's distance to target, with costs of effective abatement (Mt CO_2e)*

	ETS sector	Non-ETS sector	National emissions
Kyoto Target			60.365
Base case projections	26.350	44.904	71.254
RMUs (Forestry credits)		1.720	
Reductions @ < €10/t	2.232	1.160	
Emissions after measures	24.118	42.024	66.142
Distance to Target			5.800

Forestry Research and Development Council (COFORD) using the CARBIWARE model based on extensive Irish research and consistent with Integrated Pollution Prevention and Control (IPPC) good practice guidelines. The effect of the increased planting under the national forestry strategy will be to produce 1.72 million RMUs per year over the first commitment period. This represents almost 3% of Ireland's BSA target.

2.4 Cost of abatement assessment

The consultants estimated that Irish emissions could be reduced by 3.4 Mt per year at a cost of less that €10 per tonne, including by means of carbon taxation at this rate. As €10 was estimated to be the likely international price of allowances it was not considered prudent or efficient to undertake reductions domestically that cost more than this. The outcome of the cost assessment is summarised in Table 7.1 above.

This assessment indicated that insufficient cost-effective measures were available domestically for Ireland to meet its BSA obligations and that the purchase of additional credits via the Kyoto Protocol's flexible mechanisms would be the most efficient method of closing the remaining compliance gap.

This determination did not resolve the issue of *who* should be responsible for the purchase of such credits; the state or the private sector participants in the ETS. The aim of the consultancy process was to determine responsibility by continuing the equi-marginal assessment process, i.e. attribute responsibility to the sector with the next cheapest domestic emission reduction opportunity. However the reliability of this process at costs in excess of €10 per tonne became questionable.

2.5 Consultants' recommendation

Attempts to distribute the burden on the basis of marginal cost analysis as originally planned proved to be difficult given the increasing uncertainty with marginal cost analysis at beyond €20 per tonne. This led to the eventual recommendation that the burden be divided on a pro-rata basis according to share of national emissions. The state would therefore be required to purchase 3.7 million allowances on behalf of the emission sources not directly participating in the ETS, while the recommended allocation to the direct ETS participants was reduced by 2.1 million per year. With this split made for the first commitment period the recommended total quantity for the pilot phase was based on the BAU emissions less reductions available domestically at a cost of less that €10 per tonne. The recommended allocation was therefore 22.148 million per year, or 96.5% of base case emissions.

2.6 Finalising the decision

This proposal was first brought into the Cabinet process in January 2004, the first month of Ireland's Presidency of the European Council. Two key issues arose at this point: Department of Finance concerns over the commitment to purchase 3.7 million allowances per year for five years, and the longer-term Exchequer implications of this. As this commitment was being made in an indicative sense only these objections were addressed.

However when the final proposal was presented to the Cabinet sub-committee separate projections based an industry survey were juxtaposed with the consultants' projections and were the cause of substantial concern that the recommended allocation of 96.5% was in fact substantially less than this. The final government decision was therefore to increase the total quantity to be allocated to 22.5 million per year.

2.7 Other decisions

In addition to the decision on the total quantity of allowances the government gave the EPA general policy direction in relation to distribution. This included a direction to retain 1–2% of total allowances for new entrants, not to make allocations in subsequent years to installations that close and to auction up to 1% of the total quantity to cover the costs of administering the scheme.

3 Distribution of allowances

3.1 Background

The EPA was first established on a statutory basis in 1992, with responsibility for environmental monitoring, IPPC and waste licensing and enforcement. The EPA is a independent body, whose decisions are made through a board of four executive directors and an executive director-general. The board does not have any non-executive members. In line with Ireland's social partnership model an advisory group, with representation from the social partners, assists the board. This group does not have decision-making powers.

When the EPA was given the task of distributing allowances there was some concern in administrative circles regarding the EPA's record in dealing with industry concerns and the need to integrate emission trading with other policy areas. For this reason the government appointed a National Allocation Advisory Group (NAAG) to assist the EPA. The NAAG contained the CEOs/DGs of the Commission for Energy Regulation, Sustainable Energy Ireland, Forfás (the State Enterprise Group), the National Treasury Management Agency, the EPA, and independent chairperson Dr Ed Walsh, president emeritus of the University of Limerick. Dr Walsh had substantial experience in chairing consultation and advisory groups on behalf of the government in the areas of science, research and enterprise policy. All of these members were from the public sector and the absence of direct industry representation was criticised by industry lobby groups. However given the sensitive nature of the task the government considered it inappropriate for industry to be directly involved in an advisory group.

3.2 Consultants

The EPA appointed Indecon Economic Consultants and Enviros to advise to a basic distribution metric. The steering committee for this consultancy included members of staff from the EPA and members of the NAAG. In preparing their recommendations the consultants assessed the use of historic emissions, recent emissions, projected emissions, production output and input capacities. The consultants also assessed the competitiveness impacts of the ETS. A problem, unique in Ireland due to its magnitude, was the existence of large developments due to occur just before or during the pilot phase of the ETS. In particular two peat-fired power stations were due to be completed in mid to late 2004 and the Commission for Energy Regulation was in the closing stages of a tender process to award contracts for the construction of two new gas-fired power stations. These developments were additional to the ongoing expansion in manufacturing industry, particularly in the pharmachem and semiconductor sectors. Also a number of older peat-fired power plants were scheduled to be decommissioned. An early commitment had been given that such large developments would be allocated under the NAP rather than face the uncertainty of an application as new entrants.

The consultants' recommendations were to use a two-stage distribution based on recent historic emissions, first allocating to economic sectors and then to installations, to allocate to known planned developments (KPD) from within the sectoral allocation and not to make adjustments for competitiveness impacts or early action. Recent years had the advantage of accounting for recent economic growth and provided transparency through verification, something that could not be achieved by the use of projections.

3.3 Executive

The EPA adopted the basic recommendations of the consultants, and developed a full distributional methodology. The two-stage process was maintained with all sectoral allocations being calculated on a pro-rata basis from emissions from installations in that sector in the years 2002 and 2003. However unlike the consultants' recommendation to use economic sectors it was considered more appropriate to conduct the sectoral breakdown on the basis of the activities (Annex I of the Directive) for which the installation had been permitted. The only installations included in this calculation would be those issued with a GHG emission permit. This specifically excluded the peat-fired power stations, which would not be operational in 2005, and other industries that closed, including a large fertiliser manufacturing plant. A further adjustment was made to the historic emissions from the power generation sector to account for the increased penetration of renewable energy.

The installation level distribution in each sector was calculated on a pro-rata basis from 2002 and 2003 emissions. In the case of installations where this was less than 90% the three highest years of 2000–2003 the latter could be used. In cases where full historic emissions were not available installations' base emissions were calculated on the available months or, in certain cases, on projections. The proposed allocations in the NAP submitted to the European Commission were based solely on unverified company estimates of their historic emissions, as supplied in their GHG emission permit application forms. An interesting element of this permit application process was the large number of installations with less than 20 MW installed capacity seeking to opt in to the ETS. These opt-in applications were driven by two factors. First, the expectation was that a carbon tax would be introduced in late 2004 and that those in the ETS would receive an exemption from the tax. Second, those companies with some installations included in the scheme on a mandatory basis wanted to include all of their installations in a single regulatory regime. As the total quantity of allowances available was fixed exogenously the EPA decided not to accept applications for opt-in.

The historic emissions data supplied by companies was verified by inspectors from the EPA throughout the summer of 2004, resulting in changes to the proposed distribution of allowances.

The fact that the EPA was also responsible for issuing GHG emission permits allowed the allocation plan to be integrated with the permitting process. Because of this new entrants were defined as those who received a permit, or an updated permit for new capacity, after the submission of the NAP to the European Commission. While this definition served to simplify the processing of new entrants it resulted in a major complication to the calculation of baseline emissions for installations that had made increases in installed capacity during the base years. While this was relatively straightforward for combustion installations as the capacity of new boilers could be readily verified, complications began to occur as the EPA discovered that process industries, such as cement and lime, might have made alterations to their installed capacity during the base years. The Directive defined the installed capacity of such installations as being an achievable production level measured in tonnes per day. The actual historic output of each installation on a daily basis could be measured with sufficient reliability but attributing changes in actual output to changes in capacity proved more difficult. After all an increase in output could be for market reasons and changes to the installation might just realise rather than increase existing capacity. Conducting this assessment proved to be a classic case of information asymmetry where an operator will always have a more detailed understanding of their installation and its processes than a regulator.

Ultimately this issue required the EPA to hire outside expertise to resolve, and once each change to each installation had been assessed and its impact on installed capacity determined the EPA had to determine whether or not the changes could be treated as part of the NAP or whether they were new entrants, i.e. whether or changes in capacity levels had been permitted before 31 March 2004, when the NAP was submitted to the European Commission.

Within the same sector, cement and lime, multiple submissions were also received regarding the treatment of process emissions. These submissions focused on the irreducibility of such emissions and the treatment of this in other Member States. The EPA's consideration of such submissions, and ultimate treatment of process emissions was based on three points:

(1) The aim of the ETS Directive was to provide a dynamic incentive for the use of low-carbon technologies. Therefore zero and low emission end use alternatives had to be considered in addition to the possibility or otherwise of engineering solutions to site specific reductions.

(2) The government's consultancy had identified significant potential for emission reduction in the cement sector by means of reducing the clinker content of finished cement.

(3) It would not have been equitable to employ the 'technological potential' criterion to this sector alone.

3.4 Other elements

The EPA decided to retain 1.5% of the total quantity for new entrants, to retain 0.75% for auction and to retain just less than half a million allowances for allocation to new combined heat and power (CHP) plants.

3.5 Consultation and finalisation

The first draft of Ireland's NAP was released for public consultation on 23 February 2004. The key issues, relating to distribution, during this period of consultation was on the treatment of early action, of those subject to international competition and the use of historic emissions rather than projections. Throughout the period that followed however no concrete suggestions as to how these issues could be addressed were presented to the EPA. The main change arising from this consultation was an alteration to the treatment of anomalous years during the base period, allowing installations to use the highest three of the four base years.

A second round of consultation begun on 30 September following the European Commission's assessment and approval of the plan. Again no concrete proposals for change to the distributional methods were presented, though numerous participants did express dissatisfaction. Following this consultation the EPA released a final national allocation methodology, which set out the set of rules by which allocations would be made to individual installations.

Following an assurance from the European Commission that this methodology was within the scope of the NAP approval, the EPA sent a final proposed allocation decision to each participant. This proposed final decision was the first publication to provoke a written threat of legal action in relation to the ETS in Ireland. Following comments received on this proposed allocation, and with corrections to mistakes found by installations the EPA took a final allocation decision under Article 11(1) of the Directive on 8 March 2005, just over five months beyond the legal deadline.

4 Coordination and harmonisation

Given the time invested by Irish officials in commenting on drafts of the European Commission's guidance document, and in attending official and informal meetings of Working Group 3 of the Monitoring Mechanism Committee it is surprising how little attention was given either to this guidance or to the likely outcomes in other Member States. The guidance document was however used as a basis for determining how to explain and defend decisions taken at a national level.

The granting of free allowances to new entrants was always going to occur from the moment the Directive was finalised. This was an imperative of both the Department of Industry and the Department of Energy and lobby groups. The economic rationale of allowing the market to determine new entrant choices was never going to win out against this practical opposition. While Irish policy-makers feared that Ireland's early decision (Ireland's was the third NAP to be released for public consultation and the new entrants decision was announced just days after the United Kingdom released its draft NAP) could lead to a race to the bottom in terms of attracting new investment with CO_2 allowances, and knew that Ireland could not compete with the new Member States in this regard, they also recognised that Ireland could not expect to force a scenario where all Member States would agree not to give allowances to new entrants for free.

The treatment of closures was a decision taken quite late in the process, as discussed in more detail below, and was taken entirely for Irish policy purposes, as was the decision to have a special set-aside for new CHP plants.

The magnitude of the gap between the total allocation and BAU needs was an area where politicians were interested in the decisions of other Member States; however, given the size of Ireland's gap to its Kyoto target it was recognised that Ireland's allocation would be less generous than that of major trading partners. Ultimately due to Ireland's Presidency of the European Council the Department of the Taoiseach was eager to ensure that Ireland met the legal deadline for submission of the NAP to the Commission. This required a decision on the total quantity to be taken in early February and every effort was made by officials in the Taoiseach's office to allow this to happen.

4.1 The Commission's assessment

Within a month of submitting the NAP, the Commission responded with a list of ten questions seeking further information to assist them in making a determination on the Irish NAP. This is effect meant that the Commission viewed the NAP as incomplete and that the three-month assessment process had yet to begin. Similar questions were asked of all Member States to extend the assessment process.

The assessment process also included a presentation to all Member States, through the Monitoring Mechanism Committee, whose opinion on the NAP becomes part of the Commission's assessment. The key issues arising from the Committee's assessment were the quality of the projections, the implementation status of policies in the non-ETS sectors, the potential for *ex-post* adjustment to certain allocations and the comparison of the allocation to historic emissions. This latter issue arose due to the historic emissions of installations due to close in 2004 not being included in the NAP assessment resulting in a perception of over-allocation. Additional data resolved this issue relatively quickly.

The first informal communication of the Commission's decision on Ireland's NAP was received in mid-June 2004 when many of the relevant officials were chairing EU expert groups at the meeting of the subsidiary bodies to the Conference of the Parkes (COP) in Bonn. DG Environment were proposing that the Commission reject the NAP on two grounds; first due to the potential for *ex-post* adjustment to certain allocations and second due to an over-allocation resulting for unsubstantiated policies on the purchase of Kyoto credits by the Irish government.

The first of these issues was not of great concern to Ireland. The adjustments (return unused KPD and CHP allocations to incumbents) had been introduced following the first round of public consultation. The Commission were taking a strong line against *ex-post* adjustments due to their potential to disrupt the allowance market and while Ireland did not believe that its proposed adjustments fell into this category the EPA did make the change to secure the Commission's approval.

The second issue was the cause of much anger within administrative circles in Ireland. Officials believed that the allocation was sufficiently below need for the pilot phase and did not accept the Commission's rationale for rejection. The Commission's argument appeared to be based on the premise that Ireland had not yet bought something in 2004 that it did not require to buy until 2013, and given that the Russian Federation had yet to ratify the Protocol might never have to buy. Officials also felt that the assessment was based on the administrative procedures of a larger Member State and did not take account of the fact that a small Member State such as Ireland could take and enact decisions such as that to purchase Kyoto credits relatively quickly and that Irish public finances were in relatively good shape in comparison to the rest of the EU. The calculation of the amount to be reduced also appeared to be entirely arbitrary and based on a lack of understanding of how the Irish NAP had been prepared. Ultimately being in the Presidency necessitated compromise by Ireland and the Minister for the Environment announced a reduction in the total quantity of allowances from 67.5 million to 66.96 million. The EPA applied this cut on a pro-rata basis across the board to all allocations and set-asides.

A final footnote to the Irish NAP, and its assessment was the decision in August 2004 by the then Minister for Finance, Charlie McCreevy, to announce that Ireland would not be proceeding with the implementation of a carbon tax. Most of the officials involved in the preparation of the NAP and earlier taxation proposals had been moved to new posts directly after the Presidency so the potential impact of this announcement on the Commission's approval did not seem to have been considered. Ultimately the DG Environment decided not to reopen the assessment process due to the large number of NAPs still to be assessed and the precedent that such a move might have made.

5 Issues deserving special mention

As the preparation of the allocation plan was separated into two tasks, this section will address the special issues in the Irish NAP firstly at the macro level and then at the micro level.

5.1 Macro level

In addition to a decision on the total quantity the Minister for the Environment, Heritage and Local Government retained the right to give policy direction to the EPA in relation to the distribution of allowances. This policy direction focused on three areas: the treatment of new entrants, the treatment of closures and the use of auctioning as a distribution mechanism.

The treatment of new entrants was particularly important from the perspective of the Department of Enterprise, Trade and Employment, who wanted to ensure that the introduction of the ETS would not act as a barrier to future foreign direct investment. The Department of Communications, Marine and Natural Resources (responsible for energy policy) was also keen to ensure that the ETS was implemented in a manner that did not obstruct the ongoing liberalisation of the electricity market and the need for new generating capacity caused by concerns over generation adequacy. Therefore the consultants' recommendation that new entrants buy all of their allowances on the market, while accepted as having a sound long-term economic rationale, was always unlikely to be accepted in practice. However, these departments were also keen not to reduce the amount of allowances available to incumbents, nor to be overly prescriptive in direction to the EPA. The outcome of these discussions was to direct the EPA to retain between 1% and 2% of the total quantity of allowances for new entrants.

The consultants' recommendation that installations which closed should be allowed to keep their allocation was accepted in principle by officials. To some the idea that closures retain their allocation while new entrants receive new allowances was a recipe for gaming. However, within administrative circles it was considered the best long-term signal regarding a hands-off regulatory approach to the ETS. The proposal that closed installations retain their allocation was contained in the first draft of the policy direction prepared for government approval. Ministers however considered this to be the equivalent of allowing companies to retain establishment grants after they closed, a precedent that they did not want to set. In addition a number of the installations foreseen as likely to close were older plants in the power generation sector. Given concerns at the time regarding the potential windfall gains to power producers arising from the free allocation of allowances it became inevitable that the Cabinet would not support the policy direction as proposed. The direction was therefore altered to instruct the EPA to not allocate to installations in the year preceding a closure, and to auction these allowances for the benefit of the Exchequer. During the public consultation phases this issue was raised by a large number of companies, in particular in light of the planned rationalisation of production in the dairy sector. The matter was therefore reconsidered by departments and by government on two occasions but the original direction was reaffirmed on both. This caused substantial annoyance in some sections of industry particularly in light of transfer rules in other Member States such as Germany and the fact the rationalisation programme had been backed by the government as being a necessary development.

The initial staffing and funding of the newly established Emission Trading Unit in the EPA was resourced by cuts in the six relevant government departments. This was not a sustainable long-term solution; however, the Department of Finance, as a matter of policy, was not able to sanction increased direct funding for the EPA. It was therefore decided that participants in the ETS would pay the costs of administering the scheme. It was considered that payment on the basis of allowances allocated might not stand up to a legal challenge under Article 10 of the Directive, and as GHG emission permits were already being applied for without national transposition of the Directive charges could not reasonably be imposed for these (nor would it have been possible to do so in an equitable manner given that permits were not dependent on the level of emissions). Therefore the EPA was instructed to auction up to 1% of the total quantity of allowances to cover the costs of administering the scheme.

At the time of the government decision on the total quantity (late January/early February 2004) it was anticipated that the Commission for Energy Regulation (CER) would introduce a new wholesale market arrangement for electricity throughout 2005. This fully liberalised wholesale market would result in power companies passing on the market cost of free allowances to electricity consumers resulting in substantial gains to electricity companies. While the benefits of this in terms of attracting much-needed new generating capacity was recognised the concept of continued increases in retail electricity prices was unwelcome. In the fourteen years before the market had opened up, when ministers controlled the tariffs charged by the state-owned electricity producer, no price rises had been sanctioned. This had led to underinvestment in the national grid and had meant that liberalisation in this sector had led to increases rather than reductions in consumer prices. The CER proposed that the increases would be allowed in the wholesale market but then captured by the regulator and offset against the increasing network charges resulting from the ongoing investment in the grid infrastructure. This would require new legislation to allow the CER to take funds from private generators. Ultimately this proposal was unnecessary as the introduction of the new market arrangements was suspended due to preparations for the creation of an All-Island energy market, i.e. a fully integrated market with Northern Ireland. At present the electricity market remains a regulated one with only the value of purchased allowances being passed on to consumers; the long-term effects of this on new market entrants could well exacerbate existing concerns regarding generation adequacy.

5.2 Micro level

Following assessment of the potential for new entrants the EPA decided to retain 1.5% of the total quantity of allowances for such installations, in a specially designated New Entrants Set-Aside (NESA). This set-aside was divided into annual amounts in a ratio 1 : 2 : 3. Allocations from this set-aside would be determined on a first-come–first-served basis. Following the second round of public consultation in late 2004 it was decided to reproportion NESA in the ratio 1 : 2 : 9, to provide for the increased uncertainties in assessing new entrants further into the future, and also due to the likely commissioning of a new combined cycle gas turbine power plant in mid 2007.

While the EPA had decided to reward early action or existing clean technology in the allocation metric the effects of the ETS on the development of new CHP plants was a matter of concern. Well-designed CHP provides increased efficiency in the use of energy by producing both heat and electricity. Even the best traditional electricity generating units lose 45% of the latent energy in fuel during transformation, with older plants losing up to 65%. A CHP plant on the other hand will lose as little as 20%; therefore if an efficient use can be made of both the heat and electrical output a CHP installation will result in substantial energy savings, and consequentially a reduction in national CO_2 emissions. However the micro level effect of a CHP plant will be to increase on-site CO_2 emissions (as imported electricity is CO_2 neutral to the consumer) and the benefit of a reduced emissions liability accrues to the electricity generator not the CHP developer. If the full cost of allowances was being passed on in the retail price of electricity this problem would be addressed by the market. In the absence of such a market solution in Ireland the EPA decided to retain 148,800 allowances per year for newly constructed CHP plants. These allowances were subtracted from the power generation sector allocation, as this sector would have a reduced emission liability if new CHP plants were constructed.

The EPA decided to auction 0.75% of the total quantity of allowances, less than permitted by the Government's policy direction. This was deemed sufficient to cover the costs of administering the scheme. The current price of CO_2 allowances would seem to bear this out, though no state auction has yet been conducted.

A final point worth noting in relation to the Irish NAP is the ability to coordinate all aspects of the ETS implementation: distribution, permitting, enforcement and the registry in one agency. This has allowed decisions in each of these areas to be integrated into a coherent implementation. Necessary coordination with other relevant state agencies has been facilitated by their membership of the NAAG.

6 Conclusion

6.1 Timelines

The ETS Directive was only finalised in mid-2003, with the first NAPs due in March 2004. This timeline was extremely challenging by any standards, exemplified by the fact that only five countries met this deadline, and even fewer met the subsequent deadlines for finalising installation level allocations and issuing allowances. The need to meet the first deadline was heightened in Ireland by the pressures of the Presidency, with the Department of the Taoiseach maintaining strict records of compliance with EU legislative deadlines in all areas.

Certain government departments in Ireland had serious concerns regarding the advantages to Member States of delaying the submission of their NAPs to watch the Commission's assessment of the first wave of Plans, though equally it could be argued that it was better to have the NAP assessed early when the Commission themselves were in a learning phase.

6.2 Commission assessment

The Commission assessment process, and in particular early drafts of the Commission's decision on the Irish NAP, was the cause of considerable concern in Irish administrative circles. Certain government departments viewed the Commission's determination as a 'one size fits all' approach to areas that were the preserve of Member States (e.g. the purchase of Kyoto credits by governments). The reopening of the decision regarding the total quantity of allowances to be allocated was particularly difficult for the administrative process in Ireland due to a combination of the technical nature of the decision, the difficulty in achieving the initial compromise and the stability of the government collation at the time of the Commission's decision, following the outcome of local and European parliament elections. However such concerns were dissipated by the Commission's consistent treatment of later NAPs and by the close working relationships developed during the Irish Presidency of the European Council between Irish and Commission officials.

While the consistency of the Commission's assessments of the macro level plans was well regarded concerns remained at the level of scrutiny given to micro level issues. The aim of the Commission's assessment appeared to be ensuring the effective implementation of the ETS Directive as a stand-alone instrument without recourse to implication on national climate strategies and/or the operation of the single market. The assessment of such issues in the second series of NAPs will be watched with close interest.

6.3 Separation of tasks and coordination of functions

A final point worth considering in relation to the Irish NAP is the success or otherwise of its most unique feature: the complete separation of macro and micro level plans. The advantage of this structure was that it allowed for complete focus on individual tasks with each being entirely separate from the other. This also eased the adjustments necessary after the Commission's assessment and the baseline verification process, as individual allocations were simply a portion of an arbitrary total, and could be determined transparently and mathematically whatever change was made to the inputs.

However, the coordination of functions may become more difficult over time. Many of the areas of overlap – basic treatment of new entrants, known planned development etc. – were predetermined. This will not be the case for future NAPs, a fact that may cause conflict between the need for separation of tasks and coordination of outcomes.

6.4 Author's concluding remarks

The expectations of market players and eventual outcomes in relation to micro level allocation plans for the second phase will have considerable impact on the functioning of the market for CO_2 allowances and therefore the success or failure of the scheme. The updating of allocation plans required by the Directive takes away an underlying assumption of most economic assessments of emission trading schemes, i.e. that distribution does not affect efficiency unless transaction costs are high. Updated allocation plans have the effect of altering the abatement and production choices made by participants. Where participants believe that the base years in future allocation plans will be updated there is an incentive to increase emissions. The uncertainty as to whether or not this will happen in the second NAPs is likely to result in increased volatility and illiquidity (wider bid–offer spreads) than might be expected. Of at least as much concern however is the impact of the resolution of this uncertainty in mid-2006. Should some Member States opt to update base years and others do not the ETS could begin to act as a form of production subsidies; the macro economic impact of this may be small, given that all Member States already have different operating environments. However, the effect on the ETS is potentially significant as it could lead to reductions taking place where they do not cost the least, eroding the economic advantages of using a market mechanism. The potential for this outcome remains a concern for the author.

Reference

Fitz Gerald, J., Bergin, A., Kearney, I., Barret, A., Duffy, D., Garrett, S., and McCarthy, Y. 2005. *Medium-Term Review 2005–2012*, Economic and Social Research Institute, Dublin ICF Consulting. Integrated Power Model, Proprietary Model of European Electricity Market. www.icfconsulting.com.

8 | Spain

PABLO DEL RÍO

1 Introductory background and context

1.1 Climate policy in Spain

An integrated climate policy has been lacking in Spain, at least until 2004 when the National Climate Change Strategy was approved.[1] Several scattered, sector-specific measures indirectly leading to CO_2 emissions have been applied or recently approved, but they were mostly implemented to reach other goals (e.g. to improve employment and regional development opportunities, to reduce foreign energy dependency or to increase the competitiveness of national industry by raising its energy efficiency). The following two are worth mentioning:

- RES-E Promotion. The National Plan for the Promotion of Renewable Energy Sources (NPPES) sets a 12% energy consumption target coming from renewables in 2010 (29% of renewable electricity). This Plan aims to reduce emissions by 28 Mt CO_2e annually up to 2010 by substituting renewable energy for conventional energy sources with several measures (particularly the granting of feed-in tariffs to renewable generators).

- Energy Efficiency and Energy Savings Strategy (2004–2012) aiming at the control of energy-related CO_2 emissions by partly subsidising energy-efficient technological change in industrial firms. With the Strategy, CO_2 emissions would increase by 58% in 2012 compared to 1990 (instead of a 78% increase). This represents an accumulated reduction of 190 Mt CO_2 in 2004–2012.

These policies mainly focus on the industry and energy sectors. Measures tackling emissions from other sectors are even more limited. In general, fiscal deductions and exemptions, tax relief, direct

[1] For a more complete overview of Spanish climate policy in the past, see Hernández et al. (2004) and Tábara (2003).

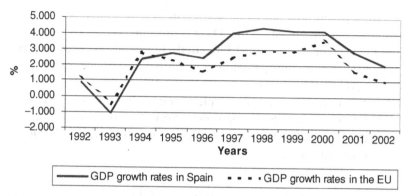

Figure 8.1. Comparative GDP growth rates (%) in Spain and the EU in the period 1991–2002.
Source: Own elaboration from Banco de España (2004).

subsidies and, to a lesser extent, voluntary agreements are applied in all sectors. As non-coercive instruments they are, therefore, relatively easy to implement.

Climate policy has been based on 'soft', politically easy measures which avoid confrontation with the affected industries and are not unpopular among voters, even though their burden falls on the great majority of uninformed and unorganised consumers. The lack of an integrated climate change policy is related to other problems considered more urgent by Spanish voters. Employment creation and real economic convergence have been the top priorities in the past. Spain traditionally had one of the highest unemployment rates of the OECD countries and the reduction of this rate has been a top goal of successive governments in the last decades.

This lack of effective measures is one of the two major factors behind the fast CO_2 emissions growth in the last decade (40% above the 1990 level in 2003), putting compliance with the Burden Sharing Agreement (BSA) target of +15% at risk. The other major factor is the fast economic growth experienced by the country (2.67% average annual growth rate of the Spanish economy in the 1994–2002 period, versus 1.92 for the EU) (see Figure 8.1). These high growth rates are expected to continue in the near term.

High growth rates have allowed the country to partly close its GDP per capita gap differentials with the EU and to reduce its high unemployment rate, one of the most crucial problems for the Spanish populace. Spain's intensity of CO_2 emissions per unit of GDP reached the EU average in the early 1990s. Reflecting the lower income level, Spain's CO_2 emissions per capita remain below, but are rapidly converging to, the EU average.

In this context, pressures to implement greenhouse gas (GHG) mitigation measures have had a top-down character in that they originated from international agreements, the Kyoto Protocol, and the EU Directive on Emissions Trading.

1.2 Theoretical framework and methodology

Although no explicit theoretical framework has been developed for analysing the National Allocation Plan (NAP) process in Spain, insights from the *public choice* and *institutional path dependency* approaches have been used in order to structure, organise and interpret the available information.

Emissions trading is a radical innovation in environmental policy in Spain, whose introduction was fraught with difficulties. Institutional path dependence theory (see Woerdman 2004) provides useful insights to interpret decisions, choices and conflicts when institutional rigidities exist. On the other hand, a public choice approach is useful for analysing the interaction between the public administration and other actors in the allocation process and some specific choices made (see Schneider and Volkert 1999; Svendsen 2000; del Río 2006).

Data sources used for this paper are multiple: own experience, interviews with key actors and experts, analysis of position statements from sectors, official documents from the Spanish Ministry of Environment (including different NAP versions), articles in the media and other written sources.[2]

---

[2] The two most relevant international and national emissions-trading specialised websites have been visited (Point Carbon and Canalmedioambiental) and relevant documents on the NAP process collected. In addition, structured questionnaires were sent to the individual associations of all the covered sectors, most of which replied in written form. Personal interviews were undertaken with several energy and climate change experts. For reasons of confidentiality of the information requested by some of the interviewed sectors, their responses are integrated in a general manner in this paper, unless their opinion has been made public otherwise (e.g. publication in journals, press releases etc.).

1.3 The elaboration of the Spanish NAP: phases, timetables, actors and legislation

The initial responsibility to carry out the NAP fell on the Ministry of Environment (MINAM) and, more specifically, on the Spanish Climate Change Office (OECC), where internal groups for the elaboration of the NAP were created. Initially, the development of the NAP was seen as a technical problem involving decisions on the criteria to be considered, installations covered, identification of emissions per sector and per installation etc. In this initial stage, the OECC felt that it did not have sufficient command of the economic aspects of the allocation and required the external assistance of a research centre (Klein Institute). During 2003, the Klein Institute worked on the criteria to elaborate the NAP and maintained constant contacts with the OECC to exchange information and data. Most of the sectors were invited to give their views on the specific features that should have been considered for the sectoral allocation.

Feeling that the allocation could have negative effects on their core businesses, the sectors started to make complaints that were transmitted to the government and published in economic journals and the general press. The process then became much more political. In autumn 2003, it was decided that the Ministry of Economics (MINECO) should take the lead because of the potential economic implications involved in allocation and the belief that the NAP should deal with issues, such as energy, that were under the competence of MINECO. An Interministerial NAP Commission made up of MINECO and MINAM (and led by the former) was set up. This group met with the relevant sectors in January–February 2004 to be informed of their official requests for allowances. Work on the NAP was carried out, but there was a widespread view that the 31 March 2004 deadline would not be met because of the upcoming national election in mid-March 2004 and the government's increasingly sceptical attitude towards the EU ETS.

As industrial groups began to realise the implications of the NAP process, an opinion developed that the BSA had been badly negotiated and the government was subjected to continuous lobbying to renegotiate Spain's BSA target. The then incumbent (conservative) government was unwilling to renegotiate because of the implication that it had indeed badly negotiated the target in 1998. Instead, the government

decided to challenge the EU ETS by stating that it would have very
negative impacts on the competitiveness of European industry in general and of Spanish industry in particular. The lack of ratification by
Russia was also used to question whether the EU should continue with
the EU ETS.

The change from the conservative to the socialist government as a
result of the general election in March 2004 had a significant impact
on the NAP elaboration. The first tangible effect was that it provided
a major justification for the delay in submitting the NAP. Following
a request of the new government on 3 May 2004 that 'special circumstances' be considered, the European Commission gave the new
government more time to finalise the NAP by setting a new deadline
of 1 August 2004.

Despite initial uncertainties about the new government's approach
to allocation, the process accelerated considerably following the election. The socialist party had been very supportive of compliance with
the Kyoto Protocol targets during the electoral campaign, but it was
difficult to predict how this general support would affect the specifics
of the NAP process. This uncertainty was resolved by the publication
of a first NAP version within barely three months of the change of
government. This speediness was an accomplishment in itself, but it
did not start from zero. Although the basic principles of the NAP had
not been made public by the outgoing government, useful information
had already been gathered and the new government made full use of
it in addition to holding new meetings with the sectors. The new government also had the political legitimacy and commitment to make
allocation a priority. In addition, the failure to meet the Commission
deadline also gave a sense of urgency for elaborating the NAP. Another
factor contributing to speeding up the process was the concern of companies (and the accompanying risk of legal action) that too much delay
in issuing the allowances would keep them from trading.

After the election, the role of different ministries changed again.
The new government decided that the MINAM should lead the process and have a central role in elaborating the NAP. The staff at
MINECO believed that it would be very busy dealing with purely
macroeconomic issues and that the allocation was above all an environmental issue (with economic implications, however). In May 2004,
the new government also set up a new, high-profile Interministerial
Group on Climate Change (IGCC) in May 2004 to replace the previous

Interministerial Commission. This new group was under the presidency of the General Secretary for Pollution Prevention and Climate Change (MINAM) and made up of representatives from MINECO and other ministries (Public Works, Industry, Tourism and Trade, Labour and Social Affairs, Agriculture and Housing). After the review of all the information and studies from the earlier Commission and after meetings with sectors in May 2004, the new IGCC elaborated several BSA compliance scenarios and submitted a proposal with the NAP criteria, which the government approved on 17 June 2004. On 8 July, it published a draft version of the NAP and solicited public consultation and comments until 19 July.

The responses led to corrections in this first version and the publication of a second version in early September 2004.

Neither the July nor the September version included the installation-level allocation, even though installations were required by 30 September 2004 to request the granting of the emission permit and an allocation of allowances. The actual installation-level allocation (as opposed to principles and formulae) was first published on 26 November 2004 and it led to some minor changes and corrections which, together with the corrections made after the European Commission Decision of 27 December 2004, were included in the final NAP version, published on 21 January 2005.

2 The macro decision concerning the aggregate total

The elaboration of the Spanish NAP has been mostly a top-down process (from macro to meso and to micro), with several stages and little interdependence between levels. The total amount of allowances was decided very early in the process, when data at the installation level were still incomplete.

Two major factors were considered when deciding on the allocation of the total quantity of allowances: the distance to the BSA target (+15%) and the preservation of the competitiveness of Spanish industry. The two factors work in opposite directions and the government tried to strike a balance between the two. The final resolution was to try to stop the increasing emissions trend in the first period and to leave the main reductions for the second period. This was as logical a way as any to proceed given the long distance-to-target indicator, the expected generous allocation of other countries, the newness of

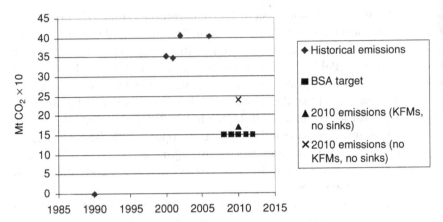

Figure 8.2. Trends and targets set in Spain's NAP. KFMs, Kyoto Flexible Mechanisms.
Source: NAP (RDL 1866/2004).

the ETS instrument and the inherent incentive for the policy-makers to leave the bulk of the efforts for the future.

The total quantity of allowances was mainly based on the independent assessment of the Klein Institute (with some adjustments). Three scenarios were considered: stringent, intermediate and lax. The macro allocation was based on the latter (constant emissions 2005–2007 in relation to base year levels and reduction in 2008–2012), with some adjustments (data updating and changes in the expected contribution of sinks and Kyoto credits). The NAP assumes a stabilisation of total emissions (covered and non-covered sectors) in 2005–2007 at average 2000–2002 levels (emissions in 2002: 401.34 Mt CO_2). The great reduction effort would be left for the period 2008–2012, when emissions are expected to be 24% over base year emissions. The 9% shortfall over the BSA target would be made up by buying credits and allowances in the international market (7%) and by sink projects (2%) (see Figure 8.2).

For the sectors in the trading scheme, the goal is that the share of total CO_2 emissions in 2002 (40%) remains constant in 2005–2007 for an annual allocation of 160.28 Mt CO_2. This total is 2.5% less than the emissions levels of the covered sectors in 2002 and it includes a new entrants reserve (NER) of 2.994 Mt CO_2 per year. In addition, 13.9 Mt CO_2 per year are allocated to non-Annex I cogeneration and mixed installations (plus a 0.364 Mt CO_2 NER for non-Annex I cogeneration installations).[3] Measures in the non-covered sectors are expected to reduce by 52 Mt CO_2e in 2005–2007. The total allocation to covered sectors was based on four variables: expected emissions trends of the covered sectors, past emissions reduction efforts of all sectors, available emissions reduction potentials and emissions trends of non-covered sectors in 1990–2002.

3 The micro decision concerning distribution to installations

The installation-level allocation was a two-step process. First, allowances were allocated per sector, and then, per installation within each sector. The sector-level allocation was based on the information collected through questionnaires sent to the sectors in January–February 2004 and from the meetings between different interministerial groups and the sectors' associations.

3.1 Meso aspects: allocating allowances per sector

Two different scenarios were considered: one for the electricity sector and another for the industrial sectors. Allocations were based on emissions projections, historical emissions and the emissions reduction potential of each sector. The general consensus is that industrial sectors have been allocated as many allowances as they will probably need and that the bulk of reductions has fallen on the electricity sector.

3.1.1 Electricity sector
The allocation to the electricity sector was partly based on historical emissions, projections of electricity demand and expected trends of the generation mix (with and without carbon constraints).

This sector receives 86.4 Mt CO_2 per year (including the NER), compared to emissions of 95.95 Mt CO_2 in 2002 (88 Mt CO_2 per year average in 2000–2002) and an expectation that emissions will average 94 Mt CO_2 annually during 2005–07. Apart from ceramics and tiles and bricks, this is the only sector that has been allocated fewer

---

[3] In the rest of this chapter, the covered sectors will be also called 'Annex I' sectors (corresponding to the activities included in Annex I of the Directive). Non-covered sectors will be called 'non-Annex I sectors'.

allowances in 2005–2007 than emissions in 2002. Therefore, the bulk of emissions reductions falls on the electricity sector, due to relatively abundant reduction alternatives, low marginal abatement costs, low degree of international competition, high profit margins and a maximum regulated annual electricity price increase of 2%. Notwithstanding electricity companies' claims in 2003 that the electricity bill 'should increase by 15%' for them to recover the costs of Kyoto compliance, both the former and the new government have rejected these demands and have argued that the 2% maximum increase allows firms to absorb the costs of the Directive and Kyoto compliance without major disruptions. This stance has mitigated the fear by industry sectors that if electricity companies passed their abatement costs on via prices, they would be excessively burdened.

3.1.2 Industrial sectors
The allocation to non-electricity sectors was based on emissions projections (2001–2006), by extrapolating the historical emissions trends per sector (annual growth rates between 1990 and 2001). The allocations were then adjusted according to the emissions abatement potentials in each sector. When combined, the industrial sectors were allocated 73.88 Mt CO_2, which was the sum of their estimated business-as-usual (BAU) emissions. This compares with 68.37 Mt CO_2 annual emissions in 2002 and average 2000-2002 emissions of 66.87 Mt CO_2.

Two adjustments were made to this general methodology:
(1) *Steel.* Since this sector had mostly exhausted its emissions reduction technological potential, given the 'remarkable reduction in emissions since 1990', an ad-hoc projection is applied, which considers the expected evolution of the sector. The allocation to the steel sector includes an additional 1.6 Mt, corresponding to electricity generation using steel gases.
(2) *Ceramics.* Several installations were finally excluded since their production/capacity thresholds were below the required minimum. The sector-level allocation was reduced accordingly. Despite increasing recent emissions and relatively high expected demand, ceramics and tiles and bricks were the only industrial sectors to be allocated fewer allowances than their 2002 emissions (5.625 Mt CO_2 and 4.75 Mt CO_2, respectively, as compared to 2002 emisisons of 6.61 Mt CO_2 and 5.29 Mt CO_2 and average 2000–2002 emissions of 6.38 Mt CO_2 and 5.92 Mt CO_2). The weaker lobbying

position of these sectors because of the predominance of small firms may account for this result.

3.2 Micro aspects: distribution to installations

The installation-level allocation faced some difficulties, due to data availability problems. These difficulties even led the government to think initially of setting the amount of allowances per sector, allowing the respective sector association to distribute the allowances per installation.

3.2.1 Basic principle

Allocation was based on capacity and some benchmark or historical emissions or output shares using a baseline period of 2000–2002, depending on the sector and certain conditions. However calculated, the shares are applied to the sectoral totals resulting from the meso level distribution.

Adjustments to the basic principle were made as follows.
(1) In the *electricity sector*, the basic allocation principle was adjusted according to geographical and technological criteria:
  • Geographical criteria. Deductions are first made from the sectoral total of 88 Mt CO_2 for electricity generators in non-peninsular Spain, i.e., the two archipelagos (Baleares and the Canary Islands) and Spanish cities in North Africa, which will receive 100% of their request for allowances, that is, expected emissions.
  • Technological criteria. After deduction for expected emissions in non-peninsular Spain, as well as for new entrants, and blast-furnace or other CO_2-emitting generating fuels, the remaining allowances were then split into subtotals for coal- and natural-gas-fired generation before making the installation-level allocations. These subtotals are based on projected generation from these two principal sources of electricity generation and on the application of emission rate benchmarks that approximately reflect the average emission rates from each type of generation (0.9421 tons CO_2/GWh for coal and 0.365 tons/GWh for combined cycle gas turbines (CCGT) units). Since coal-based generation is expected to go down and the use of CCGTs to increase, proportionally more allowances are given to the latter than was their share during the baseline period. Allowances are then

allocated to individual CCGT installations according to the expected available capacity during the year.

The allocation to coal-based installations is more complex. It is based on the following formula, which takes into account the historical emissions and emissions rate of each installation (2000–2002):

$$AD = ADPC \times x_i \times \frac{1/y_i^2}{\Sigma x_i/y_i^2}$$

where:

AD = allocation to coal-based generation installations (national coal)

$ADPC$ = overall allocation to coal-based installations (national coal)

x = historical emissions of each installation (average)

y = specific emissions factor of each installation

i = installation.

This formula effectively allocates total emissions for coal-based generation according to historical baseline output and what might be called a modified historical sectoral emission rate benchmark. The modification consists of the quadratic term which has the effect of allocating more allowances to coal-fired installations with a below-average emission rate during the baseline period and fewer allowances to units with an above-average emission rate. Coal-fired installations are further guaranteed that the unit allocation will not be less than 55% of the unit's emissions during the baseline period.

(2) The allocation to *industrial installations* is based on sectoral emissions in the period 2000–2002 and on the emissions share of the installation in the sector's emissions in that period. This basic allocation principle is adjusted to accommodate several issues. Where individual unit data are not adequate, the installation receives its baseline capacity share using an average emissions benchmark for comparable facilities. Special provision is also made for cogeneration and process emissions, special circumstances during 2000–2002 (abnormal stops due to maintenance, breakdowns, changes in the installation, strikes, adverse weather etc.), and capacity increases after 1 July 2001 leading to an increase in emissions in excess of 20%.

Box 8.1 summarises the installation-level allocation steps and formulae.

Box 8.1 Detailed steps and formulae involved in the Spanish installation-level allocation

(1) First, a distinction is made between process emissions (E_i^{proc}) and emissions from combustion (E_i^{comb}) for all installations with adequate reference period (2000–2002) emissions data. Estimates are made using the following formulae for installations that started operating after 31 December 2002 or for which emissions data in 2000–2002 were deemed to be non-representative:

$$E_i^{proc} = FE_s^{proc} \times C_i \times U_s$$
$$E_i^{comb} = FE_s^{comb} \times C_i \times U_s$$

where:

FE_s^{proc}, FE_s^{comb} are, respectively, process-related and combustion-related emissions factors based on historical data provided by installations with similar characteristics to the ones for which data do not exist

C_i = production capacity of installation i during the baseline period or after post-2002, as applicable

U_s = load factor (average capacity utilisation rate) of each sector.

After this estimation, all installations have emissions data (process and combustion) on which allocation can be based.

(2) Allocation to cogeneration installations (A_i^{cog}) based on:

$$A_i^{cog} = PE_s \times E_i^{cog}$$

where:

PE_s = rate of emission change (of the sector where the cogeneration plant is installed) between the reference period and 2006

E_i^{cog} = emissions of the installation in the reference period.

Therefore, the right-hand side of the equation represents the emissions expected in 2005–2007, assuming that the emissions of cogeneration installation evolve according to the emissions trend of the sector where the cogeneration plant is used. This means that cogeneration installations are given as many allowances as they will probably need.

(3) Allowances reserved for process emissions:

$$A_i^{proc} = PE_s \times E_i^{proc}$$

where:

PE_s = rate of emission change of the sector (s) between the reference period and 2006

E_i^{proc} = estimated process emissions of the installation in the reference period as calculated in step 1.

In granting the growth factor, the NAP recognises the 'problems to reduce these emissions', and aims to grant these installations as many allowances as they are likely to need.

(4) The sector-level allocation is recalculated by subtracting process-related emissions and emissions from cogeneration plants:

$$A'_s = A_s - \Sigma_i A_i^{cog} - \Sigma_i A_i^{proc}$$

where A_s is the sector cap determined in the meso level projections.

(5) Finally, the rest of allowances are allocated according to the share of the installation's combustion-related emissions in total combustion emissions:

$$A_i^{comb} = A'_s \times \frac{E_i^{comb}}{E_s}$$

where E'_s are the aggregate combustion emissions: $E'_s = \Sigma_i E_i^{comb}$.

Source: NAP (RDL 1866/2004).

Allocation to cogeneration installations in non-Annex I sectors was based on emissions in 2002 (if they are 'meaningful', i.e. within ±10% of average emissions in 2000–2002) and an expected growth rate of 18% between 2002 and 2006.

3.3 *Other relevant elements of the micro allocation*

3.3.1 Treatment of CHP and 'mixed' installations

'Given their environmental advantages' cogeneration plants were given favourable treatment, since they received as many allowances as they will probably need. In addition, pooling of cogeneration installations is allowed, which also represents a special treatment in relation to other combustion installations. As with process-related emissions, the expected emissions trends from cogeneration installations between 2000–2002 and 2006 were calculated, assuming that they evolve according to the trend of the sector where the cogeneration plant is used. The NAP states that, within each sector, the favourable allocation to cogeneration installations should not lead to 'an excessive penalisation' of those installations which do not use this technology (such as non-CHP installations in sectors with a large share of CHP installations). Therefore, a binding constraint is introduced whereby 'non cogeneration installations in sectors where $PE_s > 1$ (increasing projected sector emissions) will receive allowances equivalent to at least 95% of E_i (average emissions of the installation in 2000–2002)' (RDL 1866/2004, p. 30633). It is not entirely clear, however, where the additional allowances to meet this condition will come from.

CHP installations in non-Annex I sectors, which were not considered in the July NAP version, were finally included later based on 2002 emissions and an 18% growth rate expected between 2002 and 2005–2007. In contrast, Annex I cogeneration installations were considered from the start. 'Mixed installations' (i.e. installations that operate partly as CCGT and partly as cogeneration installations) were also not initially considered. This was corrected in the final version of the NAP, with allowances being allocated to them.

3.3.2 New entrants reserve (NER)

A free NER, allocated on a first-come–first-served basis, was set at 3.5% of total emissions in the reference scenario (5.42 Mt CO_2 per year). The rationale for the free allocation is 'guaranteeing the access to allowances by new entrants and respecting the principle of equality of treatment and the internal market legislation'. Other factors were claims on the loss of competitiveness, the high growth expected for the sectors covered and rumours from other countries that would also allocate these allowances freely.

The NER was initially broken down into separate allocations for the electricity generation sector and the industrial sectors. In the latter case, an allocation of 3.58 Mt CO_2 per year for all the industrial sectors and a 'guiding' distribution per sector was initially envisaged. However, following the recommendations of the European Commission, the disaggregated industrial NER was turned to an aggregate industrial NER.

The allocation of allowances to the NER has been controversial, with at least one industrial sector claiming that the allowances to new

Table 8.1 *Changes to the initial draft allocation made by the September version*

Issue	Allegation	Answer/corrections
List of installations	Requests for inclusion and exclusion of installations	Allegations considered and changes made; inclusion rejected for installations below thresholds
Number of allowances allocated to installations	Insufficient allowance allocation Allocation not following the methodology described	Changes in the allocation if justified
Sector emissions	Emissions data corresponding to lime, tile and brick manufacturing not updated	Use of new data
Installation-level allocation	Not well explained	Improved writing
Sector growth in installation-level allocation	Growth rate of the sector uniformly applied to all installations within the sector	Projections per installation deemed pertinent
Methodology for the individual allocation	Insufficient treatment of early action and cleaner technologies 'Benchmarking' considered more accurate	The methodology has advantages in terms of verification and indirectly considers early action and cleaner technologies by favourably treating cogeneration and process emissions and by considering best available technologies (BATs) in the allocation to new entrants
Industrial sectors with a large share of cogeneration installations	The methodology used penalises these installations	A special clause was introduced which avoids an excessive 'penalisation' of those installations (see section 3.3.1)
Reference period 2000–2002	Unrepresentative reference period for some installations due to 'exceptional circumstances'	Safeguard clause to avoid this
Non-Annex I cogeneration installations	Allocation to these installations unclear	Clarified in the new version
Non-Annex I cogeneration installations	Inclusion of these installations based on 2002 data Installations starting operation after 2002 not included	Corrections made
Ceramics sector	'Or' should substitute for 'and' in the determination of the minimum threshold[a]	Accepted. Only very small installations affected. Their low emissions do not justify their (relatively high) transaction costs
Quantity of allowances allocated to specific sectors	Claims for an increased allocation, based on emissions projections	Change in the quantity of allowances allocated to some sectors

[a]'Installations producing ceramic products by means of firing, particularly tiles, bricks, refractory bricks, glazed tiles, stoneware or porcelain articles, with a production capacity in excess of 75 tonnes per day and a firing capacity of over 4 m^3 and over 300 kg/m^3 load density per kilo.'

Source: Own elaboration from RD1866/2004 and MINAM (2004b).

installations are 'insufficient' in relation to the sectors' growth and at least one sector stating that too many allowances have been reserved for new entrants and that these allowances should have been allocated to existing installations. Of course, the different growth rates expected in these sectors are behind these claims. The NER has also been controversial in the electricity sector. The relatively abundant number of allowances left for new entrants will facilitate the installation of CCGT plants (which will be the ones experiencing the highest growth rates). This is criticised by firms with a relatively low share of CCGT installations and a higher share of coal-fired plants, which would have preferred that most of these allowances reserved for new entrants had been allocated to existing installations.

Allowances in the NER not allocated by 30 June 2007 will be auctioned. The allocation to new entrants will be made according to emissions projections using the average production capacity of existing installations within the sector, best available technology, and the reductions of the sector in which the new installation is to operate.

3.3.3 *Ex-post* adjustments

Two major categories of factors and issues led to changes in the initial allocation:

(1) European Commission decision on the Spanish NAP (27 December 2004). According to this decision, combustion installations with a rated thermal input of more than 20 MW not connected to the electricity grid were not included in the NAP (European Commission 2004). This was later corrected and these installations are allowed to enter the market from 1 January 2006 (Article 33, RDL 5/2005).

(2) Counterclaims from sectors and installations and requests for allowances by installations. Several changes were made during the allocation process, as observed in the major documents and legislation (laws and Royal Decrees (RDs)).

Changes from first draft (6 June 2004) to the September version of the NAP (RD1866/2004). More than 300 allegations were made in response to the first NAP draft (6 July). Table 8.1 summarises the most relevant allegations, the answers given to the allegations and the corrections made.

To sum up, the two most relevant adjustments were: (1) exclusion of small ceramic installations (allocation reduced from 11.12 to 9.6 Mt CO_2); (2) allocation to existing industrial installations reduced (parallel increase in NER), which was protested against by some sectors, arguing that it was not justified.

Changes from the September version to the final version of the NAP (RD60/2005). The requests for allowances by installations and the public information procedure (512 allegations) led to changes in the final allowance allocation. The most relevant was the final decision on the contentious micro allocation issue between coal and gas fired plants. This adjustment led to an increase in the allowances given to coal-fired plants at the expense of CCGT plants. Table 8.2 summarises the changes made to the September version of the NAP and to the installation-based allocation published on 26 November 2004.

4 Issues of coordination and harmonisation

4.1 *Influence of EU guidance and review*

The general feeling is that EU guidance has been useful and facilitated the elaboration of the NAP. However, some experts consider that in their documents the European Commission left too much freedom (for example, concerning the definition of 'combustion installations'). EU guidelines are believed to have been developed after some of the problems encountered had already been solved by the Member States (e.g. the treatment of CHP) and that they were sometimes 'imprecise'.

4.2 *Influence of signals and rumours from other Member States*

There is anecdotal evidence of the influence of other countries' NAPs. Some of this evidence is contradictory, according to the views of interviewees. It is a highly subjective, complicated issue. The real influences are difficult to discern although, a priori, the fact that Spain was one of the latest countries to send its NAP may suggest that some elements or even the general philosophy of other NAPs was taken into account.

According to some experts, MINAM checked the other NAPs to analyse the treatment of different issues. It is unknown whether

Table 8.2 *Changes to the September version made by the final NAP*

Issue	Explanation
Increase in allocation to existing installations at the expense of NER	This was due to some installations, expected to be new entrants, that were really existing installations
Removal of NER desegregation for industrial sectors	Following the recommendation of the European Commission, a single reserve for all the industrial sectors was created, rather than one 'indicative' NER for each sector. There are now only three NERs: one for the electricity generation sector, another for industrial sectors included in Annex I and one for non-Annex I cogeneration and mixed installations
Inclusion of 'mixed' installations	These were partly operating as CCGT and partly as cogeneration installations. Two allocation methodologies were used, one for each part
Revision of allocation to coal-fired and CCGT installations	The allocation was adjusted according to improved production data, expected emissions and the negative impact of new desulphurisation installations on CO_2 emissions. More allowances (4.53 Mt CO_2) were given to coal-fired generation plants based on technical arguments, reviewed by technicians from the General Secretary of Energy. Also, corrections were made when emissions in 2000–2002 were not considered to be representative. Finally, a minimum limit was set on the allocation to coal-fired installations with a high emissions factor. They will not obtain an allowance allocation below 55% of their historical emissions in 2000–2002. As a result, the allowances given to CCGT installations were reduced (from 67.53 Mt CO_2 to 66.44 Mt CO_2) and those to coal-fired installations increased (from 148.68 to 153.21). In addition, some CCGT installations, expecting to be new entrants initially, were really existing installations, and adjustments were made accordingly
Increase in the total number of installations covered (from 926 to 957)	Some installations, initially considered as new entrants, were existing installations.[a] In addition, several requests for allowances were not considered in the initial allocation because they were submitted to the Autonomous Communities (AACC), rather than to the central government
Adjustments made in the allocation to specific sectors	As a result of checking the final list of covered installations with the AACC

[a] Notwithstanding, an analysis of the data provided by MINAM (2005b) leads to the conclusion that 119 installations did not finally receive allowances for the following reasons: installations whose activity was not included in Annex I (45 installations), installations below the production or capacity thresholds (43), new entrants (30) and installations no longer in operation (1).

specific elements of other NAPs were considered. The following NAPs may have been considered: Ireland (also a cohesion country, fast economic growth, and long distance-to-target indicator), Italy (transport measures), United Kingdom (2000–2002 reference period), Portugal (future implementation of the Iberian electricity market).

Other interviewees believe that the United Kingdom's NAP was the most influential since it was the first one and its approach (generous allocation to industrial sectors, leaving the bulk of reductions to the electricity sector) was followed by the Spanish NAP, leaving serious doubt as to whether other NAPs have had any influence at all.

4.3 Influence of other EU or MS policies

The following domestic legislation is worth mentioning in the Spanish context:

- *Public promotion of renewable energy and cogeneration*, mainly based on the aforementioned feed-in tariff system, is probably the main instrument to abate GHG emissions. RES-E targets, diffusion of cogeneration and expected trends in electricity demand have all been taken into account when carrying out emissions projections. In 2003, the National Plan of Electricity and Gas infrastructures in Spain established an RES-E capacity target for 2011 (including large hydro) of 35,733 MW, 38% of expected total electricity generation of 92,959 GWh. The new Renewable Energy Plan updates these figures for 2010. Renewable energy (including bio-fuels) is expected to reduce CO_2 emissions by 27.5 Mt CO_2 in 2010. The EU Renewable Directive's indicative target for Spain is 29.4% of electricity consumption in 2010 coming from renewable energy sources (17.5% if large hydro is excluded). These ambitious goals will probably be achieved, according to the European Commission.
- *The Spanish Energy Efficiency Strategy 2004–2012 (E4)* was also considered when making projections. The new government envisaged the preparation of a '2004–2007 Action Plan which would include complementary measures to the E4'. On 8 July 2005 the government approved the Energy Savings Plan aimed at reducing energy consumption by 12 million tonnes of oil equivalent in the next three years. GHG emissions would be reduced by 32.5 Mt CO_2e. Total investments are expected to reach €800 million, €700 million of which will be provided by the public administration.

Concerning European Community legislation, several Directives were considered, particularly for the macro allocation. Some of these Directives increase and others reduce GHG emissions, and adaptations were made accordingly (either in the emissions projections or in the installation-level allocation):

- *Directive 2002/91/EC* on the energy efficiency of buildings. A recently approved law aimed at increasing the solar energy integration and the energy efficiency in new buildings responds to the Directive's guidelines and was considered to calculate energy savings from new buildings for the allocation between the covered and non-covered sectors.
- *Directive 1996/61/EC* on Integrated Pollution Prevention and Control (IPPC), and the incineration of waste. The cement sector proposed that waste combustion (other than biomass) be considered. A factor of 1 was used for this purpose (Spanish NAP, p. 50).
- *Directive 1999/32/EC* on the reduction of the sulphur content of several liquid fuels and *Directive 2003/17/EC*. The latter obliges the refining sector to reduce the sulphur content of fuels, increasing energy consumption and CO_2 emissions.
- *Directive 2001/80/EC* on the limitation of emissions to the atmosphere of several pollutants from large combustion installations. Its application to existing installations will take place from 1 January 2008. However, some installations have already taken measures to comply with it, increasing CO_2 emissions. The NAP states that they 'will not be penalised' for this increase.
- *Directive 1999/13/EC* concerning the limitation of emissions of volatile organic compounds (VOCs) caused by the use of organic solvents. Some VOC emissions reduction technologies (thermal oxidation) may increase energy consumption and, thus, CO_2 emissions. The NAP states that alternative technologies should be considered in the medium term, but not how this legislation was taken into account.

Other Directives seem to have played a relevant role in the elaboration of the NAP, although no specific details on how they affected allocation are provided, such as the Directive 2001/81/EC, which sets national emissions ceilings for several atmospheric pollutants (SO_2, NO_x, VOC and NH_3), and the Directive 2003/96/EC on the restructuring of the EC framework for the taxation of energy products and electricity.

Finally, other Community legislation is mentioned in the NAP when an analysis of the emissions projections in non-covered sectors is carried out.

5 Issues deserving special mention

Although issues, methods, rules and criteria are interrelated, some issues can be singled out and analysed, always taking into account that they are elements within a system.

5.1 Auction

Sectors argued against the possibility of having to buy allowances, claiming that this would lead to extra costs, negatively affecting the competitiveness of their products in international markets, and leading to a loss of jobs and delocalisation. In particular, the Spanish Confederation of Business Organisations (CEOE) issued, in autumn 2003, a very influential position paper on the future design of the NAP that argued strongly against this possibility. The auction was not so attractive for the government, even if it would have provided the latter with an opportunity to collect revenues to cover the administrative costs of the system, considering that the government aims at minimising the conflicts with the covered sectors and firms, whereas the administrative costs of the scheme are spread across the silent majority of taxpayers.

5.2 Early action and clean energy, technological potential

Some of the objections to the NAP criticised the allocation methodology used, claiming that a 'benchmarking' approach would be more appropriate and that there had not been a special treatment of cleaner technologies and early action. The short time available to carry out the allocation would have made it difficult to elaborate accurate benchmarks. It would also have been politically difficult to apply benchmarking because it would have been rejected by firms in sectors where it was technically feasible (e.g. the electricity sector), given the opposing views on allocation between firms.

The response to the allegations on cleaner technology and early action was that the methodology used showed advantages over others 'concerning the possibilities for verification and the favourable treatment of cogeneration, a cleaner technology'.

Obviously, there is an inherent incentive to argue that early measures have been taken to abate emissions. It is very difficult (or costly) for the government to verify those claims and to differentiate early actions from normal business practice activities. The NAP does not explicitly recognise early action, although it states that 'early action has been considered in an implicit manner', because 'allowances are allocated to cover all process emissions'.[4] The NAP has opted for simplicity in this regard because if an earlier reference period had been chosen to accommodate early action, then a problem with the quality of data might have resulted.

Finally, an estimate of the technological potential to reduce emissions was made for each sector and included in the NAP.

5.3 Pooling

In one of his first public declarations after being appointed, the new Secretary General for Climate Change supported pooling 'because it provides flexibility to meet our commitments and Spain needs as many flexibility tools as possible' (Point Carbon 2004). Pooling is thus allowed 'to reduce transaction costs and to increase the negotiation power in the market', except in the electricity generation sector. The government justifies this exclusion by the market concentration of the Spanish electricity sector on the grounds that pooling would reduce liquidity and transparency in the allowance market, affecting effective competition, and the diffusion of less carbon-intensive technologies. Iberdrola favoured this exclusion while other firms have criticised it arguing that the concentration of allowances in an eventual Spanish electricity pool would be small compared to the total amount of allowances and to the allowances received by some (large) European firms. The creation of pools between non-electricity firms has been slow due to difficulties in obtaining the authorisations for the association between firms (the authorisation of the Autonomous Communities (AACC) and the central government is necessary) and reaching agreements between competitors.

---

[4] The reasoning is that installations which invested in energy efficiency measures before 2000–2002 will have comparatively higher process emissions and, thus, a larger amount of emissions enjoying a preferential treatment (MINAM 2004b).

5.4 Data problems and influence of a decentralised administrative structure

Several data sets were used to carry out the allocation:

(1) National GHG emissions inventory. Data on the emissions of some of the covered installations (obtained through questionnaires) were included in a database on 'Major Focal Points'. Despite the cooperation between sectors and the public administration, some 'inconsistencies' between the administration and the sector emissions data were detected in the first NAP version, leading to homogenisation of both data sets.

(2) Questionnaires sent by industry associations to installations. These were especially important in determining the exact amount of emissions in some sectors and, particularly, in the bricks and tiles sector.

(3) The Spanish version of the European Pollutant Emissions Register (EPER). The AACC have the legal competence to receive information on emissions from industrial complexes. After validating this information, AACC sent it to the MINAM. Only the emissions data for 2001 were available from this source.

(4) Registry of electricity generation installations.

(5) Working group Central State Administration–CEOE, which carried out an analysis of the emissions reduction alternatives in different sectors.

(6) Collaboration agreement between the MINAM and the Klein Institute.

The availability of data at the installation level was a problem in the initial stages of the NAP elaboration, making it difficult to simultaneously apply top-down and bottom-up methodologies, as suggested in the non-paper. It was certainly an obstacle in the realisation of previous technical studies.

This problem is partly related to the decentralisation of the territorial/administrative structure in Spain. The central government is in charge of handling the allowances and the national registry, while the regional authorities are responsible for issuing these permits. Firms send the data to the AACC which then transfer them to the central administration. This is a slow process, which should be institutionally coordinated, planned and scheduled. This problem was aggravated by the short time period available to elaborate the NAP. Apart from this, the decentralisation of the Spanish administrative structure has not played a major role in the elaboration of the NAP for this period.

5.5 Involvement of the public and transparency of the NAP

The transparency of the system is believed to have increased significantly after March 2004. The covered sectors generally state that the public administration has taken their opinion into account when elaborating the NAP, informing the sectors of the steps taken, although for others this transparency was limited because the sectors were not told how many allowances would be allocated to them and the allocation criteria. Virtually all sectors declare that the relationship between the sectors associations and the public administration have been 'cordial and free-flowing' from the start. Most generally they believe that their claims have been considered in the NAP elaboration, although not all of them and, of course, they all regret that they have received fewer allowances than 'needed'. Some regret that their opinion was not requested when negotiating the BSA targets. While some would argue that this is a result of the change of government, others would interpret the increase in transparency and visibility as a logical process because secrecy about the options being considered facilitates the technical and political decisions in the NAP, while publicity is a natural (and required) step in the process afterwards. Some actors (e.g. environment non-governmental organisations (NGOs)) were initially discontent, claiming that they were not consulted. However, after publication of the NAP, many actors expressed their opinion and even NGOs such as Greenpeace admitted that the public information procedure 'has been accurate and the NAP clearly states how comments from the public were considered'.

5.6 Special interest lobbying

As expected, the zero-sum game involved in allowances allocation led to a close interaction between firms and the public authorities and intense lobbying activity. It was a source of conflict between actors, with discussions at various levels:
- meta macro level (calls for renegotiation of the BSA targets)
- macro level (split between covered and non-covered sectors)

- meso level (opposing claims between electricity and non-electricity sectors)
- micro level (major disagreement between firms within the electricity sector on crucial design elements and treatment of cogeneration).

This lobbying activity led to the creation of coalitions (both formal and informal) at these different levels to push for a greater allocation.

A major lobbying activity was to show that compliance with the BSA and EU ETS would entail high costs and negative impacts on competitiveness. The CEOE published an influential report in October 2003 defending the renegotiation of the BSA targets because Spain was believed to have been unfairly treated, calling for the EU to reconsider the implementation of the Kyoto commitments. As mentioned above, the previous government channelled this discontent by questioning the implementation of the EU ETS, rather than the renegotiation of the BSA targets.

Some claims from industry were sector-specific while others were common to all sectors (early action, exhausting reduction opportunities, limited reduction potential, high growth rates expected etc.)(see del Río (2006) for a further analysis of the strategies and arguments of different sectors in the allocation process). Obviously, the lobbying power of different sectors was different. Some sectors (the most powerful, concentrated and well-organised) carried out lobbying activities very early in the process, while others did so at a later stage. Sectors where small firms predominate had probably less capacity to influence their allocation.

The electricity sector deserves a special mention. The three major producers (Endesa, Iberdrola and Unión Fenosa) account for 90% of the market. Iberdrola is the company with the lowest associated emissions per kWh, given their investments in CCGTs and renewables. In contrast, Unión Fenosa has the highest associated emissions of the 'big three', with Endesa in the middle. The allocation process in this sector led to a dispute between Iberdrola and the rest of the sector. According to the other firms, Iberdrola regarded the NAP as an opportunity to gain market quota. Advocating scarcity, it asked for the sector to be given 70 Mt CO_2, arguing that 'this would be enough for the sector to comply with'. In contrast, the rest of the firms asked for 96 Mt CO_2, claiming that, otherwise, coal-thermal installations would be severely affected. Endesa further claimed that, when it was a public company, it had to invest in the exploitation and burning of coal in depressed areas.

Table 8.3 *Coverage of sectors' demand for allowances*

Sector	Coverage (%)[a]
Electricity generation	93.51 (89.4 most firms, or 121, Iberdrola)
Oil refining	96.39 (90 according to the sector)
Steel	97.07
Cement	94.73
Lime	89.15 (85 according to the sector)
Ceramics	
Bricks	94.23
Tiles	97.18 (91.6 according to the sector)
Total	94.72
Glass	97.13
Pulp and paper	95.54
Subtotal industrial sectors	95.41
All covered sectors	94.38

[a] Coverage = share of the initial requests that were awarded in the final allocation.
Source: RD1866/2004.

While Iberdrola asked for an allocation based on expected production, the rest favoured an allocation based on historical emissions. The final allocation reflects a middle path between both positions.

Finally, cogeneration also led to intensive lobbying. Seven industrial sectors (paper, chemicals, ceramics, bricks and tiles, oil refining, textile and electricity autoproducers) presented a common position to the government, requiring that cogeneration receive a special treatment in the NAP and be considered a 'separate activity'. An intensive consultation process with the covered sectors was undertaken in January–February 2004. After the change of government there was a new round of meetings with the involved sectors which reactivated the lobbying activity. According to the government, the NAP awarded at least 90% of the request for allowances from the sectors (Table 8.3). Some sectors (e.g. cement) claimed that real coverage has been lower, however.

The fact that loud voices were not heard against the allocation could be interpreted as meaning that the NAP had struck a reasonable balance between all the interests and actors involved. Consumers and taxpayers have been the most relevant absentees from the NAP discussion, however. Policy-makers have an inherent incentive to reduce the conflict with visible and noisy interest groups (firms) and to shift the costs to uninformed and unorganised taxpayers and consumers, unlikely to be very noisy against any NAP decision.

6 Concluding remarks

6.1 Important political dimensions facilitating or complicating agreement

The elaboration of the NAP in Spain has been an arduous process complicated by several factors, which included among others: the lack of a climate change policy tradition (and of economic instruments in environmental policy), the long distance to BSA target, problems with the data (solved during the NAP process), the change of government, the territorial/administrative structure of the Spanish state, the priority given to economic development, the expected dynamism of the Spanish economy (and particularly, the high future growth rates of the sectors covered) and the opposing views between firms in the electricity sector.

In contrast, other political factors may have played a positive role in the elaboration of the NAP, such as the political and social legitimacy given to legislation coming from Brussels (Directive), the fact that Spain would significantly lose from climate change (and from non-compliance with the BSA target) and the renewed impetus and political legitimacy of the new government.

With a delay, and in spite of all those factors, the government was able to prepare the NAP, an accomplishment in itself. The NAP strikes a reasonable balance between two major concerns: safeguarding the competitiveness of the covered sectors and positioning the country in a Kyoto path. It could be argued whether this was an achievement on the part of the government or a logical way to proceed. The fact that there were not very loud voices against the final NAP version and that most actors praised the transparency of the NAP elaboration process might not have been expected at the beginning of the process.

What was to be expected, however, was intensive lobbying activity by the covered sectors, the costs falling to the greatest extent possible on the large silent majority of consumers and taxpayers and leaving/postponing some difficult decisions (e.g. the treatment of emissions from diffused sectors) and the largest reductions for the medium to longer term.

6.2 Lessons learned

- The elaboration of the NAP is a political process, whose outcome may not be 'rational' from an economist's perspective, but the result of the interaction between several actors and the balancing of conflicting interactive variables and criteria. Policy-makers have an incentive to minimise short-term political costs, which is not necessarily optimal in a longer-term perspective.
- The elaboration of the Spanish NAP confirms that it is politically difficult to tackle the emissions from non-covered sectors. This is a negative aspect of the NAP. The treatment of the policy measures to be applied in these sectors is superficial, which makes the overall objective of GHG emissions not exceeding 24% of base year levels by 2008–2012 a weak one. A wide array of measures is envisaged but no budget provision, no estimate of specific emissions reductions and no time schedule for the implementation of these measures are mentioned at the time of writing this chapter. There seems to be a sense of resignation that the emissions from diffused sectors will increase and that Kyoto units will be bought to compensate for them. There is no detailed and integrated strategy tackling the causes behind GHG emissions from these sectors. It is a difficult task politically, with (uncertain) long-term rewards (probably exceeding the government's term in office).
- The burden of reductions was put on sectors with relatively abundant low-cost reduction alternatives and which do not face international competition (electricity sector), while industrial sectors, which have more difficulties in abating emissions, are given as many allowances as they are likely to need. This implicit adaptation to the abatement costs of different sectors is key in mitigating the conflicts with sectors and in keeping the costs of the scheme low. In turn, this facilitates the elaboration of the NAP and the implementation of the ETS in Spain.
- The coordination between different administrative levels concerning installation-level data should be improved, maybe through the implementation of specific rules and schedules for transferring the data.

- Smooth communication between the government and the sectors affected is also key to ensuring the successful elaboration of the NAP. Clear criteria by the incoming government, transparency and simplicity (although not taken to the extreme) were crucial in mitigating conflicts with the affected sectors in the preparation of the NAP. Claims for 'special treatment' due to 'special circumstances' were carefully addressed and, if possible, integrated within the general architecture without causing major changes. When this was not possible, the legitimacy of these claims was put in the balance with the possible disruptions an exception to the rule could bring (including increases in transaction costs). The government maintained a continuous dialogue with the affected sectors, explaining what the government's position was and making clear what was not negotiable. This focuses and restricts lobbying activity and prevents the questioning of everything.
- The government also showed considerable leadership and authority in its relationship with sectors. The existence of skilful people, involvement of different ministerial departments and attachment of a high political profile to those negotiating was also important in facilitating the negotiations with the sectors. Requiring independent technical expertise at the start of the process helps substantiating the government's position. In turn, communication with sectors and the elaboration of the NAP itself is easier when firms within a sector maintain a homogeneous position which was, for example, not the case in the electricity sector.
- Policy-makers face trade-offs and conflicting objectives when deciding on certain NAP choices. It is then very difficult to satisfy simultaneously all criteria, variables and actors. Policy-makers have an incentive to minimise short-term conflicts at the expense of leaving the decision on critical issues for the long term. Therefore, it is likely that a gradual approach to reduce emissions will be adopted, that a proportionally higher share of the effort will be put on the least powerful/able to negotiate and that the costs will fall on the large silent majority of consumers/taxpayers.
- Auction was excluded from the start in the Spanish NAP to avoid a negative impact on competitiveness and to minimise the potential conflict with the affected sectors as a result of the NAP. Free allocation contributed to ensure the support of affected (and potentially conflicting) actors and was key to facilitating the successful

implementation of the ETS in Spain, although it is less optimal than auctioning from a social welfare perspective. Once the system has been implemented, auction could be introduced in future compliance periods, although there might be a risk of institutional lock-in.

- Free allocation had another indirect positive effect: it improved existing data on emissions by requiring closer cooperation between different actors and administrative levels. This was so because small firms whose emissions were not initially identified requested an allocation (ceramics sector).

- A regulated limitation on final electricity price increases (in oligopolistic electricity markets, with an incomplete liberalisation process and without major international interconnections such as the Spanish case) may mitigate concerns that industry and households will inequitably bear much of the costs of the ETS through higher prices. However, it discourages consumers from changing electricity consumption patterns.

- Decentralisation of allocation has its pros and its cons. Governments have an inherent incentive to reduce the impact on the competitiveness of the sectors that are most exposed to international competition by allocating them a higher share of allowances. If not harmonised at an EU level, this may cause severe distortions between the same sector in different Member States whose NAP may treat the same sector differently. ·

References

Banco de España 2004. *Annual Review of Statistics*, Madrid.

Del Río, P. 2006. 'Implementing the EU emissions trading directive in Spain: a public choice perspective', in R. Antes, B. Hansjürgens and P. Letmathe (eds.) *Business and Emission Trading*. Editorial Springer/Phisica, Heidelberg (Germany): 293–312.

European Commission 2004. 'Commission Decision of 27 December 2004 concerning the National Allocation Plan for the Allocation of Greenhouse Gas Emission Allowances notified by Spain in accordance with Directive 2003/87/EC of the European Parliament and of the Council', C(2004) 5285, Brussels.

Hernández, F., Del Río, P., Gual, M.A. and Caparrós, A. 2004. 'Energy sustainability and global warming in Spain', *Energy Policy* 32: 383–94.

MINAM 2004a. 'Aplicación del comercio de derechos de emisión', Nota explicativa, OECC-MINAM, Madrid.

MINAM 2004b. 'Acuerdo por el que se aprueba la asignación individual de derechos de emisión a las instalaciones incluidas en el ámbito de aplicación del real decreto Ley 5/2004', Madrid.

MINAM 2004c. 'Propuesta de asignación individual de derechos de emisión para el periodo 2005–2007', MINAM y Ministerio de Industria, Turismo y Comercio, Madrid.

MINAM 2005a. 'Nota sobre las modificaciones introducidas por el real decreto ley 5/2005 en el ámbito de aplicación del comercio de derechos de emisión', Madrid. www.mma.es (in Spanish).

MINAM 2005b. 'Acuerdo del consejo de ministros de 11 de marzo por el que se desestima la solicitud de asignación gratuita de derechos de emisión a las instalaciones no incluidas en el ámbito de aplicación de la Ley 1/2005, de 9 de marzo, por la que se regula el régimen del comercio de derechos de emisión de gases de efecto invernadero', Madrid.

Point Carbon 2004. 'Carbon Market Europe', *The Carbon Market Analyst*, 27 April 2004. www.pointcarbon.com

RDL (Royal Decree Law) 1866/2004. National Allocation Plan for 2005–2007. September 7th 2004. BOE number 216: 30616–30642.

RDL (Royal Decree Law) 5/2005 on 'urgent reforms for the improvement of productivity and public contracting'. March 14th 2005. BOE number 62: 8832–8853.

Schneider, F. and Volkert, J. 1999. 'No chance for incentive-oriented environmental policies in representative democracies? A public choice analysis', *Ecological Economics* 31: 123–38.

SGPCCC 2005. 'El consejo de ministros aprueba la asignación final de derechos de emisión de CO_2', SGPCCC. www.mma.es

Svendsen, G. T. 2000. *Public Choice and Environmental Regulation: Tradable Permit Systems in the United States and CO_2 Taxation in Europe*. Cheltenham: Edward Elgar.

Tábara, J. D. 2003. 'Spain: words that succeed and climate policies that fail', *Climate Policy* 3(1): 19–30.

Woerdman, E. 2004. 'Path-dependence climate policy: the history and future of emissions trading in Europe', *European Environment* 14: 261–75.

9 | Italy

DANIELE AGOSTINI

1 Introductory background and context

Throughout its development process the Italian National Allocation Plan (NAP) required by Directive 2003/87/EC (the EU ETS Directive) had to face a complex context. The challenging emission reduction commitments taken up by Italy through the Kyoto Protocol had to be pursued vis-à-vis a number of challenging issues including: rapidly rising greenhouse gas (GHG) emissions trend, contrasting energy policy objectives of supply diversification and smooth market liberalisation, challenging time-frames, lack of data and a climate of political uncertainty over the entry into force of the Kyoto Protocol itself. Furthermore, at both the EU and national level lack of clarity existed over the role of the EU ETS as a 'policy setting' rather than a 'policy compliance' tool. All of these issues have significantly influenced the NAP development process and their understanding is critical for drawing conclusions over lessons to be learnt.

1.1 Italy's reduction target and abatement potential

Italy's GHG emissions reduction target under the Burden Sharing Agreement (BSA) appears relatively modest when compared in absolute terms to targets of other Member States. However, it becomes quite challenging when considering the country's rising emission levels and high abatement costs. Under the BSA Italy has committed to a 6.5% reduction of its 1990 GHG emissions. A commitment requiring in absolute terms a reduction from the 508 Mt CO_2e per year emitted in 1990 to an average 476 Mt CO_2e per year during the period 2008–2012.

However, a number of factors make such an apparently limited reduction extremely challenging. On the energy demand side, a review of energy pricing data shows Italy's energy consumers facing some

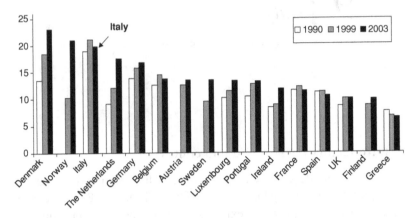

Figure 9.1. Electricity prices to the residential sector (1 January €/kWh, taxes included, 3,500 kWh per year).

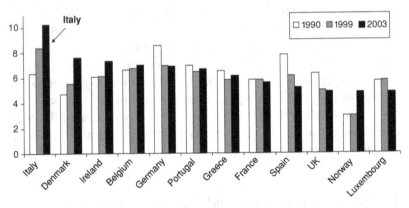

Figure 9.2. Electricity prices to industry (1 January €/kWh, taxes included, 10 GWh per year, 2,500 kW).
Source: Eurostat.

of the highest prices in the European Union. In the residential sector Italian prices rank third after Denmark and Norway. In the industrial sector Italy has the highest electricity prices, which top by more than 20% the prices of second-ranking Denmark. High electricity prices in both industrial and residential sectors have resulted in low energy intensity within the Italian economy (see Figures 9.1–9.4).

On the energy supply side the early penetration of gas in the first half of the 1990s, the limited use of coal and the full exploitation of existing hydro power have resulted in an electricity generation system

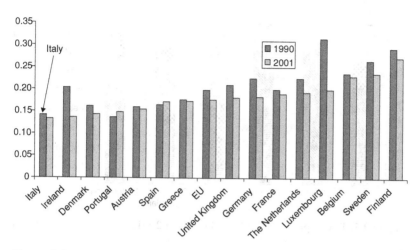

Figure 9.3. Energy intensity of GDP in the EU (TOE/$1000 1995 using purchasing power parities).
Source: *Energy Balances of OECD Countries*, 2002 edn.

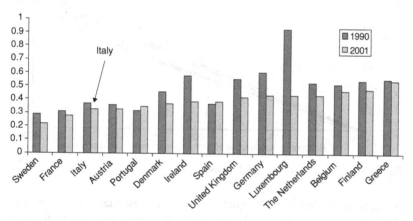

Figure 9.4. CO_2 emission per unit of GDP (kg CO_2 per $US using 1995 prices and purchasing power parities).
Source: IEA, CO_2 *Emissions from Fuel Combustion 1971–2001*, OECD, 2003 edn.

characterised by low carbon intensity if not for the absence of nuclear plants. In Italy in the early 1980s a popular vote phased out nuclear energy in light of strong environmental concerns over the risk of accidents. However, such concerns have led among other things to a non-level carbon playing field between Italy and other members of the EU

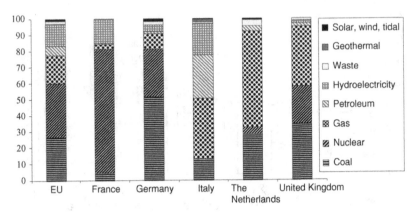

Figure 9.5. Power production (%) in Europe in 2001.

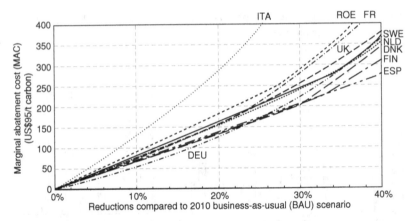

Figure 9.6. Abatement costs in Europe.

using nuclear energy (see Figure 9.5). Recent estimates indicate that the ban on nuclear has burdened Italy's BSA target with an additional 21 Mt CO_2e per year, drastically increasing its carbon abatement costs vis-à-vis other Member States.

Low energy intensity caused by high energy prices together with low carbon intensities despite the ban on nuclear energy are some of the key factors making Italy's carbon abatement costs among the highest in Europe. Analysis performed by the Massachusetts Institute of Technology (Viguier *et al.* 2003) show Italy's marginal abatement cost curve to be above that of its main EU counterparts (see Figure 9.6).

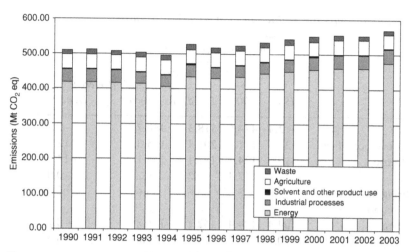

Figure 9.7. Emission trends in Italy.

1.2 Historical trends and future perspectives

A scenario of high abatement costs turned out to be even more critical when placed within a recent trend of increasing emission and forecasts of further electricity penetration[1] implying additional emissions growth. Data on emissions of GHGs from 1990 to 2003 show a reduction in total emissions falling below 500 Mt in 1994 in order to rise sharply in 1995 and then proceed with a steady increase in the period 1995–2003 (see Figure 9.7). The reduction in the early 1990s is attributed to the early penetration of gas in the Italian energy market, an effect that was exhausted in the late 1990s and more than offset by increasing electricity penetration by the year 2000. Penetration occurred despite the high power prices observed, energy demand in Italy often being characterised as quite inelastic. The medium- and long-term forecasts indicate a continuing rise in emission levels only partly mitigated by the policies and measures foreseen by the National Emission Reduction Plan (see Table 9.1 for business-as-usual (BAU) and reference scenarios).

Within a context characterised by increasing emissions and high abatement costs the recent process of electricity market liberalisation is

---

[1] Electricity penetration implies greater electricity consumption which in this case is associated with diversification of electricity uses (e.g. electronics, air conditioning) rather than with reduced energy efficiency.

Table 9.1 *National Emission Reduction Plan scenarios in Italy*

| | GHG emissions (Mt CO_2e) | | | | |
| | | | 2010 | | (Reference − Trend)/ |
	1990	2000	Trend	Reference	Trend (%)
Energy use	412.4	444.5	518.3	480.7	−7.3
Energy industries, of which:	127.6	151.6	201.3	175.3	−12.9
thermoelectric	110.5	134.2	182.1	156.1	−14.3
refineries	17.1	17.4	19.2	19.2	0.0
Manufacturing industries and construction	89.6	78.0	83.6	83.6	0.0
Transportation	104.4	124.4	142.1	136.8	−3.7
Residential and tertiary	70.7	72.9	74.1	67.8	−8.5
Agriculture	9.2	8.9	9.6	9.6	0.0
Others (fugitive, military, distribution)	10.9	8.7	7.6	7.6	0.0
Non-energy use	95.6	99.4	95.0	95.0	0.0
Industrial processes (mineral and chemical industries)	40.9	45.4	51.0	51.0	0.0
Agriculture	40.4	40.3	36.1	36.1	0.0
Waste	12.6	12.4	6.9	6.9	0.0
Others (solvents)	1.7	1.3	1.0	1.0	0.0
Total GHG	508.0	543.9	613.3	575.7	−6.1

bound to play a positive role. The introduction of power exchange trading and the further mitigation of the significant market power enjoyed by Ente Nazionale per l'Energia Elettrica (ENEL) should together contribute to a reduction in electricity prices. If on the demand side such an effect may release electricity consumption increasing further the rate of electricity penetration, on the supply side increased competition should exert upward pressures on productivity and thus increase generating efficiency and consequently reducing carbon intensity of power generation. In this perspective a stream of investments made in the early 2000s should bring on line significant new generation capacity characterised by lower emissions per unit of power through the use of combined cycle gas turbine (CCGT) technology (see Figure 9.8).

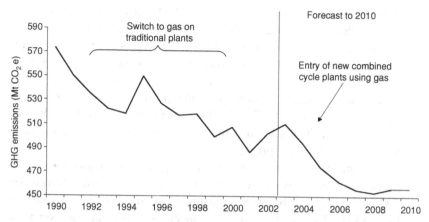

Figure 9.8. The impact of new investments on the Italian power sector.
Source: NAP.

1.3 Policy context and emission information uncertainties

In addition to the important technical factors described above, two additional elements introduced significant uncertainty in the development and pursuit through the NAP of a clear path to achieving the Kyoto target: on the one hand, before Russia's ratification in February 2005, the uncertainties associated with the coming into force of the Kyoto Protocol, while on the other the time-frame for the transposition into Italian law of the EU ETS Directive.

The key phase of the development of the NAP coincided with increasingly strong rumours regarding the likelihood of Russia's not ratifying the Kyoto Protocol and giving in to the pressure exerted by the US government. The uncertainty over the entry into force of the Kyoto Protocol led many of the industrial players opposing GHG reductions on the grounds of loss of international competitiveness to doubt whether the EU ETS Directive would in fact be implemented. Such views were made all the more credible by the difficulties experienced in enforcing the stability pact, with supporters claiming that, if the very EU stability pact could be questioned, then Decision 2002/358/EC and the BSA surely could not be considered irreversible.

The uncertainty over the entry into force of the Kyoto Protocol together with the pressure from those who opposed GHG emission reduction targets contributed to significantly slowing down an already difficult transposition of the EU ETS Directive. The December 2004 transposition deadline set by the EU ETS Directive represented a procedural challenge for many EU governments in light of both the complexity of the subject matter and the lengthy procedures required by parliamentary debates. Furthermore in Italy, as in other Member States, all EU Directive transposition measures need to be included in a single law, which required extensive formulation and debating time. The late transposition of the EU ETS Directive in spring 2005 resulted in the absence of the legal mandate for the government to drive forward the NAP development process.

The uncertainties associated with the procedural issues were only partially mitigated by an urgent government decree issued in December 2004. The goal was to ensure continued operation of plants regulated by the EU ETS Directive and the collection of the data necessary to develop national and sector-level allocations. The data collection exercise was especially important for moving forward a NAP development process stalled by the lack of plant-level data: pre-existing plant-level emission data were insufficient and, when available, often could not serve as basis for the NAP. The latter was due to the fact that Italian law does not allow the use of data collected for purposes other than the ones originally intended.

2 The macro decision concerning the aggregate total

The macro decision followed a one-step bottom-up approach and led to an allocation apparently consistent with the path to achieving the Kyoto targets. However, the method resulted in a potentially generous allocation, which could only be justified by the need for both a smooth transition and the minimisation of impacts on power prices.

2.1 The one-step sector-level bottom-up approach

In contrast to the procedures followed by other Member States and recommended by the Commission's guidelines on NAPs, the Italian allocation process did not develop through a two-step top-down approach a first decision on overall allocation based on macro level policy goals which sets the constraints for the following second decision on sector level allocations. The Italian NAP decisions were, instead, intrinsically linked to the GHG National Reduction Plan (NRP) characterised by a one-step bottom-up approach: a first-step single decision on sector-level caps determines the second-step overall allocation obtained as a simple sum of sector level allocations.

The NRP constitutes the highest policy level document on climate change in Italy. As a consequence it requires at the highest government levels validation from the Government Committee on Economic Planning (CIPE). Starting from sector-level analysis the NRP on the one the hand defines current and expected emissions scenarios, while on the other it identifies the policies and measures necessary to ensure Italy's compliance with the BSA. The bottom-up approach of the NRP made the NAP's decision on the aggregate total number of quotas to be distributed a simple sum of the sector-level emission scenarios previously approved. The CIPE working group had developed sector-level emission scenarios taking into consideration not only expected economic growth and abatement potential but also economic, energy, social and environmental policy goals. The latter included: sustained economic growth, preservation of international competitiveness, security of energy supply, energy market liberalisation and maintaining the ban on nuclear energy. Although on the one hand consideration of such diversified policy goals ensured a desirable integration of economic, energy, social and environmental policies, on the other hand it introduced new challenges to the achievement of domestic emission reductions.

2.2 The path to the Kyoto Protocol

In developing its emission scenarios the CIPE working group assumed, in line with economic theory, high energy prices to have led through high energy efficiency to the exhaustion of opportunities for CO_2 combustion emissions abatement. As for CO_2 process emissions, significant reductions were excluded as they appeared possible only by sacrificing economic growth. The analysis concluded that in Italy abatement beyond the levels defined by the United Nations Framework Convention on Climate Change (UNFCCC) Reference Scenario[2] were hard to achieve and any BSA compliance gap should be addressed through a strong contribution of the Kyoto Protocol's flexible mechanisms. The latest update of the NRP abatement integrated such results causing

---

[2] The UNFCCC reference scenario includes all policies and measures already in place.

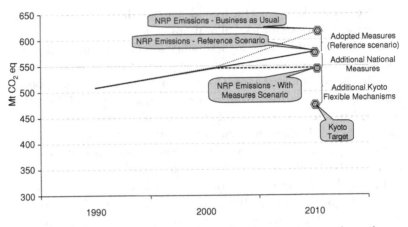

Figure 9.9. The Path to Kyoto in Italy: total national emissions of greenhouse gases.

quite a stir: the 2010 Reference Scenario emissions outlined in the NRP[3] (553 Mt per year) showed national emissions exceeding 1990 levels (508 Mt per year), which, to most of those not familiar with the Kyoto Protocol's Marrakech Accords, appeared to breach the BSA and place Italy off the path to compliance with its commitments. In reality the modalities for the use of the Kyoto Protocol flexible mechanisms imply the possibility for parties to exceed 1990 levels in those cases where BAU trend emissions are sharply rising.[4] Italy's context did exhibit such circumstances together with high abatement costs which made domestic abatement possible only through significant effort. Reliance on the flexible mechanism was thus justified and consistent with a path for positive BSA compliance (see Figure 9.9). Nevertheless, misconceptions on the subject reinforced by the national and international press led to strong pressures being exerted on the NAP development process as well as negative repercussions on its credibility.

The Italian government through the joint endorsement of both the Ministry of the Environment and the Ministry of Industry maintained

[3] UNFCCC methodologies require NRPs to define three scenarios: (1) business as usual, a no-action scenario; (2) reference scenario, with adopted policies taken into account; (3) with additional measures, taking into account measures not yet adopted.

[4] The Marrakech Accords only call upon domestic policies and measures to contribute significantly to a party's emission reduction effort; they leave significant margins for flexible mechanism to make up for any remaining compliance gap.

the approach already defined in the NRP of economic–energy–environmental policy integration. The EU ETS NAP was used as a 'policy compliance' rather than a 'policy-setting' tool: the 2005–2007 NAP allocations notified to the European Commission were aimed at 'ensuring compliance' with the long-term NRP 2010 emission Reference Scenarios rather than 'set new policy goals' in contrast with the NRP's assessments of limited abatement potential. As a result the amount of allowances allocated were generous enough to avoid significant negative impacts on economic growth with an average annual allocation of 255 Mt CO_2. The allocation was in line with maintaining constant the EU ETS sectors' share of the total national GHG emissions (estimated 51% in the year 1990, 47% expected in the year 2010).

2.3 The decision for a smooth transition through low scarcity

The result of the NAP development process was a scenario of likely abundance rather than scarcity of allowance. The Italian authorities carefully evaluated this outcome and endorsed it based on a number of considerations. The two main driving factors were on the one hand the expected impact of scarcity on Italian company's competitiveness while on the other the preference for a gradual transition from the traditional more rigid command and control to the new flexible market instrument approach to environmental policy-making.

Scarcity in allocations for Italy was deemed undesirable partly because of its impacts on Italian industry's cost structure and competitiveness. High energy prices had already internalised to a significant extent the carbon externality in the Italian economy making its carbon abatement costs among the highest in Europe. Diversity of supply issues requiring a greater role for coal and environmental policies banning nuclear energy made reducing the carbon content of energy extremely challenging. As a result scarcity, rather than inducing domestic abatement, would have led Italian companies to purchase allowances from other EU companies operating in national contexts characterised by lower abatement costs. The costs of purchasing allowances would have eroded existing product margins with possible impacts on inflation and international competitiveness. For products like cement and lime characterised by low levels of international trade the additional costs would be passed on to the consumer. For other products, such as glass and tiles, the cost of purchasing allowances would instead expose the sectors to unfair competition from companies operating in the context of lower national abatement costs either in the EU (Spain, Greece and former Eastern Europe) or outside (North Africa, Turkey). In both cases, although from an environmental economics point of view because of static efficiency considerations these are the very dynamics that emissions trading strives to introduce, from a balance of payments and social point of view the impacts are less desirable because of impacts on price and import/export. Adequate quantification of such effects within the reduced time-frame allowed by the EU ETS Directive proved to be difficult. Furthermore, in a time of recession and globalisation-driven relocation of production, the pressures to avoid further threats to national economic recovery were strong. The additional costs of purchasing allowances appeared as one of such threats and led very early on to a strong and public political commitment by the Italian authorities to provide companies with the allocation needed by avoiding allowance scarcity.

Although the dangers of price impacts and international competitiveness issues were quite vivid, most operators and some of policy-makers failed to appreciate fully the potentially significant role of the opportunity cost of allowances as well as their impacts on company balance sheets. The fact that once issued allowances could become part of the company's assets and, if optimally managed, yield positive results was not acknowledged. If on the one hand such failure could to some extent be part of a deliberate lobbying strategy aimed at amplifying negative repercussions of the EU ETS Directive, on the other it was most likely the result of senior management's inability to appreciate adequately the potential risks and opportunities associated with emissions trading (see also Section 5). It was often found difficult to understand that opportunity cost dynamics would have led to reduced production whenever the market value of a product's carbon content exceeded the product's profit margin. This basic market dynamic would occur irrespective of the level of allocation generosity or scarcity: on the one hand allocation scarcity through purchasing costs would have led to competitiveness and thus production losses; on the other hand, allocation generosity, despite the absence of allowance purchasing needs, could still lead to production losses due to opportunity cost dynamics especially in low-profit-margin sectors. The fact that opportunity cost could hamper to some extent any effort to avoid impacts on domestic production through allocation generosity was also often not recognised by some of the operators and policy-makers involved in implementing the scheme.

2.4 *The EU review process*

The strong political position at the national level to adopt a low scarcity allocation aimed at providing plants with all the allowances they needed was bound to be challenged by the Commission's review. The decision, endorsed fully by both the Ministry of the Environment and the Ministry of Productive Activities, had to undergo an EU policy evaluation process driven by different policy priorities, expectations and procedural constraints. At the Commission's level the environmental integrity of the overall EU ETS had an overriding effect on other national economic, energy and social considerations. The Commission assessed negatively the possibility of stretching the path to the Kyoto target in the first period with a low scarcity allocation in order to follow up in the second period with a tighter allocation. The decision was based on a number of factors. First, the risks associated with Italy's ending off-path on the medium to long term were deemed to be too high, especially in the light of national policy and measures considered to be not sufficiently aggressive. Second, delays associated with the inability to collect plant-level data through legally binding instruments had greatly reduced the time allowed for the review process and left negotiators with insufficient time for an in-depth review. Because of such time constraints, the process adopted a simplified two-step top-down approach rather than the more accurate sector-level one-step bottom-up approach used to develop the NAP. By looking at national energy demand and GDP growth at a macro level the top-down approach identified higher abatement potentials and led to a negotiated lower aggregate level allocation. Third, low scarcity might have led to the absence of thriving trading. To many the latter clearly would have indicated failure of an important EU initiative rather than smooth integration of a new system into existing complex economic dynamics. Fourth, aggressive environmental policy decisions in other Member States had led on the one hand to scarce allocations elsewhere in the EU, while on the other to the common perception that the Italian allocation was too generous, a judgement apparently not sufficiently appreciative of factors such as the political dimension of the allocation decision, national abatement costs and the important role of subsidiarity. The amplification through the media of all the above factors generated significant external pressures on the EU NAP review process, pressures that were partly based on incomplete understanding of the issues at play. The final result was a very significant cut in total allocation equal to approximately 10% of what originally allocated. In contrast with the original NAP one-step bottom-up approach, the reduction had to be achieved through a two-step top-down approach: the 10% cut in total allocation imposed from the top was distributed among the different sectors according to their abatement and growth potentials; within this redistribution the power sector carried most of the burden, followed by the mineral sector.

3 The micro decision concerning distribution to installations

As pointed out in the previous paragraph, the bottom-up methodology implied determining within the macro decision sector-level allocations for incumbents and existing plants in order to arrive at the total amount to be allocated. The basis of the micro decision was then the existing and new entrant allocations determined in the previous step and its goal was determining the best intra-sector installation-level allocation criterion. Very different issues characterised the macro and micro decisions: on the one hand, aggregate allocation determination had to face significant political pressures and challenging confrontations between contrasting policy goals; on the other hand, installation-level distribution and its requirements in terms of objective formulae became more of a technical exercise and was thus challenged by the need to account for both market share dynamics and production process complexities.

3.1 Allocation principles for new and existing plants

For new entrants the installation-level allocation criterion implied the issue of a number of allowances based on environmental performance per unit of output and expected production. Although the details of the procedure were to be finalised later through special decrees, the NAP outlined the basic principles: environmental performance was to be based on the top performance of existing plants through statistical formulae, while expected output would be based on plant capacity.

For existing plants the NAP used a simple allocation formula to determine the share of allowances the installation received from the sector's total allocation for existing plants. The number of quotas allocated to each of the existing plants was computed as follows:

$$Q_{t,j,n} = Q_{t,j} \times X_{n,j}$$

where:

$Q_{t,j,n}$ = allowances allocated for year t to plant n belonging to sector j

$Q_{t,j}$ = allowances allocated for year t to existing plants in sector j

$X_{n,j}$ = share of the total allowances allocated to plant n belonging to sector j.

Within the above formula the share of the total number of allowances was determined as follows:

$$X_{n,j} = L_{n,j} / \sum (i = 0, \ldots, m) L_{i,j}$$

where:

$L_{n,j}$ = level of activity of plant n within sector j

$L_{i,j}$ = level of activity of plant i within sector j

m = total number of plants in sector j.

The plant's level of activity was determined by the choice of an activity indicator and a historic reference period over which the indicator was to be quantified.

3.2 The selection of activity indicators

The choice of activity indicators implied adopting for each sector one of three possible allocation methods:

(1) *Historical production method*, which allocates on the basis of historical production of the plant and is applicable to reference activities characterised by relatively homogeneous production outputs.

(2) *Historical inputs method*, which allocates on the basis of historical consumptions of essential inputs such as fuel and is applicable to reference activities characterised by relatively homogeneous inputs.

(3) *Historical emissions method*, which allocates on the basis of historical emissions and is used for to activities for which the previous two methods are not applicable.

Among the three methods in terms of environmental and economic efficiency the historical production and historical emissions methods rank respectively first and second as they tend to reward early action more effectively and to allow for greater flexibility in abatement strategies.

Table 9.2 *Allocation criteria for existing plants in Italy*

Sector	Criterion
Energy activities	Electricity – expected production with a change of regime component
	Heat – historical emissions
	Heat and electricity – historical production
Lime	Historical production
Steel	Historical production
Pulp and paper	Historical emissions
Bricks and ceramics	Historical emissions
Cement	Historical production (clinker)
Oil refining	Historical emissions
Glass	Historical emissions

However, data availability and technical characteristics of the production process can greatly affect their level of applicability. Although the NAP development process preferred whenever applicable the most efficient of the three methods, in most cases the historical emission methods was the method of choice (see Table 9.2) more often due to technical difficulties rather than data availability.

Allocation used the historical production method in the power, lime, clinker production and steel sectors, which are all characterised by relatively homogeneous outputs. In the other sectors two main reasons ruled out allocation based on historical output. The first was that often the goods produced were not sufficiently homogeneous as in the case of paper production, oil refining and glass production. Under such circumstances allocating by volume or weight, the main output parameters, would be unfair as neither of them necessarily relates to the carbon intensity of the goods. A second reason for ruling out historical output as allocation criterion was the degree of penetration within sectors of on-site power production. In some cases operators had built on-site power production plants to satisfy partly or totally their electricity consumption needs. As a result their direct emissions per unit of output would have been greater than those competitors generating indirect emissions by buying power from the grid. In such cases allocation per unit of output would have penalised on-site power generation and would have been in conflict with the general energy policy encouraging decentralised power production. Although allocating additional emissions for on-site power generation could have compensated for the impact of on-site power generation, the procedure would have required more time and separate historical fuel-accounting records for on-site power generation. The latter in particular were very difficult to retrieve either because of historical aggregate fuel accounting or because the power plant was often found to be a combined heat and power (CHP) plant.[5]

3.3 Selecting the reference period

Both data availability and sector dynamics were the basis for selecting reference periods. In order to maximise perceived fairness, efforts were made to maintain the same historical reference period for all sectors. Because in all cases but the oil refining sector the formula foresaw an average over the historical period with the exclusion of the minimum value, a minimum four-year reference period was necessary. For all sectors the reference period selected was the period 2000–2003. Inclusion of recent 2004 data, although desirable, proved to be too difficult because of the unavailability of part of the plant-level data which was due to be submitted by the end of December 2004. Going back beyond the year 2000 was undesirable for two reasons: on the one hand increasing the length of the period would have amplified the impact of plant turn-around (old installations closing and new ones opening) on the homogeneity of the statistical sample; on the other hand, most of the plant-level data available had not been subject to any collection and archiving requirements so that often their level of reliability decreased significantly the further back one went in time.

The choice of a fixed historical reference period to describe plant production levels turned out to be inappropriate in a number of cases including plant opening, production halts longer than a year due to special circumstances (e.g. the intervention of public authorities, ownership disputes) and significant structural interventions on the installation's technical layout. Redefining the historical reference period easily solved the first two cases: for new plants the reference

---

[5] For technical reasons separation of fuel use for electricity production from fuel use for heat production in CHP plants is quite difficult.

period used to compute the average would begin from the plant's start-up date, while for production halts the period of zero production exceeding the year would be filtered out from the average. However, in the case of significant structural changes to a plant's layout (e.g. production line closures or installation of new CHP units) the solution was simple in principle but more difficult to apply. If on the one hand the beginning of the historical reference period could easily be made to coincide with the date of entry into operation of the modified configuration, on the other hand a significant structural change was hard to define. In assessing such cases the National Competent Authority in charge of implementing the EU ETS Directive used two screening criteria: the first one was an annual drop or increase in production greater than 10% to be maintained in subsequent years; the second one was a structural change implying an increase in emissions either due to expanded production capacity or adoption of new technologies.

3.4 The special circumstances of the power sector plant allocation

The power sector allocation was the only plant-level allocation in the NAP adopting as a basis expected output together with an updating component. The sector required special treatment because of both new entrant dynamics and the need to minimise power price impacts. The power sector in Italy is expected to be experiencing an intense turnaround with almost 25% of the production capacity possibly being replaced in the next four to five years. The process makes allocation in the power sector complex not only because of the difficulties associated with regulatory uncertainties affecting start-up dates and with forecasting fuel prices, but also because of the possible impacts of the newly built CCGT plants on recently established power market merit order dynamics.[6] Under such circumstances over-allocation and significant windfall gains for operators could occur: allocating to coal-powered plants based on their historical base-load merit role may lead to significant windfall gains when such plants decide to curtail production because of a return to lower price levels of fuel oil and the coming

---

[6] The merit order is the order with which power plants are called upon to produce within an electricity pooling market.

on line of the new CCGT plants; similarly, allocating to fuel-oil running plants based on their historical mid-merit ranking class could lead to over-allocation when they are forced into peak load service, their production being displaced by the new CCGT plants. These are only some of the new entrant dynamics likely to lead to a reorganisation of the individual power plants placements within the different market segments of the Italian electricity market, segments corresponding for all practical purposes to the different merit classes.

The other issue requiring special treatment in the power sector is the need to minimise the impact of power price increases. While power prices in Italy rank among the highest in Europe, opportunity cost dynamics of a cap-and-trade scheme such as the EU ETS are bound to increase power prices. Limited competition in the power sector could further amplify such effects. As a consequence the NAP development process adopted as one of its highest priority policy goals minimisation of power price increases. This goal was partly in conflict with the environmental goal of integrating carbon cost in power production, which was thought to be already addressed through fiscal policies on fossil fuels leading to some of the highest energy prices in Europe. Given the situation, the only promising strategy was the development of an allocation criterion allowing for the introduction of a relative targets trading component into the EU ETS cap-and-trade scheme. The relative target component would have at least partially removed the ability of operators to pass on opportunity costs to consumers thereby increasing power prices.

In order to manage such situations, the Italian NAP development working group explored various methods. The first method, the *merit ranking method*, foresaw allocating to plants by keeping track of the historical merit class in which they were initially placed. A change in the merit class placement due to either technological changes or variations in fuel prices would trigger the revisions of the plant's allocation. Merit-order-based market segments were defined on existing market dynamics in terms of equivalent operating hours (see Table 9.3).[7] Each installation would initially receive allowances equal to its average equivalent operating hours multiplied by capacity and by a fuel-specific emission coefficient. Should a plant change market segment (i.e. merit class) during a specific year, the number of allowances allocated to it

---

[7] For instance, annual production divided by full capacity.

Table 9.3 *Power sector market segmentation based on merit order*

Segment	Yearly hours of equivalent full capacity operation
Base	>6,900
High merit	5,700 – 6,900
Mid merit	4,100 – 5,700
Low merit	2,900 – 4,100
Peak	<3,000

the following year would change based on the new segment to which it belonged: every parameter would stay the same except for the hours of operations which would change to the average equivalent hours of operations associated with the market segment just entered.

The merit ranking method appeared to be able to effectively manage the specific situation the Italian power sector was experiencing while remaining compatible with the letter of the EU ETS Directive and maintaining the necessary trading incentives. Allocation and issue of allowances would in fact change depending on the plant's repositioning within the power market. This mechanism would allow new entrants not to be penalised compared to incumbents, making increased equity one of its main merits. The cases of repositioning, although characterised by potentially significant windfall gains, were expected to be limited and therefore be managed through the EU ETS Directive's provisions for new entrants.[8]

The merit ranking method maintained abatement incentives in terms of both emission intensity and hours of operations. Given a specific mix of fuel and technology, on the one hand relative target type of incentives would be in place, as plants reducing reducing emissions per unit of output below the allocation emission coefficient could retain the allowances in excess, while plants with emissions per unit of output above the allocation emission coefficient would have

---

[8] A plant changing merit class would be considered a new entrant and receive a new authorisation and allocation in light of the fact that it was changing market segment, such treatment being allowed by the Directive's definition of new entrant as a plant who receives a new authorisation based on changes in its functioning.

to purchase allowances. On the other hand, absolute cap-and-trade dynamics would still persist, as plants reducing emissions by reducing their hours of operations could still retain their allowance if they did not fall below the bottom limit of their merit class, while plants increasing emissions by increasing their hours of operations would still have to purchase allowances until they reached the top limit of their merit class.

The second method proposed, the *change of regime method*, was based on a closer tracking of power plant's emissions. Allocation would be based on plant capacity and on forecasts of both equivalent hours of operation and specific emission factor by technology mix (combination of technology and fuel type). More specifically allocation to plant n for the year t would be equal to the sum of two components:

$$A_{n,t} = A^1_{n,t} + A^2_{n,t}$$
$$A^1_{n,t} = \Sigma k\ \Sigma s(k) \in n(\alpha_{k,t} h *_{k,t}\ P_{s(k)})/1000$$
$$A^2_{n,t} = \Sigma k\ \Sigma s(k) \in n\ \alpha *_t\ P_{s(k)}(h_{s(k),t-1} - h*_{k,t-1})/1000$$

where:

$A^1 n, t$ = main component

$A^2 n, t$ = change of regime component ($A^2_{n,2005} = 0$)

$s(k)$ = indicates unit s belonging to category k of plant n

$\alpha_{k,t}$ = reference emission coefficient specific for category k, for the year t (g CO_2/kWh) to be used within the main component

$\alpha *_t$ = reference emission coefficient specific for the year t (g CO_2/kWh) to be used within the change of regime component

$h*_{k,t}$ = number of hours of expected operation of plants belonging to category k, relating to year t

$h_{s(k),t}$ = number of hours of actual operation (equivalent to full load) of unit $s(k)$, for year t

$P_{s(k)}$ gross efficient capacity (in MW) of unit $s(k)$, belonging to category k.

In practice the main component would be allocated based on forecast of a plant's production while the change of regime component would on the one hand partially correct forecasting errors while on the other hand rewarding or penalising less or more polluting production thus maintaining trading incentives. In the change of regime method the main component is determined using emission coefficients ranging between 500 and 2,800 g CO_2/kWh, while the change of regime component is determined using a single emission coefficient set to 550 g CO_2/kWh. On the one hand, such provision implies that plants exceeding their expected hours of operation or more polluting installations would be penalised by receiving an allocation adjustment inferior to their effective needs, while cleaner installations would receive allowances in excess of what was actually required. On the other hand, the same provision provided an incentive for more polluting plants to operate less and maintain possession of part of the allowances in excess. Compared to the merit ranking method the change of regime method tracked hours of operations more closely, leaving all abatement incentives to emission intensity rather than production and further mitigating the potential impact of opportunity costs on electricity prices.

Following the national consultations the NAP formally presented to the European Commission adopted the change of regime method, mainly because the Ministry of Productive Activities believed it to be the most suitable to the conditions of price instability observed within the recently introduced electricity pool. By removing any incentive to alter hours of operations in the light of carbon market dynamics, the method was considered to be more predictable and to interfere less with the severe price oscillations experienced on the Italian power markets. Despite the worthiness of its goals and the fact that it did maintain some trading incentive, the change of regime method had the major weakness of being hardly compatible with the letter of the Directive: systematically adjusting allocation of allowances to all plants diverging from their assigned hours of operations made it impossible to treat them as new entrants and give them the benefit of access to a new individual allocation. On the other hand the merit ranking method's strengths of maintained significant trading incentives and compatibility with the letter of the Directive were not sufficient to overcome its weaknesses in terms of apparent complexity to most policy-makers and operators. After a number of advance warnings the Commission through its review procedure formally rejected the change of regime method and required the removal of the change of regime component. The resulting allocation method in the power sector became therefore a purely *ex-ante* expected output allocation characterised by all the difficulties of forecasting in advance plant hours of operations in a sector experiencing rapid as well as dramatic changes.

4 Issues of coordination and harmonisation

During the NAP development process coordination over technical matters as well as harmonisation over political issues played an important role and were the subjects of an exponentially increasing number of meetings. Informal meetings between Member States, special seminars and conferences complemented, but at times duplicated, the regular sessions of the Climate Change Committee's Working Group 3 in charge of coordinating the technical implementation of the EU ETS Directive. The complexity of the subject and the tight time-frame did not contribute to the meetings' effectiveness. Furthermore, often the timing of NAP development was different from one Member State to another because of different progress in the legislative process, in the data collection exercises or in the political debate. All such factors affected the ability or willingness of Member States to discuss and possibly coordinate at the EU level the specifics of their NAPs.

Italy's case was made even more challenging by a number of different elements including the delay in the transposition process, the lack of data to assess fully the significance of certain issues and the absence of agreement among national stakeholders due to the many existing uncertainties. Technical issues playing an important role in terms of coordination included: assessing abatement potential, treatment of special cases such as management of blast furnace gas within integrated steel works or that of emissions from cleaner technologies such as CHP, treatment of new entrants and closures, integration in the allocation process of concerns over competitiveness from non-EU countries. There were also a number of political issues which were significant in terms of need for harmonisation including: compatibility of the BSA with a level playing field, interaction with other energy and environmental policies, scope of the EU ETS Directive. Unfortunately, all issues, although addressed, were only partially resolved.

4.1 Evaluating abatement potential

Assessing abatement potential from the different sectors was probably one of the most controversial issues discussed at the EU level. Paradoxically abatement potential should have been the least relevant issue, as one of the main virtues of emission trading is its ability to bypass the issue of asymmetrical information regarding abatement and let the markets determine the best distribution of the abatement effort. Never-theless, often because of the strong command and control tradition of all of those involved from both the operators' and policy-makers' sides, the connection between allocation and abatement potential assumed a prominent role in the EU coordination and review process. As we have seen, this was a role that turned out to be so significant as to pre-vail over what most economic textbooks indicate as the purely political exercise of setting total and sector-level caps. In determining allocation, what should have been purely technical EU coordination ended up giv-ing prominence to abatement potential over other equal if not more important policy goals. In the case of Italy these goals included energy diversification, economic development and stabilisation of prices in the power market.

Although EU coordination over abatement potential was success-ful in finding a near consensus on the fact that emissions should be reduced, it failed to identify effective methods to determine by how much industrial systems were underperforming or overperforming: the absence of homogeneous benchmarks made it easy for some countries to claim leadership (as in the case of the Netherlands), while in other cases the late penetration of gas appeared as the result of virtuous car-bon policies rather than global fuel prices dynamics (see the switches to gas in the United Kingdom and Germany). All exercises in coordinating the assessment of abatement potential did apparently fail to take into consideration energy prices, possibly the easiest indicator to monitor and the one traditionally subject to most studies: high energy prices generally ensure high efficiency through market dynamics and lead to limited carbon abatement potential. In some cases failure to acknowl-edge such evidence was due, even in this case, to the predominance of an engineering/technical approach to the problem rather than an economist's view of how the EU ETS would interact with a country's industrial system.

4.2 *Interpretation of the scope of the Directive*

The single most discussed issue of harmonisation must have been and still is the scope of the EU ETS Directive. Despite efforts by the Com-mission to use in the Directive's scope definition wording extracted from previous legislation (the Integrated Pollution Prevention and Con-trol (IPPC) Directive), the desired clarity failed to materialise. Three different interpretations of scope followed depending on a Member State's interpretation of what constituted energy activity combustion plants: the first one including only power-producing combustion plants, the second one including energy-producing combustion plants, the third one including all combustion plants. Reconstructing the legal basis supporting the three interpretations is probably not as interesting as pointing out that, surprisingly enough, most of the discussions over scope could not benefit from data relating to the number of installations and size of emissions associated with each interpretation of scope. Those data could in fact be collected only once the scope issue was resolved. As a result, in stead of focusing pragmatically on the effectiveness of the policy tool in regulating installations over a narrower or wider scope, the debate ended up deliberating over the legal letter of the EU ETS Directive, which most obviously was not sufficiently clear. The scope debate also created confusion among operators, harming the credibility of the system and its perceived fairness in terms of competitiveness. Greater time and resources might have been of benefit to the debate and allowed Member States to go through an impact assessment that, at the time the EU ETS Directive was proposed, had to necessarily be approximated due to missing installation-level data.

4.3 BSA, total allocation and the level playing field

While scope was the most discussed issue, the compatibility of the BSA target with a playing field level in terms of competitiveness were never fully debated despite it probably being the most fundamental harmonisation issue within the allocation process. Based on political grounds and commitments made by national governments, the BSA led to differentiated reduction efforts across the EU. Compared to 1990 levels Italy appeared to face a relatively moderate reduction target which, however, became quite challenging when considering BAU trends and national abatement potential. Through the Commission's NAP guidance the EU review procedure required the BSA reduction target to be passed down onto trading sectors and non-trading sectors according to their relative contribution to a country's total emissions. The mechanism implied a differentiated abatement effort for trading sectors in different Member States. Although the latter would be consistent with obligations to comply with the BSA, by requiring more from trading sectors in some Member States than from trading sectors in others, the *target pass-down* method was likely to significantly impact intra-EU competition. The BSA through the EU ETS Directive contributed to the creation of a non-level carbon playing field with its obvious repercussions on EU markets: installation from some Member States (e.g. the Netherlands and Italy) were bound to have to purchase allowances from installations from other Member States (e.g. Germany and the United Kingdom). Although the situation described was the result of Member States' autonomous decisions over national emission reduction targets, its impacts on EU single-market dynamics might have been mitigated through greater harmonisation in the modalities with which the national abatement effort could or could not be distributed within national economic systems.

4.4 Barriers to coordination and harmonisation

While coordination over abatement potential and scope interpretation attracted excessive technical attention, issues deserving greater technical coordination failed to receive it, including treatment of blast furnace gases, CHP, closures and new entrants. The absence of coordination over such issues was not due to political considerations, but mainly to the different timing with which these issues were approached by the different Member States; early movers having finalised decisions that other Member States were not ready to endorse were in turn unwilling to go back and reconsider. The lack of time, resources and effective informal channels to exchange information also inhibited technical coordination. The result of this situation often provided arguments for national lobby groups to pressure governments inaccurately claiming and/or purposely misrepresenting decisions taken in other Member States. Timely and full exchange of information on specific technical decisions would have allowed to better chance to address challenges made by trade organisations which on technical issues, rather than on strategic ones, appeared to enjoy a European-level coordination more effective than the one set up by Member States.

5 Issues deserving special mention

There a number of additional issues which deserve special mention. At the political level the hierarchy of the different policy acts was probably not sufficiently appreciated. The dependency of the NAP on the

National Reduction Plan (NRP) was clearly determined from both a policy and procedural point of view. From a policy point of view, emission trading should only serve as an effective policy tool aimed at maximising the economic efficiency of achieving predetermined environmental policy goals defined by the NRP. From a procedural point of view the NRP is the highest climate change policy instrument in Italy. As a consequence the NAP caps have necessarily to be a derivation of the emission scenarios defined in the NRP. Throughout the NAP development process there was often a failure especially at the EU level to recognise the hierarchic structure of policies: while emissions trading should have played merely the role of a compliance tool aimed at ensuring achievement of long-term policy goals set by the NPR, it became instead a means of dictating changes to the very climate policy acts and plans it should have been derived from. In many ways the EU ETS was thus often used by the EU as a way of introducing rigour into national climate plans. This was an intervention that, although it might have been justified by the failure of those plans to deliver, was outside the scope of the EU ETS Directive.

At the sector level market structure is maybe one of the most prominent issues so far only implicitly addressed. Allocation based on historical behaviour often tends to favour incumbents. As we have seen the problem was particularly dramatic in the Italian power sector where the predominance of ENEL, the former state power company, was challenged only in a limited way by other players. In such situations historical allocation may play a part in protecting incumbents and inhibiting competition. Market structure and monopoly power also plays an important role in the operator's ability to pass on to the final consumer the real or opportunity cost of the allowances. In those sectors where competition is lowest or price elasticity is limited, operators will be able to raise prices in order to accommodate the impacts of allocation.[9] In light of such circumstances, market structure and monopoly power considerations should play an important role in evaluating the economic and equity issues associated with a NAP and arriving at aggregate, sector-level and plant-level allocation rules. Special attention should be focused on ensuring that the results of the allocation process do not reinforce existing market barriers or penalise new entrants.

[9] It could of course also be easily argued that, given the market structure of such sectors, price increases would have occurred anyway. However, as is often the case, raising prices on a pretext is less prone to create a consumer backlash.

At the operator level, the trade-off between flexibility and simplicity in allocation is an important aspect which emerged throughout the allocation process. On the one hand the novelty of the policy instrument made its strength and basic logic difficult to understand for stakeholders (operators and policy-makers) and thus made the introduction of very simple allocation rules highly desirable. On the other hand, a command and control tradition of being able to negotiate with authorities led operators to request special and complex rules to handle all sorts of special situations. Although articulated rules might have added to the perceived fairness of the allocation in dealing with the complexities of industrial plants, they would have detracted from the transparency of the process as well as possibly creating problems where simple rules would have been effective.

The complex and innovative nature of the EU ETS also contributed to stakeholders' failure, in both government and the private sector, to appreciate fully the real impacts on policies and business. From the policy-makers' point of view the lack of sufficient data regarding the extent and depth of impacts resulted in possibly inadequate participation in terms of policy levels and breadth. Both at the national and EU level authorities responsible for energy, economic, financial and industrial matters joined the debate often late and not in sufficient depth. Most often the EU ETS dossier was, instead, handled by environmental authorities, which led to a predominance of environmental goals and a lack of integration with other policies.

Similarly, in the private sector the environmental affairs department of companies often managed the implementation of the EU ETS. Traditionally such departments play a marginal role within industrial organisations and are often not in a position to communicate adequately the relevance of the issues at hand to upper management. In many cases the important consequences of allocation on a company's balance sheets and investments plans were acknowledged only late in the process, once the EU ETS Directive had been published if not well into the NAP development process. Last-minute attempts by operators to affect the NAP development process became frantic, with high-level calls to the ministries involved more frequently causing misunderstanding and confusion rather than contributing to the integration of business concerns into the policy-making activity. The dialogue between government and industry as well as the integration of the EU ETS with other policies could have been made more effective if there had been more time available as well as better high-level information, the latter having failed also because of the many uncertainties surrounding the process.

6 Concluding comments

Various lessons can be learnt from the Italian NAP development process at the EU, macro allocation and micro allocation levels.

6.1 On NAP contexts

A review of the NAP development context highlights important elements including the feasibility of emission reduction targets and the role of political uncertainties. Although emission reduction targets may seem realistic based on historical data, they can become extremely difficult when put into a medium- to long-term perspective. The apparently modest 6.5% reduction compared to 1990 emissions required by the BSA transforms itself in a challenge when set into a 2010 perspective (16% reductions) and in the face of high abatement costs. The challenge becomes even greater when significant political changes occur. Aggressive environmental policies of a previous centre-left government led to a BSA target which then needed to be acted upon by a centre-right government advocating fuller integration of environmental concerns in economic policies.

The other important context element was *political uncertainties* which can play a major role in hindering full and positive participation of all parties involved in the NAP development process. Uncertainties over the entry into force of the Kyoto Protocol together with the technical challenges of its targets led many to question the commitment of EU governments to the BSA, thus weakening and slowing down the NAP development process. Strong and public commitment at the highest levels of government is necessary to ensure the credibility of the process and the full participation of all actors involved.

6.2 On aggregate allocation

Regarding the process leading to aggregate level allocation, there are two political issues playing an important role and deserving special attention. The first one is the full acknowledgment and proper management of policy integration issues. The NAP should clearly identify the way in which goals from differing policy fields may affect total and sector-level allocation. In addition to the environmental issues the trading system was designed to target, such goals may include economic growth, energy source diversification, employment in specific sectors and international competitiveness. The relative priority given to each goal within the determination of the aggregate allocation must be negotiated at different policy levels: in the case of the Italian NAP, EU-level policy concerns over environmental integrity overrode the national-level policy goal assessments and led to the need to the review of the NRP developed by the CIPE working group.

The second issue concerns the speed of approach to the predefined set of policy goals. The issue is particularly relevant in those cases, as the EU ETS, characterised by the possibility of periodically reducing the number of total allowances allocated. In this respect the Italian NAP had foreseen a more gradual approach path to the final targets: an initial period of low scarcity aggregate allocation would have allowed for the system to be tested, for companies to experiment with trading and for the myths on its negative impacts to be dispelled; a second period of scarce allocation, announced well in advance, would have instead lent credibility to the system, allowed for adjustments in industrial capital investment cycles and finally led to compliance with the BSA. However, driven by the political need to assert rapidly the credibility of the system, the EU-level goal of thriving trading through an initially scarce allocation overrode the political national decision for a path of gradual approach characterised by an initial low scarcity allocation.

6.3 On installation-level distribution

At the level of installations technical issues dominate. Despite the theoretical merits of output-based and emissions-based allocations, in practice the complexity of production processes and the diversity of plant configurations often challenge their use. On the one hand, output types produced within the same sector may not be sufficiently homogeneous and make it difficult to apply straightforward emissions per unit of volume or weight methods. On the other hand, the different penetration of on-site power production within a sector complicates the task of developing benchmarks on emissions per unit of input or output. As a result allocation based on historical emission is often easier to use despite its lower efficiency and failure to reward early action. Data availability at the installation level as well as development of sector-specific benchmarking techniques could partly overcome the problem.

Plant-level allocation in the Italian power sector deserves special attention as it provided the opportunity to explore more complex and 'sophisticated' allocation formulae. Two methods (see Section 3.4) were developed to manage some of the peculiarities experienced by the Italian power sector, namely: price instabilities in a recently developed electricity pool, significant new entry, strong historical market structure favouring incumbents. The two methods aimed at mitigating potential rises in electricity price, limiting the amount of windfall gains for incumbents and reducing the risk of entry barriers for new entrants. Despite their rejection, the approaches showed how more complex allocation formulae, while maintaining trading incentives, could better secure compliance with a variety of policy goals, integration of microeconomic dynamics and preservation of equity. More importantly, the experience pointed out the difficulty of ensuring a full understanding by all stakeholders involved, both policy-makers and operators, of allocation formulae characterised by more sophisticated dynamic mechanisms.

6.4 Final considerations on coordination and harmonisation

NAP development within a trading system such as the one set up by the EU ETS Directive did call for significant coordination and harmonisation. Great efforts were undertaken regarding technical issues, but their complexity, the short time-frame and the often limited resources mobilised did not allow for effective solutions. Furthermore, the reluctance of Member States to relinquish their sovereignty in light of a common and coordinated solution played an important role. This reluctance was often based on a lack of the data and information necessary to apprehend fully the consequences of a fragmented approach. More information, a wider time-frame, greater resources and a shared view on the relative priority of the different issues might have facilitated the coordination and harmonisation of NAP development, greatly adding to the credibility of the process among the stakeholders involved.

On the other hand, harmonisation over political issues rarely occurred. Questions regarding the interaction of such an important system as the EU ETS with energy policies, environmental policies and development policies were rarely approached. A number of reasons could be identified for such failure: the information available was insufficient to understand the full impact of trading on important parameters such as economic growth, inflation and energy dependency; the political content of the issues was too high to be discussed within the technical working groups; and the time-frame not sufficiently wide for the full dimensions of the issues to be able to rise to the highest levels of government.

6.5 On additional issues deserving attention

Some of the additional issues deserving attention are the policy role of emissions trading, market structure in the allocated sectors, the level of information of relevant stakeholders and the relative complexity of allocation formulae. From a policy point of view the most important issue is the use of emissions trading as a 'policy-setting' or 'policy compliance' tool. The difficulties experienced within the Italian NAP development process were often due to different interpretations over the policy role of the NAP. In terms of economic efficiency, the need to account for market power is an issue often not receiving sufficient attention. The Italian electricity sector's market dynamics provided an important opportunity to explore various allocation strategies aimed at safeguarding competition and avoiding impacts on prices. Finally, failure for the appropriate information to reach all relevant stakeholders in a timely manner may have significant impacts on the efficiency of the process as well as the outcome in terms of integration with other policy and company matters. Within the NAP development process late full involvement of both public and private sector high-level decision-makers across areas beyond environmental management and policy-making led to the failure of a full integrated approach to materialise.

6.6 Final considerations

Experiences from the Italian NAP development process appear to confirm the theoretical finding that initial allocation is mainly a question of equity and policy-setting rather than economic efficiency. As in the case of the electricity sector in Italy, exceptions may occur in the presence of market power. In such cases, allocation can not only affect barriers to entry and potential price impacts, but also the achievement of economically efficient solutions in terms of both static and dynamic efficiency. From an equity point of view allocation to new entrants rather than to existing installations plays an important role. However, through the NAP development process the political dimension of setting policy abatement targets predominates. The lack of clarity over the role of emission trading as a 'policy-setting' or 'policy compliance' tool can greatly hinder progress. Such difficulties become particularly challenging in the Italian context where the EU ETS is bound to have significant impacts on energy production and electricity prices, two issues that play a fundamental role within the Italian economic system.

Reference

Viguier, L. L., Babiker, M. H., and Reilly, J. M. 2003. 'The costs of the Kyoto Protocol in the European Union', *Energy Policy* 31: 459–481.

10 | *Hungary*

ISTVAN BART

1 Introductory background and context

As all the other continental new Member States of the EU,[1] Hungary is a 'country that is undergoing the process of transition to a market economy' according to the United Nations Framework Convention on Climate Change (UNFCC), and shares most of their characteristics with respect to climate policy.

Hungary's trading sector CO_2 emissions are relatively small compared to its size. The total quantity of allowances issued is about half of the trading sector emissions of Belgium, a country with an almost equal number of inhabitants. Though only half as populous as Hungary, Slovakia's trading sector emissions are almost equal to Hungary's. These relatively small emissions are attributable to the relatively small heavy industry, and an exceptionally high (40%) share of gas in the country's total primary energy supply. Significantly, unlike in most new Member States, electricity generation and distribution has been almost fully privatised, and is now owned mostly by the multinationals E.ON, RWE and EDF. The state still owns and operates the national electricity grid and the Paks nuclear power station.

Hungary, like all new Member States, has seen a steep decline in its plan-based carbon-intensive economies during the 1990s, and despite the subsequent rebound of economic activity, emissions are still well below their levels in the 1980s. As new Member States have all pledged in the Kyoto Protocol to reduce their emissions in 2008–2012 by 6–8% compared to base years between 1985 and 1990, none of them except Slovenia need any further measures in order to meet their reduction obligations. Hungary has a reduction target of 6% compared to the average of the years 1985–87. In 2003, its emissions stood at 73%

---

[1] Any further reference to the 'new Member States' should be understood as not referring to Cyprus and Malta, countries in an entirely different situation with regard to climate policy.

246

of this target, and it does not expect to surpass the target in most business-as-usual (BAU) scenarios. (Depending on the extent of decline after the collapse of the planned economy, and the speed of recovery afterwards, some new Member States are further below their Kyoto Protocol cap than others. The three Baltic states' current emissions have experienced such a sharp drop that they could under no circumstances see their emissions rise even close to their targets in 2008–12. Hungary is on the other end of the spectrum, where a stronger-than-expected growth in emissions could bring it close to its Kyoto Protocol target.)

Climate change is not an important issue for Hungarians. Hungary is a country that has still not realised that it ranks among the more developed nations of the world. Hungarians aspire to western European living standards, and they will continue to feel underdeveloped until such standards are attained. The lack of a colonial past has left the Hungarian public without much interest for the well-being of the planet at large. Environmental issues get attention only when they affect the everyday life in Hungary's territory. Given that Hungary's targets set out in the Kyoto Protocol are not restrictive, reducing emissions is largely perceived as a problem that is relevant only for the old Member States. Politicians and the general public in the old socialist countries consider that their duty in combating climate change was fulfilled by the economic restructuring of the 1990s, which, although not carried out for this purpose, had enormous social costs.

Therefore, though Hungary is an Annex I country, the approach of the general public and of many experts to climate change is rather similar to that of a developing country. It is sometimes argued that targets based on total emissions instead of per capita emissions are unfair to Hungary as these could in theory restrict economic growth before Hungary reaches western European levels of development; indeed Hungary has indeed very low per capita emission levels (Figure 10.1).

As a result, Hungary traditionally devotes only meagre resources to climate change issues in general. In a period when environmental ministries were all but overwhelmed by the tasks of EU accession, the climate change administration typically had one or two persons in 2003, when the EU ETS Directive appeared on their horizon. Though further manpower eventually became available for the implementation of emissions trading, the severe lack of human resources proved to be a constant impediment.

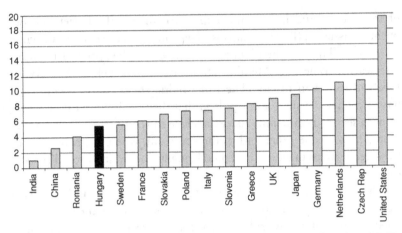

Figure 10.1. CO_2 emissions (t per capita) in selected countries (based on OECD data).

It is against this background that two distinct approaches emerged in Hungary towards the impending introduction of the EU ETS. On the one hand, there existed a not inconsiderable degree of suspicion towards the introduction of emissions trading, which was perceived by some as a scheme aimed at extorting hot air from the new Member States. The EU ETS Directive was also strongly criticised for not taking into consideration the special situation of the new Member States, thus making implementation more difficult. Indeed, for a time the Hungarian government seriously discussed starting negotiations to delay the implementation of the ETS for a few years.

On the other hand, another approach regarded emissions trading not so much as a burden on the economy but rather as an opportunity to foster foreign investment. It was understood that the governments were able to provide generous allocations to the trading sector, thus boosting their assets. Even in new Member States where hot air was less abundant, as in Hungary, it was clear that with an allocation that was not restrictive, and with reduction opportunities that were cheaper than in many old Member States, the trading sector stood to gain with the introduction of emissions trading. Businesses became excited at the possibility of receiving surplus allowances which could be resold to Western customers. Allocation even came to be viewed by as a compensation for the hardships of the past decade.

With time, however, both approaches led to disappointment. Instead of stripping the new Member States of the surplus allowances, the

ETS rather increased awareness of climate change issues and made available more government resources for research into climate change. In Hungary, the personnel involved in climate change administration has now grown from two persons to seven, a serious forecast has been made for future emissions up to 2010 and a large scientific programme has been initiated on the impacts of climate change in Hungary.

On the other hand, the 'free lunch' of selling surplus allowances never materialised in Hungary. The opportunities advertised by the Ministry of Economy and Transport came at a cost for operators. New measurement and accounting tasks had to be developed, and allowances had to be considered as a new risk factor in business planning. These burdens were not in the end accompanied by a soothingly generous allocation for all, but rather by one that was short for some and long or just enough for others, depending on each installation's fortune with the allocation rules.

2 The macro decision concerning the aggregate total

Arriving at the total quantity of allowances to be distributed constituted a twin challenge. The first task was to find out how to interpret the Directive's provisions on the total amount. After that, the total amount had to be established and negotiated with all the stakeholders.

2.1 Interpreting the Directive's criteria on the total quantity of allowances

It has been mentioned before that the ETS Directive was often criticised in (*inter alia*) Hungary for not taking into consideration the special circumstances of the new Member States, namely that they stand to fulfil their Kyoto Protocol targets and therefore there is no political basis for expecting their allocations to be restrictive. This is most apparent with respect to the rules on establishing the total amount. Annex III of the Directive, the part outlining the requirements for setting the total quantity in the National Allocation Plan (NAP), is more or less clear when applied to countries that need to reduce their emissions in order to meet their Kyoto Protocol target. However, when it came to those countries that do not need to reduce their emissions, i.e. the new Member States, the Directive lent itself to radically diverging interpretations. This lack of clarity is surely due to the difficulties in achieving a compromise between the then fifteen Member States, which also explains why there

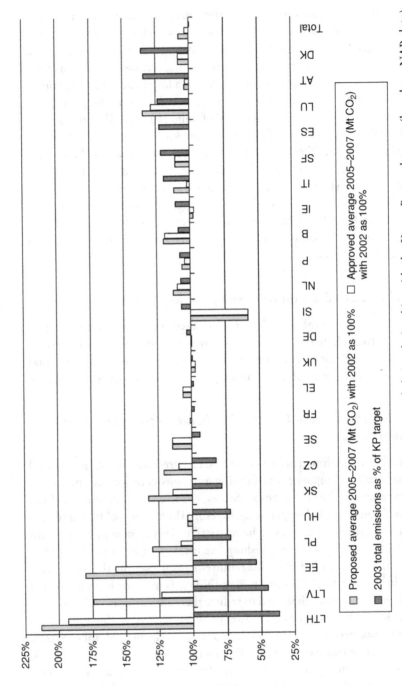

Figure 10.2. Proposed and approved allocations and their relationship with the Kyoto Protocol target (based on NAP data). Countries are ordered by their distance to the Kyoto Protocol target.

was no political room to take into account the circumstances of the then Acceding Countries.

There is no space here for a detailed legal analysis of the ambiguous wording of the Directive's relevant provisions.[2] Suffice to say here that criteria 1–3 of Annex III were interpreted by the Commission to the effect that Member States on a course to overshoot their Kyoto Protocol target in 2008–2012 (i.e. most of the old Member States and Slovenia) should not allocate more allowances than a level that still ensures compliance with the target, taking into consideration measures outside emissions trading. On the other hand, Member States that were set to fulfil their Kyoto Protocol target in 2008–2012 (i.e. all new Member States except Slovenia), were supposed to – in the interpretation of the Commission – allocate a total quantity that equalled the sum of the projected emissions of the sectors covered by the ETS.

While this interpretation was not openly challenged by any Member State, when it came to applying the above vague rules, Member States arrived at different conclusions, as demonstrated by Figure 10.2. It can be assumed that if the method of establishing the total amount had been clear in the Directive, the proposed total quantities would be comparable at least among Member States in a similar situation.

The figure shows how many of those old Member States that are currently a great distance from achieving their Kyoto Protocol target tried to allow for a significant growth in the emissions of the sectors falling under the ETS, which attempts were then restricted by Commission's decision. It also shows that, albeit on a grander scale, a similar game was played by most new Member States, significantly though with quite different justifications. It is interesting to look into these.

In the case of the new Member States below their Kyoto Protocol target, it is apparent that the proposed allocations are quite different when compared to historic emissions. Three countries – Lithuania, Hungary, and the Czech Republic – used macroeconomic analysis to forecast the total expected emissions of the ETS sector, which figure, after several revisions, was the primary input for the total amount of allowances to be created.

---

[2] The reader is invited to make an attempt at understanding the two criteria in Annex III of the Directive (see Appendix IV) that are relevant for setting the total quantity of allowances.

In Poland, Slovakia, Estonia and Latvia, macroeconomic forecasting played an ancillary role. The NAPs were drawn up primarily on the basis of the installations' production forecasts, and in effect proposed to create as many allowances as possible without compromising the fulfilment or their Kyoto Protocol target. This approach naturally led to NAPs that proposed to allocate significantly more allowances than the sectors under the ETS could have reasonably expected to emit in the first trading period. These Member States considered this method of establishing the total amount compatible with the Directive. The high total amounts were justified with strong expected economic growth, and – most strongly in the case of Poland – with the need for the economy to benefit from the EU ETS.

The Commission's decisions on the new Member States' NAPs did partially succeed in bringing the new Member States' total quantity closer to emission levels that can be reasonably expected in the course of the period 2005–2007. However, when the Hungarian NAP was approved after the Slovak and the three Baltic NAPs, there was a strong feeling among the Hungarian public that those countries who adopted the second approach and tried to set the total quantity as high as possible ended up in a better position than the other group even after cutting their quantities as requested by the Commission.

2.2 *Establishing and negotiating the total amount*

Once the method of establishing the total amount (with top-down macroeconomic prognosis instead of with bottom-up operator-level forecasts and the interpolation of Kyoto Protocol targets to the period 2005–2007) was decided, the next step was to establish the total amount.

In Hungary, macroeconomic analysis was aimed at forecasting the total amount of CO_2 emissions in the sectors under the ETS in 2005–2007. The sum of the sectoral projections would constitute the total quantity to be allocated. The study, which took about six months to prepare, was prepared by researchers at the Budapest University of Economic Studies. At the outset of the public consultation, this forecast was presented as a total quantity. For the Ministry of Environment, the ministry responsible for setting the total quantity, keeping the amount of allowances low was of paramount importance because Hungary, though below its Kyoto Protocol target, is much closer to it than other new Member States (except Slovenia). Therefore, attempting over-allocation would have been a much riskier strategy than in other countries.

As is apparent from Figure 10.2, Hungary finally proposed a total amount that is 3.8% more than its emissions in 2002. This scenario of emission growth accounted for both expected market developments and technological improvements. The proposal was accepted in December 2004 by the European Commission without any reductions in the total amount. (Only Slovenia and Hungary did not have to reduce their proposed allocations before the Commission's approval, which would suggest that the total amount proposed by Hungary was reasonable.) However, the quantity of emissions originally forecasted by the researchers and presented to the public in August 2004 amounted to only about 90% of 2002 emissions. In other words, the public consultation process in the autumn of 2004 resulted in a roughly 14% inflation of the projections. (A similar phenomenon occurred in the Czech Republic, where the total quantity increased 17% between the summer 2004 draft and the final version submitted to the European Commission.)

In Hungary, the public consultation process largely constisted of increasingly powerful industry representatives arguing for ever higher allowances before ever higher instances. In this process, the Ministry of Environment partly defended its own position, or where impossible, sought to accommodate the demands for higher projections. The Ministry of Economy assumed its role as the champion of the industry, and in particular the energy industry, and worked on convincing the Ministry of Environment to agree on increases of the projections in the sectors that clamoured the loudest. Both ministries, knowing that the proposed total quantity was not overly generous, attempted to soften the impact by discussing first the sectoral quantities, without reference to individual allocation levels. Smaller increases were agreed in the initial consultations after the publication of the sectoral figures in August 2004. However, when the proposed installation-level allocations were published in October 2004, industry attacks aimed at increasing the total quantity resumed with greater intensity. Finally, the power generation industry joined hands with the miners' labour union (Hungary's few mines are owned by the power generators), and started to threaten mass lay-offs of miners if power plants had to close down due to a lack of allowances. This persuaded Hungary's then newly appointed prime minister to ask the responsible ministries for a redraft of the total quantity.

As a result, the NAP already submitted to the European Commission was withdrawn, and a new version with a higher total quantity was submitted. The electricity generation industry's power is amply demonstrated by the fact that the increase in allowances was almost exclusively allocated to the large power generators. (The higher quantities were justified by the emergence of new, better data, an upward adjustment of GDP forecasts and an upward adjustment of domestic coal-based generation due to rising international oil prices.) However, the inflation of the total quantity is only partly attributable to the government's weakness in the face of industrial interests. Another important factor was that the Directive is unclear about the rules for setting the total quantity. The lack of clear rules and the absence of a restrictive Kyoto Protocol target made it very difficult for the government to defend its proposed quantity. The main argument was to refer to the sectoral production forecasts done by the Ministry of Environment's researchers, which were of course strongly contested by the sectors themselves.

Another argument in defence of the proposed total quantity was that this should not be too high, as an over-generous NAP would be refused by the Commission. However, when Slovakia's NAP was approved on 20 October 2004, and the decision permitted a 13% increase on 2002 levels (the proposal was a 32% increase, admittedly at projected GDP growth levels that were higher than in Hungary), the responsible ministries found their position seriously undermined, as an increase of only 1.7% was proposed in Hungary's then current draft NAP. The industry accused them of putting the Commission's interests before domestic considerations, while Slovakia 'got away' with a more generous NAP. It was hinted that a good NAP is one that is reduced by the Commission, thus ensuring that the maximum quantity was reached. (The keen interest by Hungarian stakeholders in the events of other new Member States should not be a surprise, as the same multinationals own the installations of a particular sector in all these countries. The events in neighbouring Slovakia were of particular significance due to the merger of the two national oil companies, MOL and Slovnaft.)

Though in the case of new Member States no restrictions need to be factored into the total quantity, establishing the total quantity is politically just as difficult in these countries as if there were too few allowances to distribute. This is because a reasonable emissions forecast for a given sector is by definition lower than the 'safety point' of the sector's players, i.e. the amount of allowances that would satisfy the expectations of all the installations involved. (As operators are aware of the Kyoto Protocol target and Hungary's current emission levels, they fully expect their allocations to cover their needs.) All installations expect to improve their performance in the future, or at least do not want to be constrained in their growth by the lack of allowances. Therefore, even if the total quantity equals the projected emissions on a sectoral level, many players will receive individual allocations that are restrictive. (Many of course will receive more allowances than necessary, but these installations will not complain.) If a country's Kyoto Protocol target itself is not restrictive, such frugal allocations are difficult to explain, even by referring to the general reliability of macroeconomic forecasting.

3 The micro decision concerning distribution to installations

3.1 Setting the principal allocation rules

In Hungary, the establishment of the rules on distributing the allowances among installations was the responsibility of the Ministry of Economy and Transport. The figures for the total quantity and the sectoral quantities were provided by the Ministry of Environment and Water, and since Hungary does not need to reduce its emissions in order to reach its Kyoto Protocol target, the quantity distributed in each sector was equal to the forecast volume of emissions in 2005–2007. Thus, the establishment of distribution rules did not entail a decision on how to distribute the burden of emission reductions among sectors. In fact, there was no significant interplay between the two issues of setting the total amount and establishing the allocation rules.

Thus, the process aimed exclusively at reaching the fairest possible solution. However, it was clear from the start that whatever the outcome, some operators would be disadvantaged in comparison to others. The Ministry of Economy and Transport sought therefore to minimise the numbers of those disadvantaged, and simultaneously, to prevent any participant from receiving too many windfall allowances. This was to be ensured by involving the operators to the greatest possible extent in the process of establishing the distribution rules.

With Hungary being a relatively small economy with only around 250 operators, this was done through convening a meeting for each industrial sector, to which every operator of the sector was invited. Such sectoral meetings were only not possible with the populous and diverse group of non-energy sector combustion installations (hospitals, schools, smaller factories and the like.) At these meetings, the participants of the sector were presented with the possible options for the distribution rules, and were asked to give ideas on how to distribute allowances in the most equitable way. (As a longer consultation document was circulated a few months before these meetings, operators had prior knowledge of the possible distribution design options.)

Besides the sectoral meetings, the Ministry of Economy and Transport held countless bilateral meetings with practically every operator who wished to have such a consultation. Notwithstanding that, the Ministry of Economy and Transport always made clear that only sectors as a group may submit proposals on allocation rules. Of course, in the case of sectors where there was only one operator (oil, steel, coking), the only possibility was to hold individual negotiations. In such sectors, the allocation was equivalent to the projected emissions, and no allocation rules were necessary. Therefore in such cases discussions were almost exclusively about the emission projections, which meant that the Ministry of Environment, the ministry responsible for projections, was the main negotiating partner.

Although the possibility of benchmarking was not excluded outright by the ministry, it did not seem viable in most sectors due to the diversity of technological processes and the sheer lack of time. Discussions therefore centered on 'grandfathering'. Eventually, the primary allocation method to emerge was to take each installation's share of the total emissions of its sector during the base period (for most installations 1998–2003) and use this percentage share to distribute the total sectoral emissions projected for 2005–2007. In the case of longer base periods, the allocation rules allowed the omission of the year with the lowest emissions.

In a few special situations, however, the adopted method of calculating the individual allocation was different from the general rules. In all these instances, it was the operators of the sector who proposed to depart from the general grandfathering method to distribute the sectoral quantity. When adopting such a proposal, the government ensured that the solution proposed by the industry is accepted by all participants. Special rules were invariably proposed in sectors with rather few participants using very similar technology. The following special rules can be found in the NAP:

(1) in the cement sector (two operators) the installation-level allocation was established through estimating the installation's share in total cement production in 2004, multiplied by the CO_2 emission factor of the installation

(2) in the lime sector (two operators) the installation-level allocation was established by multiplying the sectoral allocation by the installation's share of the sector's 2003 limestone consumption

(3) in the paper sector (five operators), future planned production volumes multiplied by past specific emission factors was the basis for calculating an installation's share of the sector's allocation

(4) the sugar industry (four operators) was to see a substantial increase in production after joining the EU, due to the EU sugar production quota system. Therefore, future production data with past installation-level emission factors were used to establish the allocation.

3.2 Setting of the base period

As there was no reason to try to enforce a single base period for all participants, the sectors were invited to come up with a proposal for a base year, or base period. Depending on the level of concentration in the sector and the relative importance of emissions trading for the participants, this strategy had various results. The cement and lime sectors have only two or three players of equal size and both industries are highly exposed to carbon costs. These sectors quickly agreed on a base period and eventually even proposed complex equations for the distribution for the total quantity. At the other end of the spectrum, the brick sector has over thirty small players, half of whom could not even make it to the initial meeting.

In sectors where no single voice could be found, the base year had to be decided by the ministry. More recent base years or periods were preferred to earlier ones as these would be closer to current trends, and fewer installations with changes or upgrades after the base period would need to be excepted from the general rule. Longer base periods were preferred to shorter ones as these would be better at ironing out annual irregularities. In conclusion, the period 1998–2003 was set as

Table 10.1 *Base periods in various Hungarian sectors*

Base period data used for allocation	Sector
2003 emissions	Large-scale power production
1998–2003 emissions	District heating
1998–2003 emissions	Industrial and other combustion installations
EU production quota of company	Combustion installations in the sugar sector
1998–2003 emissions	Mineral oil refinery (one installation)
1998–2003 emissions	Coking and steel (one installation)
2004 planned production	Cement production
2003 lime consumption	Lime production
1998–2003 emissions	Glass manufacturing
2001–2003 emissions	Production of bricks
1998–2003 emissions	Production of ceramics other than bricks
1998–2003 emissions	Production of pulp, paper and board

a default base period, but the ministry was ready to change this if the sector so demanded. The result is a rather eclectic assembly of base periods, as demonstrated by Table 10.1.

An interesting situation emerged in the large (over 50 MW) power plants sector, responsible for two-thirds of all trading sector emissions. The overwhelming majority of operators supported a long base period between 1998 and 2002. However, a lignite-fired power plant, which is also by far the biggest, responsible for one-third of the sector's emissions, supported 2003 as a single base year. This was the year in which nuclear power plant had to be partially shut down, and its generation was substituted by this lignite-fired power plant. Defying the majority of the sector, the ministry eventually decided on 2003 as a base year for the sector, because this year was overall more representative of the contemporary situation then earlier years. However, using 2003 as a base year was unfavourable to the newest and most efficient combined cycle gas turbines (CCGTs), as their percentage in total fossil generation in 2003 was lower than usual (because more fossil-fuel-based electricity was generated due to the partial shutdown of the nuclear power plant). In order to avoid disadvantaging the cleanest power plants, a small amount of allowances was redistributed as compensation to these installations. This was done by multiplying base period averages by 1.032 for installations using CCGT technology.

3.3 Other design features

Hungary's NAP was among the last plans to be published and approved, which allowed the drafters to build on the structure and contents of many other NAPs. While Hungary's NAP does have a few original solutions, most of its ideas are recycled from several other NAPs. This 'free-riding' technique greatly reduced the amount of resources needed for the NAP while avoiding pitfalls already identified through the adoption of other NAPs. (For example, *ex-post* amendments of allocations, very popular among installations, were dropped at an early stage of drafting when the Commission made it clear in connection with other plans that such a solution would not be allowed.)

An interesting feature of Hungary's NAP is the early action reserve, interesting because of the reasons that gave birth to it: 2005 was not only the year of the start of emissions trading but also the year when Hungary's temporary exemptions on the limits on sulphur dioxide emissions from large combustion plants expired. There were several large power stations in Hungary that were exceeding these limits, and would therefore have had to close down. Two of these operators responded to this challenge in an original way, and converted some of their furnaces to biomass combustion through a Joint Implementation (JI) project. As biomass combustion was only introduced in late 2004, these installations had very high emissions in the base periods. It was considered unfair to allow them to receive surplus allowances due to their high past emissions given that they would not have been able to continue operating on their previous emission levels due to the constraints on sulphur dioxide emissions. Moreover, the drop in fossil-fuel emissions due to their switch to biomass was also factored into the sectoral forecasts, which meant that the distribution would have become even more distorted.

In order to maintain fairness, the NAP has a rule that the base period is only applicable if the technology is the same as the one used on 1 January 2005. If no base period with the current technology exists, than installation-level projections need to be used, as in the case of the power plants that switched to biomass. On the other hand, it was considered unjust that these operators should receive no reward for their investment in environmentally friendly technologies. Therefore, a small amount of allowances was set aside as an early action reserve. The early action reserve is to be distributed on the basis of applications. Thus, operators who can prove that there was a reduction in their emissions per unit of production may apply for allowances from the early action reserve. The allowances in the early action reserve are distributed among the applicants in proportion to the size of the reductions achieved. Though the early action reserve is not restricted to the power plants that have switched to biomass, their sheer size (i.e. their share in the total early action reductions) ensures that most of the allowances in the early action reserve will be awarded to them.

When discussing the new entrants reserve, two problems in its future operation became clear. First, it will be difficult to verify the new entrants' claims about future emissions, as these depend mainly on production levels. Second, all new entrants would seek to apply for new entrant allocations as early as possible, which would make the verification of claims even more difficult. The first problem of verifying operators' production forecasts was eventually not resolved in the NAP, but a unique method is employed to remedy the second difficulty. According to this, all requests for allowances from the new entrants reserve are collected during the course of a calendar year, and the annual tranche of the new entrants reserve is distributed proportionally among the applicants. If the quantities applied for exceed the quantities available, the amounts to be distributed are proportionally reduced. If the sum of requests is smaller than the available quantity of allowances, the remaining amount accrues to the reserves of later years. As new entrants request allowances for the entire period, the new entrants reserve is divided into progressively decreasing tranches, i.e. half of the reserve is distributed for new entrants in 2005, one-third for new entrants in 2007, and one-sixth for those in 2007. The advantage of this method is that it prevents operators rushing for allowances, and distributes evenly the risk of the new entrant reserve being too small.

Auctioning is only applied in four Member States, among them Hungary, which auctions 2.5% of the total quantity of allowances. Originally, 1% of the total quantity was intended to be auctioned, with the justification of recouping the costs associated with the introduction of the ETS. Another reason was to introduce auctioning early in order to be able to increase the total quantity to be auctioned as soon as possible. The proposition was very unpopular with the operators, who perceived it as a tax on their allocations (to which they considered themselves entitled).

Both the government and the operators realised that even if auctioning were restricted to operators, the allowances would be auctioned off at a price that is close to the current market prices, because no buyer could be prevented from selling on allowances to others. (Clearly, when auctioned allowances are only a fraction of all allowances in the EU market, auctioning is no different in its effects than if the government sold its allowances to a single trader on the market.)

Operators' protests against auctioning grew louder when at a very late stage, during the interministerial negotiations on the NAP the Ministry of Economy and Transport proposed raising the level of auctioning to 5% (with a view to obtaining higher revenues for the state). The ministries responsible for drafting the NAP opposed this last-minute increase, fearing a backlash from operators. After tough negotiations, it was agreed that 2.5% was to be auctioned, and a government resolution was adopted stating that revenues should be spent on action against climate change. (It must be noted, however, that such resolutions are not guaranteed to be automatically transferred into the budget. It is possible that the Ministry of Finance will eventually decide to pour the revenues into the common pot.) An important lesson from these events is that if policy-makers intend to increase the share of auctioning in the future, a likely and strong ally in these efforts is the Ministry of Economy and Transport.

Overall, the setting of distribution rules went relatively smoothly in comparison to the establishment of the total quantity of allowances. (It was mentioned before that a sectoral cap which is based on a macroeconomic forecast is necessarily too tight to make all operators satisfied.) By the end of the allocation procedure, the drawbacks of grandfathering based on past emission data also became clear. Though simpler to administer then benchmarking, grandfathering is perceived as unfair by those who are disadvantaged by it, as it presumes that emissions, and consequently production levels, will in the future be similar to those in the past. In every sector there are operators who are increasing their production and market share, and others who are shrinking. When allowances are distributed on past emission data, the well-performing, growing operators are punished and the laggards are rewarded. These punishments and rewards are not connected to environmental performance, but rather to the change of performance in the future, compared to the past. These distortions were accentuated in the Hungarian NAP, which distributed allowances on the basis of an operator's percentage share of sectoral emissions in the base period.

4 Issues of coordination and harmonisation

There are differences between the NAPs of small and large countries. In countries of Hungary's size, a few dozen installations account for the bulk of emissions. These operators usually have a level of access to the government which in larger countries is available only to the largest of the large operators. Therefore, NAPs of smaller countries tend to have more specialised rules or solutions tailored for a particular industry's needs than the NAPs in large Member States.

4.1 Allocation methods in the future

The difficulties with grandfathering were discussed above. Hungary, like many other countries, came to the conclusion that apart from some smaller sectors, benchmarking is not a viable alternative, as there is not enough time to develop the benchmarks that would need to be used. It was also often implied that in later years, this may be a possibility. In my opinion, due to the 2006 deadline for the submission of the NAPs for the 2008–2012 period, no Member States will have found the time to develop a comprehensive technology-based benchmark system. There is great potential in benchmark systems that use the actual average emissions (by unit of output or input) of a sector to distribute allowances. In such a system, the principal question would be the delineation of sectors, or sub-sectors with installations using the same technology. Setting industry averages is of course a problem in the case of sectors like oil or steel, where often there is only one installation in a country. However, a great advantage of a system based on industrial averages is that it could set the stage for a single EU-wide allocation system in the period after 2012.

The treatment of sectors where only one player is present is a particular difficulty of drafting NAPs in small countries. Hungary for example has only one oil refinery and one steel plant. As MOL's Hungarian oil refinery is part of a larger refinery network in central Europe, the future emission levels of this refinery depend not so much on future fuel consumption in Hungary, but rather on MOL's operational decisions in organising production. At the time of drafting the NAP, the steel plant was undergoing privatisation and not even the management had a clear idea of future production. Moreover, the steel market is international, and as such, to a great extent outside the scope of the macroeconomic forecasting carried out in preparation to the NAP.

Therefore, the forecasts and as a result the allocations in the oil and steel sector had to rely heavily on the operators' own business plans, which is not a satisfactory method in the long term. Moreover, these forecasts were very difficult for the European Commission to verify. In future NAPs, further harmonisation could be envisaged in the treatment of these large industries, either through a consolidated method of forecasting, or through the establishment of EU-wide benchmarks.

4.2 Scope of installations

A well-known problem of the ETS is that the scope of installations under the system is too wide. This was observed in Hungary as well, where more than half of all allowances were issued to the ten largest installations, while the smallest one-third of the installations received only around two per cent of all the allowances. The small installations are of course disproportionately burdened by participating in the scheme due to the fixed costs associated with monitoring emissions and participating in trading. While reducing the scope of the system might be a needless reversal of environmental policies, the possibility of opting these installations out (and placing them under other emission reduction policies) should be maintained in later trading periods.

4.3 Evaluation by the Commission

The Commission did very well in the primary objective of ensuring overall scarcity, as demonstrated by Figure 10.2. However, this was not accompanied by the harmonised and consistent method of evaluating NAPs. The evaluation of NAPs were done through bilateral negotiations, and other Member States had difficulty in finding out about most changes, or the reasons for the requested changes, apart from the proposed total quantity being 'too much'.

This lack of transparency on behalf of the Commission can well be attributed to the vagueness of the Directive's relevant provisions, as discussed earlier. Notwithstanding that, there is a clear albeit unintentional pattern in the total quantities proposed by and eventually allowed to new Member States. This pattern shows that the greater growth of allowances (compared to past emissions) a new Member

State proposed to issue, the greater the final quantity was, despite the reductions requested by the Commission. The pattern did not go unnoticed by Hungarian operators, and soon, the Commission's evaluation procedure came to be viewed as a raw bargaining exercise where, if you want to get the most allowances possible, you have to start with a very high number, and descend from there, until the Commission agrees.

In the future, more transparency is needed with regard to the evaluation carried out by the Commission. For both old and new Member States, the evaluation criteria for the total quantity need to be more evident. There will always be Member States who consider the reduction of emissions a greater priority than others. If these Member States cannot be sure that other Member States with other priorities will not be permitted to interpret the Directive broadly when setting the total amount, they will think twice about endangering the competitiveness of their industry by ambitious allocations. (And let's not forget that industry pressure for more allowances can only be expected to be stronger and more effective after the lessons of the first allocation process.)

Furthermore, the difficulties encountered with using forecasts as a primary basis for allocations in the new Member States can expected to re-emerge in the preparation of the second NAP. Under the current rules, the 2008–2012 NAPs in new Member States would have to rely on forecasts made in 2006 for a period that is two to six years from then, well beyond a reasonable time-span necessary for the predictability that is required when distributing valuable assets. (The outcome of the public consultations on forecasts in Hungary was discussed above in detail.) The proposed total quantities will therefore be based more on the outcome of a political process that decides on the level of emissions that is considered acceptable. It was mentioned before that reducing CO_2 emissions is not a primary political objective in countries that are still far from their Kyoto Protocol targets. We should not therefore expect very ambitious proposals.

5 Issues deserving special mention

5.1 Institutional structure

An important factor in the NAP procedure was the active participation of the Ministry of Economy and Transport in the operation. Due to the strong personal interest in the concept of emissions trading by the then leadership of the Ministry of Economy and Transport, the ministry fought hard from early 2003 to get a stake in the allocation process, and eventually it was entrusted with setting up and negotiating the allocation rules, while the establishment of the total quantity remained with the Ministry of Environment. The Ministry of Economy and Transport, traditionally on better terms with industry than the Ministry of Environment, was able to gain the trust of operators and involve them in the NAP process, thereby making it more acceptable in the end. As the NAP was a truly common project of the two ministries, in the sense that some chapters were written by the one, and some by the other ministry. Hungary managed to avoid the last-minute battles between economic and environmental administrations, thus considerably simplifying the approval of the NAP. (The leadership of the Ministry of Economy and Transport changed by early 2005, and consequently the ministry scaled back its activities in the climate change field, leaving the field to the Ministry of Environment.)

5.2 Data collection

The collection of proper data proved to be a greater task then initially thought. Data collected on the basis of previous legislation turned out to be less useful than originally thought, and there was no time to adopt legislation to enable the gathering of new data. In Hungary, the entire NAP was drafted on the basis of 'soft data', i.e. data not gathered for the purpose of emissions trading. On the other hand, this is not expected to have significantly influenced the overall outcome of the allocations, as good-quality data were available for all the large installations.

5.3 Auctions

Beyond setting the percentage to be auctioned, auctioning posed another dilemma for the drafters of the NAP. Auctioning was recognised as a distribution method better than grandfathering or benchmarking, as administrative costs are low, revenues are generated and operators are discouraged from inflating their allowance needs. However, auctioning as it is currently envisaged in the EU ETS also has a significant drawback for new Member States. This lies in the difficulty of exploiting the advantages of auctioning as a distribution tool without simultaneously importing the single market price of allowances, which is presumably higher than the price would be if trading was restricted to only Hungarian operators. As trading of allowances is unrestricted, any bidder in any auction would want to bid until the price reaches the price level on the international markets. Large-scale auctioning would therefore mean placing the burden on Hungarian operators of having to buy allowances for compliance at the international market price, which is much higher than the price level that could develop if allowances could only be traded within Hungary, where scarcity is much smaller than in new Member States (because the Kyoto Protocol target is still far away). Notwithstanding that this is compensated by that fact that allowances given for free would have a greater value on a common market than on a restricted Hungarian market, the perception of unfairness due to high EU-wide prices would constitute a political difficulty.

A technique was invented in the draft NAP to circumvent this, where operators would have been allowed to buy the remaining 2.5% of their allocations (the quantity that was designated to be auctioned) at government-set prices. Ironically, the Ministry of Economy and Transport opposed the idea because it would have resulted in lower government revenues than a simple auction. However, if auctioning is to be introduced on a significant scale, the issue of keeping prices low in countries with low domestic allowance demand (as compared to domestic supply) could be important in making auctioning acceptable. This could be done through an auction where only domestic operators are allowed to participate and all participants have an individual ceiling of allowance purchases. However, such a solution would still be quite complex, and would certainly bring less revenue than a simple auction.

6 Concluding comments

According to all observers and participants, the allocation process was even more difficult than initially expected, and there is no reason to expect that the next allocation process is to be any easier. Nevertheless, the first period should be able to provide us with some useful insights for the future.

6.1 The political process of setting the total quantity

It was discussed before that climate change is a back-burner issue in new Member States for several reasons. This fact had its impact not only on the allocation of resources for the implementation of emissions trading but also on the allocation process.

The government was in a weak position throughout the process as it could not point to any Kyoto Protocol requirements that were yet to be fulfilled. Further, there was no home-grown political urge to do something about climate change if not required by greater powers. The implementation of the ETS was largely seen as just another obligation in the long march to the EU. As a result, the process of setting the NAP happened largely outside the attention of the general public. It was essentially a 'private' affair of the government and industry, an exercise in the allocation of resources.

Environmental considerations rarely emerged in the talks on the NAP. This does not mean that industry or the government were climate change sceptics or opposed to environmental protection, it simply means that the allocation process never extended to the issue of why emissions trading is necessary. Tellingly, the government's standard argument against demands for more allowances was that they would not be accepted by the European Commission.

The lack of a genuine debate on climate change and the perception of emissions trading as a simple question of resource allocation can be expected to continue until 2012, after which Hungary might receive a restrictive target. By then hopefully the public will also become more concerned about climate change.

6.2 Rules on setting the total quantity

I have mentioned several times the difficulties that arose from the many issues left open-ended by the vague wording of the ETS Directive. Most important of all these is the establishment of the total quantity. With due acknowledgement of the political realities prevalent at the time of the adoption of the Directive, we need to remind ourselves that any given quota system should be able to set the total quantity of allowances in legislation. This was the case of the Kyoto Protocol, and this is the case with the countless EU rules on various agricultural quotas.

It is difficult enough to allocate even if we know the total quantity beforehand. In the EU ETS however, the Member States need to set the total quantity simultaneously with the domestic discussions on the rules on the distribution of allowances which makes the process very exposed to the public and to interest groups. For Hungary (and for other Member States with emissions well below targets), the current rules provide much greater leeway in setting the total quantity than for old Member States, again making the whole exercise more difficult.

Simpler and clearer rules are therefore needed for the establishment of the total quantity. Preferable would be a situation where there is already an EU-wide political agreement on the total quantities for the trading sector even before the domestic discussion on allocation rules starts.

6.3 *Simplicity and harmonisation*

Hungary's NAP is not among the most complex NAPs, but it has some solutions that are somewhat unwieldy to implement. This is true in particular with respect to the evaluation of the planned emissions of new entrants, and with respect to the distribution of the early action reserve. The development of implementation provisions for these rules requires much more time and effort than would by justified by their significance. Smaller Member States in particular cannot afford to spend too many resources on the development of rules. An important lesson is to strive for simpler allocation rules wherever possible.

In the quest for more simplicity, further harmonisation of at least some allocation rules would be of great assistance for Member States with fewer resources. As mentioned, the smaller the Member State, the greater the pressure to introduce rules tailored for certain large individual installations. More harmonised allocation rules would provide a useful counterbalance against this complexity-generating tendency.

11 Czech Republic

TOMAS CHMELIK

1 Introductory background and context

1.1 Czech industry in the context of economic development

The Czech Republic is a relatively industrialised country with a higher share of industrial activity than its closest neighbours. The carbon consequences are aggravated by the relatively poor quality of the available coal which is used in the power sector. Higher-quality black coal, which is also available in smaller quantities, is mostly used for the production of coke and for the blast furnaces in the steel industry. These factors explain the relatively high carbon intensity of the economy expressed in carbon intensity per capita or per unit of GDP.

During the past decade and a half, Czech industry has undergone a significant transformation that shaped the conditions in which the Czech National Allocation Plan (NAP) was prepared and the attitudes towards carbon emissions trading. The main factors were:

- the transformation from a directive-based economy to a market-based one along with economy-wide privatisation and an increased opening to external influences
- the breakdown of stable Eastern European markets after political changes in the former Eastern bloc required an orientation to more demanding Western markets
- the relatively quick change from non-existing environmental legislation in the communist period to implementation of economy-wide environmental legislation that required massive investment for air and water protection
- Entry into the European Union (EU) in May 2005 and the further strengthening of environmental legislation due to implementation of the environmental *acquis* in order to fulfil conditions of a single market.

The first two factors from the above list were the main reasons for the economic recession in the early 1990s (similarly to other Eastern European countries), which was followed, in the Czech Republic, by relatively quick economic growth driven in substantial part by industry. The first peak of growth was reached in 2000, when the annual GDP growth amounted to 3.9% (in fixed prices). Thereafter, growth slowed down until the end of 2002 and then resumed in 2003.

All these factors contribute to the relatively high level of industrial CO_2 emissions in the Czech Republic compared to the EU average; and they need to be taken into account when considering a constraint on CO_2 emissions. That the Czech Republic has in fact no domestic sources of natural gas or oil only exacerbates the problem. These higher-quality fuels have to be imported and they have a negative influence on the balance of trade. All these factors are behind a relatively high share of greenhouse gas (GHG) emissions covered by the trading sectors (60%), whose share is expected to grow even further (to 65%).

1.2 Climate change policy in the Czech Republic

The Czech Republic is a signatory to both the United Nations Framework Convention on Climate Change (UNFCCC) and its Kyoto Protocol (signed on 23 November 1998 and ratified on 15 November 2001, respectively).

Climate change policy in the Czech Republic is profoundly influenced by the absence of problems in fulfilling the country's quantitative Kyoto target. In the early 1990s, because of structural changes in the economy, the implementation of environmental legislation and technological improvements, GHG emissions dropped well below the target. Despite economic growth in recent years, GHG emissions remain safely below this value due to the so-called decoupling of emissions from economic growth. The Czech Republic is not part of the Burden Sharing Agreement (BSA) and therefore the Kyoto target is the only relevant international quantitative target in GHG emissions until 2012. Nevertheless, the domestic climate policy, as defined by the National Programme on Abating Effects of Climate Change, adopted by the government in 2003, clearly states that the good position regarding the Kyoto target is not a reason for inactivity. That Programme sets a domestic GHG emissions reduction target of 20% from the year 2000 to be achieved by 2020.

Still, the ease in achieving the Kyoto target is the main reason for the weak implementation of climate change policy. For an economy that has undergone such significant transformation during a relatively short time period, the main challenges remain the attainment of high and stable economic growth, especially if climate change (and sometimes more generally all environmental) measures are perceived as a drawback for the development of the economy. Without a wider political consensus, it is difficult to implement proactive climate change policy. As a result, climate-related policies are quite often interpreted as declaratory documents whose contents are purely indicative and not essential for the economic life of the country. The main aim remains the stable growth of the economy together with the improvement of the main economic indicators.

1.3 Experience with market-based instruments

Environmental regulation in the Czech Republic is based on a mix of administrative and economic instruments. However, emissions of CO_2 have never been subject to regulation or even direct monitoring. Lack of experience with emissions trading is one of the factors explaining the relatively reluctant position of industry to the EU Emissions Trading Scheme (EU ETS). It is also worth noting that the usual industry partners negotiating the implementation of the EU ETS with the government were the same persons who were responsible for fulfilling the obligations resulting from other environmental legislation and for reporting data or who were connected with the production process, not those taking a more forward-looking view of the investment opportunities made available by flexible approaches such as emissions trading. The position of industry was therefore rather negative in viewing the EU ETS as another regulatory instrument, overrating its negative aspects, and systematically underrating the incentive side of the instrument.

2 The macro decision concerning the aggregate total

The assessment of the NAP by the European Commission (EC) effectively split the NAP process in the Czech Republic into two phases: the first to reach a national consensus before submission to the EC and the second to respond to the changes required to gain EC approval. In

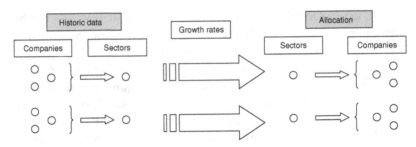

Figure 11.1. Digram of the Czech allocation process.

the case of the Czech Republic, the second phase resulted in a methodological change in the allocation mechanism and the preparation of a 'new' version of the allocation. To clearly divide the two parts of this process, both are described separately.

The allocation process can be illustrated by the schematic drawing above (Figure 11.1).

2.1 Original version of the NAP (before assessment by the EC)

2.1.1 Allocation vs. Kyoto target

The crucial feature influencing the allocation process in most of the new Member States is the problem-free position in relation to the Kyoto Protocol. Like other accession countries, the Czech Republic is not part of the BSA so that the only relevant absolute target limiting the emissions is the Kyoto target: aggregated GHG emissions during 2008–2012 should be 8% below the 1990 level. Current emissions are approximately 17% below this reduction target. From the standpoint of the Kyoto Protocol, therefore, no further emission reductions are needed, which is something that environmental non-governmental organisations (NGOs) do not accept as an argument. The Czech situation is described by Table 11.1 and Figure 11.2.

The absence of a 'Kyoto problem' was also the basis for a simple idea that the allocation should be 'strictly' based on the Kyoto target as mentioned in the Commission guidance, i.e. the allocation was expected to include some of the Kyoto surplus. This idea was supported by arguments that the drop in emissions, which had benefited the country in meeting its Kyoto target, was a burden to industry, and that an allocation of this surplus would represent an excellent payback option. While this argument has some merit, it is neither correct nor in

Table 11.1 *Greenhouse gas emissions from 1990 to 2004 excluding bunkers in the Czech Republic (Gt CO$_2$e)*

	CO_2 total[a]	CO_2 LULUCF[b]	CH_4	N_2O	HFCs	PFCs	SF_6	Total emissions Incl. LULUCF	Total emissions Excl. LULUCF
1990	165,060	163,281	18,590	12,604	n.a.	n.a.	n.a.	194,474	196,253
1991	155,261	145,254	17,012	10,853				173,119	183,126
1992	140,160	130,466	15,881	9,611				155,958	165,653
1993	136,704	127,781	14,809	8,580				151,170	160,093
1994	131,242	122,908	13,914	8,417				145,239	153,573
1995	132,125	124,314	13,580	8,724	1	0	75	146,694	154,505
1996	133,506	123,012	13,470	8,260	101	4	78	144,924	155,418
1997	138,032	133,035	12,716	8,469	245	1	95	154,562	159,559
1998	129,188	125,560	12,258	8,416	317	1	64	146,615	150,243
1999	122,099	117,227	11,553	8,069	268	3	77	137,195	142,068
2000	129,017	122,136	11,531	8,258	263	9	141	142,338	149,218
2001	129,033	121,960	11,458	8,491	393	12	168	142,483	149,556
2002	124,040	117,875	11,434	8,204	391	14	67	137,984	144,149
2003	128,075	122,326	11,109	7,744	590	25	100	141,894	147,644
2004	127,297	122,427	10,895	8,318	600	17	50	142,306	147,177
Per cent[c]	−22.9%	−25.0%	−41.4%	−34.0%	817 – times	141 – times	−34.1%	−26.8%	−25.0%

[a] CO$_2$ emissions excluding LULUCF (Land Use, Land Use Change, and Forestry) sector.
[b] CO$_2$ emissions from LULUCF sector.
[c] Relative to base year.

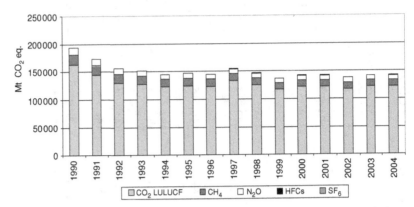

Figure 11.2. Greenhouse gas (GHG) emissions from 1990 to 2004 excluding bunkers in the Czech Republic (Gt CO_2eq.). (Data from Czech Hydrometeorological Institute.)

line with the EU legislation as a whole (in particular with regard to state aid rules). Once this was pointed out, an opinion gained ground among industry that the whole system was intended to solve the problems of EU15 and to represent a drawback to new Member States.

When it became evident that an allocation 'according to Kyoto' or 'on the basis of installed capacity' was not possible, the whole discussion became focused on determining the level of 'needed' emissions as the key factor setting a national ceiling of CO_2 emissions for the industry sectors covered by the EU ETS. It quickly became clear, however, that a bottom-up approach based only on company-level expectations would result in an allocation not consistent even with Kyoto itself. Accordingly, some top-down constraints became necessary.[1] Thereafter, the preparation of the allocation plan was driven from both ends – with criteria limiting allocation on a macroeconomic level, but with a pressure from the bottom on the basis of company-level data and expected development.

2.1.2 Individual data collection

From the very beginning, it was clear that the only realistic way to allocate would be to use historical emission data, or 'grandfathering'.

---

[1] The Ministry of Industry and Trade initially asked the fifty biggest emitters to present their allowance requirements. When the sum of these 'needs' came in at almost 150 million allowances annually, well over the Kyoto target, this approach proved not to be the right way forward. Companies seemed not to distinguish between 'as needed' and 'as wanted'.

A baseline period of 1999–2001 was chosen in order to permit cross-checking of emissions data with data previously and independently reported concerning fuel consumption for purposes of air protection. Although this cross-check was possible for only a limited number of installations, these facilities accounted for a majority of emissions (large and medium sources of air pollution, mainly from the energy sector).

Using available emission databases and supplementary data provided by industrial associations, it was possible to identify most of the installations potentially falling under the scope of the EU ETS. On the basis of this identification, operators of installations received a specifically prepared questionnaire, based on a draft version of the Monitoring and Reporting guidelines, asking for emission data for the baseline years. The response rate of the questionnaire was not perfect, but the coverage in terms of emissions was somewhere between eighty and ninety per cent. Surprisingly, no intentional 'adjustments' of data were identified and therefore the match to already available information was very good. This is probably one of the advantages of technical people being involved in EU ETS implementation at the company level.

2.1.3 Sectoral growth rates

Once historic installation-level data were available, the emissions of concrete sectors were aggregated and the discussion could move on to the identification of growth factors for each sector as the main variable of the NAP. In fact, in the absence of conflict with other allocation criteria, expected growth can be used as a variable for the determination of the total amount. Growth rates therefore represent a key variable in the allocation process.

Once growth factors were determined, it was possible to calculate the so-called 'basic allocation'. This basic allocation represents the vast majority of allowances to be allocated; other factors influencing the total were relatively minor. The allocation plan included adjustments to reflect early action and combined heat and power (CHP) (see below). As this adjustment had to be made 'within' the projected ('expected') emissions, it was necessary to deduct allowances for these bonuses from this projected allocation. In total 4.5% was deducted (3% early action, 1.5% CHP). Early action and CHP bonuses were allocated back to companies on the basis of separate criteria, influencing allocation at sectoral and company level, but not the total.

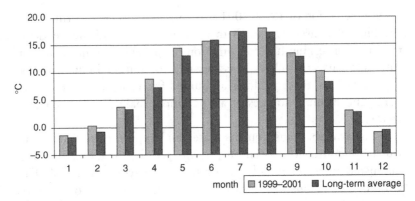

Figure 11.3. Average temperature in the Czech Republic. (Data from Czech Hydrometeorological Institute.)

2.1.4 Other factors influencing the total allocation

Total allocation was further adjusted by the so-called *CZT reserve* (CZT is the abbreviation in Czech for 'district (centralised) heating'). In this context the term is used for installations providing heat as an output (or partial output) for heating either as a primary function or as a by-product. The reason for this reserve was an approximately seven per cent higher level of temperatures in the reference period, resulting in lower emissions (less heat was needed) and therefore in a lower allocation. As temperatures often have a cyclic course, it is assumed that in the period 2005–2007 the level of temperatures most probably will not be above average as in the reference period. The district heating sector is a very specific phenomenon of the Czech Republic, based on the one hand on the specific climatic conditions of an inland country and on the other hand on a very developed sector of municipal (public) heating systems, replacing individual heating sources and bringing a number of environmental benefits especially in local air quality. The potential problems of this sector also depend on their strong social background (heat supply to households, public services etc.) and were one of the very sensitive elements of the plan.

Reasons for this reserve can be illustrated by Figure 11.3.

Additional allowances were added to the total allocation for the so-called *individual corrections*. Individual corrections were an element added to the plan on the request of the Ministry of Industry and Trade (MIT) to deal with cases when the allocation formula resulted in an allocation that had significantly adverse effects on a concrete installation. The main aim was to cover non-typical levels of emissions in the historic period used for allocation or to cover changes in production levels or other changes at the installation that cannot due to its nature be covered by the allocation mechanism and thus damaging the situation of a concrete company. For such purposes 1 million allowances were added to projection levels. The allocation among companies was supposed to be based on individual applications. Had the approved number of corrections exceeded 1 million, the corrections would have been cut on a pro-rata basis.

The final factor influencing the allocation was the *new entrants reserve*. As the projected emissions were based on micro data, the allowances for newly commissioned installations were added on top of these emissions. The size of the reserve was set to provide a safe space for the further development of the economy, as only limited information about the planned installations to be put into operation during the first trading period was available, based on the indications of companies. No specific research for this purpose was made.

2.1.5 Allocation plan as notified to European Commission

The basic figures about total allocation can be described as follows:

Total allocation = base period emissions × growth factor	(separately for sectors, added together; early action and CHP bonuses are made within this figure)	
	103.88 million	ΣE_i
Reserves added on top of total allocation:		
CZT correction (in fact 673,468, remaining to be cancelled)	1 million	$R(CZT)$
Individual corrections	1 million	$R(IJ)$
New entrants reserve	2 million	$R(NE)$
Total NAP	*107.88 million*	

In mathematical terms, allocation can be expressed as follows:

$$PC = \Sigma E_i + R(CZT) + R(NE) + R(IJ).$$

The NAP was officially notified to the European Commission on 12 October 2004. Part of the work was over.

2.2 NAP after assessment by the European Commission

2.2.1 Differences in national projections

After examination of the information provided by the Czech Republic, the Commission concluded the assessment by preparing a draft decision on allocation, permitting an allocation of no more than 90.16 million allowances annually. The draft decision was unofficially provided to Czech officials to create time for reaction and potential discussions. The origin of this figure is official Czech projections of national GHG emissions, where using the expected share of EU ETS sets a number of allowances to cover the expected reality (i.e. needs). The original proposal of allocation (107.88 million) and the draft decision (90.16 million) thus determined the interval for potential adjustments and negotiations. Using national projections for setting the allocation predetermined the reaction of the Czech side, based on describing the differences between projections and reality (the projections were from 2001 and some assumptions were no longer valid) with an attempt to quantify them and so provide arguments for an allocation higher than the one proposed by Commission.

The arguments focused on the following:

(1) *Increase in electricity export.* This argument was based on the different level of electricity exports assumed in the projections and on the actual development in recent years. The original projections expected a 10 TWh export level in 2005, with recent expectations of 18 TWh. This difference was quantified using the emission factor ($1\,TWh = 1.05\ Mt\ CO_2$). This argument proved to be the most difficult one for the Commission to accept and only a partial adjustment of the projections was made.

(2) *GDP growth rate.* The projected growth rate was slightly higher than recent predictions, so in this step the projections were downgraded for consistency with the recent data set. The adjustment, however, was relatively small (minus 0.28 Mt).

(3) *Increase in industrial production.* The originally predicted industrial production was lower than what can be assumed on the basis of recent development (between 2000 and 2004). The difference in industrial growth per cent was quantified on the basis of macroeconomic data about GDP growth and GHG emissions growth.

(4) *Utilisation of natural gas and renewables.* The projections were adjusted to accommodate more recent forecasts on the share of

natural gas and renewables on the total primary energy supply which is expected to be lower than originally predicted. Again, this effect was quantified.

All the above factors were added to the original projections to produce 'updated' projections, which were presented to the Commission for consideration in the calculation of the total allocation. Detailed discussions with Commission experts took place during the first week of April, and were followed by an update of the NAP by the Czech Republic that reflected the discussions with the Commission and allowed a 'clean' final decision to be reached (in the sense that the Commission agreed with the text with no exceptions). A revised allocation plan was adopted by the European Commission by decision K(2005)1083 on 12 April 2005.

2.2.2 NAP as agreed with the Commission after negotiations

As described above, the discussions with the European Commission resulted in an updated plan for the Czech Republic. The number of allowances was reduced to 97.6 million. This figure included all the allowances to be distributed including the new entrants reserve. This led to the reconsideration of the existence and size of some types of reserves or corrections as follows:

Total allocation = base period emissions × growth factor	(separately for sectors, added together; early action and CHP bonuses and individual corrections are made within this figure)	
	96.43 million	ΣE_i
Reserves added on top of total allocation as approved by the European Commission:		
CZT correction (673,468)	0.67 million	*R(CZT)*
New entrants reserve	0.5 million	*R(NE)*
Total NAP	*97.6 million*	

In mathematical terms, the allocation can be expressed as follows:

$$PC = \Sigma E_i + R(CZT) + R(NE).$$

Consultations with the European Commission resulted in an update of the total allocation figure with minor adjustments for those elements that were not accepted by the Commissions (mainly *ex-post* adjustments). However, a micro economic decision on the allocation to individual installations was not explicitly made and was the subject of the second round of discussions at the domestic level. In general terms, adjustments of the allocation methodology were allowed, but they had to respect the relevant Commission decision.

3 The micro decision concerning distribution to installations

3.1 Key elements of the Czech allocation plan

3.1.1 Basic allocation
Basic allocation represents a key step in the allocation process. The number of allowances to be assigned in basic allocation is calculated as total allocation minus allowances for reserves that are made within the total allocation. Basic allocation is distributed among installations on the basis of emissions in a so-called reference period. The figure for the average reference period of emissions is calculated as an average of two years with the highest emissions selected from the period 1999–2001 for each installation. In specific cases, this number can be replaced by a different figure to reflect the specific situation of an actual installation (absence of historical data or abnormal growth, for example). In fact, this figure represents a 'share' of an installation in the emissions of a sector:

$$ZA_j = RE_j / \Sigma\, RE_k \times PA_i$$

where:

ZA_j = basic allocation for installation j
RE_j = average reference emissions for installation j
PA_i = allowances for basic allocations in sector i
$\Sigma\, RE_k$ = sum of average reference emissions from all installations that fall within sector i including emissions of installation j.

After reaching a consensus with the European Commission, on the total number of allowances to be allocated, one of the problems to be solved was how to 'cut' allocation to remain within these figures and how to be fair at the same time. The advantage was that at this time (spring 2005) preliminary data about 2004 emissions were available. They were taken from permit applications that operators were obliged to send to the Ministry of Environment (MoE) no later than mid-February. After some careful consideration, the following formula on the level of installations was used (common for all variants).

If in 2004 the emissions of an installation were higher than the average reference emissions (the two highest years from the period 1999–2001), the 2004 emissions were used instead. If the emissions were lower, the average from these values was calculated. The aim was to identify the countries whose emissions were growing and those whose emissions were stagnant, and to adjust allocation accordingly. Since the total number of allowances was lower than in the original plan, the aspect of fairness, in terms of avoiding under-allocation, became the most sensitive element of the plan. The situation was complicated by the fact that the 2004 data were not available for some installations; the original average reference emissions were therefore used for the sake of consistency with the allocation formula. The 2004 data were used once more – for individual corrections (see below).

3.1.2 Distribution to sectors

As previously explained, the allocation for the Czech Republic was made in two steps. For the bottom-up calculation of the emissions, the installations were grouped into separate sectors (see Table 11.2). Generally speaking, the sectoral division follows the criteria included in the Annex to Directive 2003/87/EC, however, the energy production sector was split into three separate sectors to reflect the difference in the factors influencing the growth of each sector. In terms of size, the main sector was the *public energy production* sector representing the most important sector in the whole NAP, while the *corporate energy production* sector was separated to group the installations supplying most of their production to the industrial sectors not covered by the NAP (such as production of automobiles, food etc.). The energy installations that were supplying most of their production to the sector covered by a separate category in the NAP (i.e. paper, iron and steel) were included in that sector. Another category created for the purposes of a separate growth factor was the *chemical* sector grouping energy installations in chemical plants.

Table 11.2 *The Czech NAP by sector (share of emissions, number of installations)*

	Share of total emissions	Number of installations
Public energy production	66.59%	139
Corporate energy production	3.53%	135
Oil refineries	1.10%	4
Chemicals	5.28%	17
Coke	0.26%	2
Production and processing of metals	16.22%	19
Cement	2.95%	6
Lime	1.34%	5
Glass	0.84%	21
Ceramics	0.78%	60
Pulp	0.19%	2
Paper and board	0.91%	16
Total		426

Source: Internal calculations by the Ministry of the Environment based on allocation algorithm.

3.1.3 Adjustment of average reference period

The sub-section on basic allocation briefly mentioned that these reference emissions could be modified under specific circumstances for two reasons. First, the two years with the highest emissions in the period 2000–2002 were used for installations that had become operational in 1999, while for installations that had become operational in 2000, the corresponding two years in the period 2001–2003 were used. The year with the highest emissions in the period 2002–2003 was used for installations that had become operational in 2001. For such installations, RE_j therefore equals the emissions in that year. For installations that had become operational in 2002, RE_j is equal to the emissions in 2003. For installations that had become operational in 2003 and 2004, allowances were allocated in accordance with the rules for new entrants from the reserve for new entrants.

Second, whenever the installation underwent a more dynamic growth than was anticipated for the whole sector, the reference period could be adjusted on the basis of the application form filled out by the operator of the installation and submitted to the MoE. For such purposes Excel tables with automatic calculation formulas were prepared and made publicly available via the Internet. In general terms, the growth of the installation had to be higher than the growth of the sector and if the difference was greater than 10%, the allocation was adjusted. A 10% threshold was introduced to avoid a large number of small-scale changes.

3.1.4 Bonuses

The Czech allocation plan was, to a certain extent, specific for the number of bonuses it included. The three different bonuses can be split into the following groups.

The first group includes the bonuses that adjust allocation to take into account past investments in environmental measures (early action) and investments in more efficient technologies (CHP production). Their main purpose is to deal with the issues resulting from a grandfathering approach to allocation. For the applications, specific questionnaires using automatic calculations (Excel) were prepared and made available via the Internet.

In case of *early action*, a reserve equal to 3% of the total emissions projection was deducted from the projected emissions and distributed to the installations based on the assessment of questionnaires (operators of the installations had to fulfil certain conditions to be eligible for bonus allocation). The early action bonus $B(EA)$ for one year was calculated as follows:

$$B(EA) = (EF_{original} - EF_{reference}) \times V$$

where:

$EF_{original}$ = the emission factor for two successive years (selected from the period 1990–1998)

$EF_{reference}$ = the emission factor for the two years selected from the period 1999–2001 for the basic allocation

V = lower of the values for the average annual production (in relevant units – e.g. TJ heat for boilers, tons of bricks, etc.) in the years selected for the calculation of the emission factor $EF_{original}$ and in the years selected for the calculation of the emission factor $EF_{reference}$.

In case of *CHP* production, this factor was reflected in the allocation by assigning 430 allowances for every GWh of electricity produced by CHP in 2003. This 'benchmark' was based on an estimate of the displaced electricity production from a coal-fired plant and its sole purpose was to recognise the contribution of CHP in the allocation at the installation level. Eligibility was based on a written document stating how many GWh were supplied to the electricity grid or on the statement of a statutory representative in case of own consumption (electricity consumed inside the installation, not supplied to the grid). For the purposes of this bonus, a reserve of 1.5% of the total projected emissions was deducted.

In case of both the above-mentioned bonuses, the reserves were fixed, depending on total allocation only (3% and 1.5% respectively). If the approved number of allowances to be allocated on the basis of these bonuses was higher than the number available in reserves, the allocations were cut on a pro-rata basis. Both bonuses were cut, as expected; early action bonuses had to be cut by 46% and CHP bonuses by more than 65%.

The third bonus – bonus for *CZT* – was a specific one, because only specific types of installations could apply for it (those supplying the so-called centralised heat) and because the number of allowances to be allocated was added on top of the allocation. The reasons for the implementation of this adjustment to allocation have already been described (Section 2). On the basis of the difference in temperatures and heat produced in 2003, it was decided that the bonus was equal to 7 allowances per TJ of heat sold in 2003. Before the applications were complete, it was decided that 1 million allowances would be sufficient for this purpose. After the assessment of all the applications only 673,468 allowances were allocated and the remaining ones up to 1 million were not used according to the allocation plan rules.

Bonus distribution to the sectors in the allocation plan is described in Table 11.3.

The table clearly shows that some sectors are rather 'net' recipients of bonuses (more bonuses are allocated back than deducted), others are in the exactly opposite situation (such as the refineries and coke sectors). CZT bonus distribution naturally adjusts allocation to the energy sector; however, for CZT purposes heat from the installations in other sectors is used as well, but to a limited extent.

Table 11.3 *Allocation of bonuses in the Czech Republic (number of allowances)*

Sector	EA	CHP	CZT	Total
Public energy production	996,991	1,001,262	601,548	2,599,801
Corporate energy production	244,519	56,884	18,714	320,117
Oil refineries	0	0	0	0
Chemicals	315,593	109,163	6,492	431,248
Coke	0	0	0	0
Production and processing of metals	818,193	206,050	45,974	1,070,217
Cement	223,598	0	0	223,598
Lime	15,694	0	0	15,694
Glass	7,449	70	453	7,972
Ceramics	31,122	0	0	31,122
Pulp	82,229	4,851	0	87,080
Paper and board	157,408	68,118	287	225,813
Total	2,892,796	1,446,398	673,468	5,012,662

Source: Ministry of the Environment (2005). Internal calculations by the Ministry of Environment based on allocation algorithm.

3.1.5 Correction for system services in the electricity sector

A specific correction was introduced for the auxiliary services of the producers in the public energy sub-sector, whose purpose is to partly reduce the risk of an inadequate number of allowances, caused by the provision of auxiliary services and an increase in power electricity production. A maximum of 200,000 allowances, determined as the number of MWh of auxiliary services related to an increase in the output (positive regulation, dispatching reserve, quick start) provided in 2001 (data provided by ČEPS a.s., electricity network provider) multiplied by a coefficient of 0.6 is added to the basic allocation. Because this adjustment is made 'inside' the sector (no allowances were deducted or added for such purpose), adding these 'auxiliary services allowances' to the basic allocation leads to an increase of the allowances to be allocated for that sector, and the allocations had to be reweighted back to the original number of allowances before this correction. Only three companies received this correction. Any bonuses for early action, CHP and CZT corrections were then added to this reweighted basic allocation, as appropriate. The coefficient of 0.6 is based on the ratio 5240/8760, where 5240 corresponds to the average number of hours of production of power electricity mainly in coal-fired power plants in the Czech Republic (55 TWh of produced electricity corresponding to 10.5 GW of installed capacity) and 8760 is the number of hours of auxiliary services per annum.

3.1.6 Individual corrections

As requested by the MIT, another element was added to the plan, this time allowing those operators who despite all possible corrections were still facing an inadequate number of allowances – and thus felt themselves to be discriminated against – to have their allocation 'corrected'. The main reason was the very specific development and course of emissions in the reference period, so that the allocation formula did not reflect the latest reality or expected future.

The operators who felt discriminated against by the allocation, once the plan this was made publicly available, had to apply in writing for a correction. These applications were analysed in due time and – if approved – the corrections were made. However, during the assessment phase both ministries – MIT and MoE – faced the problem of analysing extremely complicated data, with often weak or incomparable documentation serving as a justification for correction. Given the limited time and human resources available, as well as the extreme sensitivity of an allocation based on individual negotiations, this approach was abandoned, even though in principle such corrections were acceptable to the European Commission, so long as they did not change the approved total figure. All the operators were extremely careful to monitor any other deduction from their total allocation, as this would have meant giving up some of their allowances for such purposes; this served as another argument against this relatively unclear element of the plan.

Individual corrections, however, are used in the latest version of the allocation plan, but in a completely different form. The modification of allocation methodology includes respecting emissions in 2004 that were available as preliminary figures before NAP completion. The allocation formula includes a condition in which should the allocation calculated be lower than the 2004 emissions for a specific installation, the allocation is adjusted to reach this level by deducting a corresponding number of allowances from the others. This means that for such purposes a specific reserve is made of the exact size needed for such a correction during the calculation process. This approach is less controversial than individual negotiations and it is generally accepted although it affects the allocations to others.

3.2 Comparison of sectoral allocation variants presented to the government

The allocation plan was finally presented to the government in three variants, which differed by the sectoral allocations and also by the method and rationale according to which the sectoral shares were determined. An important element of the ensuing discussion was finding a criterion of 'fairness' – even if there is nothing like a fair allocation.

The first variant was based on the strict use of preliminary data for 2004 and the application of the same methodology that had been used in determining installation-level shares of the sectoral totals. If the sector's 2004 emissions were higher than those for the reference period, the 2004 value was used; otherwise, it was the arithmetic average of these two values. Each sector's 'share' of the sum of these newly calculated sectoral values determined the 'share' of the national total already agreed with the European Commission (97.6 Mt) minus bonuses and reserves. This methodology is illustrated in Figure 11.4.

From the methodological point of view, this variant was 'the cleanest' – trying to be apolitical and relying on data alone. In fact, it represented a new allocation plan in which the previously negotiated sectoral allocations based on projected growth were not taken into account at all. In effect, it reallocated allowances to the sectors that had grown the most, as of 2004, relative to the earlier projections, at the expense of the sectors whose growth was lower than expected. The smaller national total implied a limited reduction for the public energy sector and a greater reduction for the manufacturing sectors. It was thus called the 'pro-energy' variant.

The second variant, which was based on a request from manufacturing sector representatives, granted sectors their 2004 emissions and then allocated the 'growth surplus' (the national cap less reserves less 2004 emissions) according to the shares of the previously projected total, as follows:

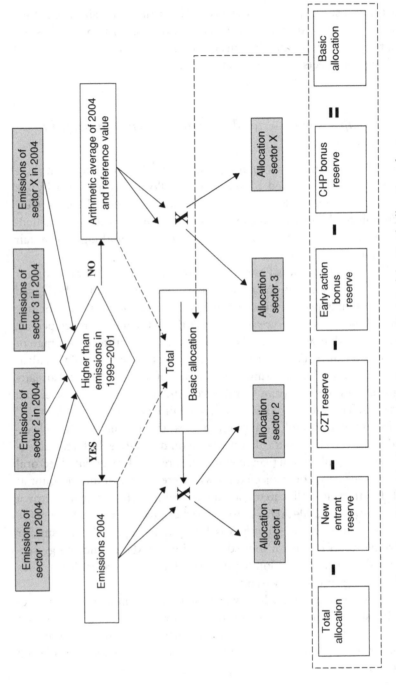

Figure 11.4. An illustration of the methodology used to prepare variant I of the Czech allocation plan.

Total number of allowances used for surplus determination	96,426,532 t/year
Total emissions of sectors in 2004	90,438,342 t/year
Surplus for allocation	5,988,190 t/year

This variant was more friendly to the manufacturing sectors by 'conserving' at least partially the expected growth in emissions included in the previous allocation plan. Naturally this approach was subject of critique from the public energy sector.

The third and last variant was added at the request of industry and on the basis of a suggestion from the Association of Industry and Transport, when the approaches using 2004 data were refused as not reflecting the real needs of the sectors (despite the initial support from major stakeholders for using the latest data). Allocation was then determined on the basis of discussions with sector representatives. This variant was called 'pro-chemical', because it visibly increased the allocation of the chemical sector (including the refinery sector).

All three variants are compared in Table 11.4.

The comparison clearly shows that an 'ideal' variant does not exist. Each sector will of course prefer the variant with the highest number of allowances for itself. Consequently, decision-making tended to be influenced by the political strength of the individual sectors rather than by objective criteria. The choice of the variant was therefore expected to end up in a conflict between the energy sector and big manufacturing sectors (mainly steel and chemical) with the smaller sectors having a very limited chance to influence the result. Discussion about variants ended on 20 July 2005, when the government decided to choose Variant III during its last regular session before the summer holidays. Another part of the work was over.

4 Issues of coordination and harmonisation

4.1 Institutional set-up

From the institutional point of view, the MoE was responsible for the implementation of the EU ETS directive. In the Czech Republic, the responsible body is decided according to the nature of the issue; however, if another ministry feels that the legislation to be implemented represents a cross-cutting issue, it takes part in the implementation

Table 11.4 *Comparison of allocation variants in the Czech Republic (number of allowances)*

	Variant I	Variant II	Variant III
Public energy production	63,992,006	63,295,778	63,458,493
Corporate energy production	3,797,092	3,710,687	3,766,771
Oil refineries	1,030,907	1,220,874	1,370,498
Chemicals	5,175,252	5,457,386	5,574,288
Coke	242,483	268,244	249,827
Production and processing of metals	15,662,314	15,723,109	15,455,479
Cement	3,063,841	3,073,444	3,047,260
Lime	1,232,322	1,423,424	1,341,085
Glass	783,758	846,638	827,848
Ceramics	809,007	799,547	808,166
Pulp	260,213	280,781	251,899
Paper and board	1,050,805	1,000,090	948,384
Total	97,100,000	97,100,000	97,100,000

Source: Ministry of the Environment (2005). Internal calculations by the Ministry of Environment based on allocation algorithm.

as a cooperative body. In the end, the assignment of responsibilities must respect the Competence Act, which defines the responsibilities of different members of the government and their ministries. The final decision is made by the Office of the Prime Minister which can also solve disputes should the ministries be unable to find a consensus. In the case of the EU ETS, no other ministry expressed an interest in taking part in the implementation of the Directive when the implementation started (October 2003). The MIT became involved at a later stage when pushed by the industry to find an alternative to the expected 'green' approach of the MoE on the issue of allocation. Experts from the MIT did not take part in expert meetings in Brussels or anywhere else in the preparatory phase, which was, together with the MIT's late involvement and confrontational approach, one of the factors of inefficient cooperation. A pragmatic approach was difficult to find at this stage, especially when the MIT almost automatically supported industry's proposals for allocation – including the ones that were totally in conflict with Directive rules, such as allocation strictly to Kyoto level.

The situation was further complicated by two other circumstances. First, due to EU entry in May 2004, the allocation plan for the first period had to be prepared in parallel with legislation. The legal basis for collecting the information needed for allocation did not exist and cooperation with industry was made on a more or less voluntary basis. Industrial representatives proved to be pragmatic. While a warning voice was raised from time to time saying that almost everything that the allocation was based on could be legally questioned, a serious challenge was never made.

Second, the progress of allocation activity was significantly delayed by a governmental crisis, which resulted in a slightly restructured government in the summer 2004. Since both the allocation plan and the act itself were very controversial, there was no political will to make a decision at this particular time. Both issues were therefore delayed by several months before the political situation stabilised.

Another very important disadvantage was caused by systematic capacity constraints at the MoE. Despite the fact that climate change issues including the EU ETS had been centralised in a separate Climate Change Unit since May 2004, there were only two people from this unit involved in NAP preparation. Cooperating institutions, such as the Czech Hydrometeorological Institute (involved in data collection and preparation) and the Czech Environmental Inspectorate (involved in some specific cases concerning the definition of installations and their identification) were also available to help, but conceptual work and the lead responsibility rested with the Climate Change Unit.

The role of consultants was surprisingly very limited and focused mainly on the principles of allocation or on some specific issues. Since the information at company level was considered confidential, the detailed set of data used for allocation was available to governmental representatives only. For the broader purposes of the EU ETS implementation, the MoE was also the recipient of assistance provided by the Netherlands. A consortium of experts selected by the Netherlands was involved in assisting Czech experts on a number of issues, including preparation of national legislation, implementing the monitoring and reporting provisions and setting up the system of permit issuance. Due to the sensitive nature of company data, this assistance was intentionally limited in relation to the preparation of the NAP.

To improve communication with industry representatives, a working group was established by the MoE, consisting of representatives of all major stakeholders from industry, which was very helpful especially during the initial stage of EU ETS implementation. In the final stages of NAP preparation, contacts tended to be more direct with associations and companies rather than through the organised working group.

Local environmental NGOs took a very active role when total allocation was decided, proposing of course a decrease in the number of allowances to be allocated; however, once the total allowances were fixed, their involvement in discussions about the distribution of the allowances to the installations significantly decreased.

As the work on the NAP progressed, and after an initial inactive stage, the MIT became more heavily involved. This led to an adjustment in the originally proposed legislation during its approval process in the Czech parliament. As a result, this ministry was granted competence to participate in the NAP preparation. There was also a proposal to relieve MoE of the responsibility and to give it entirely to the MIT, but this option was abandoned as part of a compromise. Involvement without sufficient cooperation in the early stages led the MIT to make its own proposal for a NAP, to be presented to the government as a counter-proposal to the MoE proposal, but political considerations pushed towards a consensus from both ministries. The final stage of NAP preparation before the final approval by the government and during negotiations with the European Commission was coordinated by the Vice-Prime Minister for Economic Affairs, serving as a consensual third party with a political rather than technical role. After inefficient cooperation at the beginning of the process, both responsible ministries started to cooperate much more efficiently in the later stages and they eventually defended the proposal together at both the domestic and EU levels.

4.2 Commission guidance

During the preparation of the allocation plan, the Czech Republic, similarly to other Member States, used as guidance a document prepared by the European Commission.[2] The aim was to develop a clear and transparent allocation plan draft for discussion at both the domestic and EU level. As the structure of the guidance document was generally complied with, the allocation plan was comparable to other allocation plans (for example the Irish one).

---

[2] COM(2003) 830.

The guidance document, however, was used only to a certain extent. As described above, the key issue in the development of the Czech allocation plan was setting the total number of allowances in a situation where the country was below its quantitative Kyoto target. The result was that some of the Directive's criteria were not applied at all, while the consistency of the relation between allocation and projections and state aid rules, which was considered most important, was not discussed in sufficient detail in the document and was a partial source of misunderstanding. The question is whether a more clear interpretation would have helped to avoid these discussions on a domestic level, or if at least it would have helped the argumentation of those who were responsible for the preparation of the allocation plan.

There is another point that should be mentioned now – total misunderstanding or misinterpretation of the relation of allocation to the Kyoto target in a situation where the Czech industry's emissions are below this target. A number of industry representatives believed that the 'Kyoto path' mentioned in the allocation criteria could be read as a way to allocate part of the Kyoto surplus. The negative position of industry was further strengthened by the fact that the new Member States were not taking part in the negotiations of the Directive during the legislative process on the EU level, and criticised the scheme as an instrument prepared to fit EU15 which did not consider the specific situation of the new Member States. Concerns about the behaviour of the multinational companies who were expected to 'protect' homeland facilities by adjusting the production in the newly invested branches in new Member States, potential threats from non-EU competition, negative influence on economic development including slowing down incoming investments in the country were constantly present in the comments from industry.

4.3 Poor harmonisation

Due to the differences in approach of the EU15 and the new Member States and also due to the delays in the preparation of allocation plans, issues of harmonisation were not the first priority. Even if some elements from other allocation plans were inspiring, the Czech allocation plan was built around the key methodical pillars that were based on our own experience and availability of data and information. A relatively important advantage was that the information about most of the companies included in the plan was already available, so that work on the allocation plan could be started on a relatively solid basis.

It must also be stated that allocation plans were not written in a unified structure, so for the purposes of harmonisation it was difficult to make any kind of comparison or assessment within the preparation of the Czech allocation plan, especially if the human resources working on this issue were relatively limited. Also – as another point – notified allocation plans were often first published in the mother tongue or non-English language.

Harmonisation was also difficult due to the relatively long period of time devoted to the assessment of allocation plans. Opinions vary whether it was better to be one of the first movers or to submit allocation plans at a later stage; according to the Czech allocation plan, it was difficult to change or adjust the methodology on the basis of the successful implementation of a particular element in someone else's allocation plan – it seemed that Member States went mostly their own way.

4.4 Assessment of allocation plans

At this stage it is also difficult to evaluate the results of the Commission's assessment, especially in the interpretation of the important criterion of the 'Kyoto path'. The guidance document does not specify what should be understood under this 'path' and the practical interpretation by various Member States showed quite a lot of creativity. The big question is, what effects will the experience with the first round of allocation plans have on the second round? The European Commission has been declaring for quite a long time that its willingness to be 'flexible' will be much more limited than it was at the beginning of the scheme. Allocation plans should be notified in time, and a unified summary table was prepared on the basis of the difficult understanding of key elements in the allocation plans. However, the short period between the finalisation of the first NAP and the preparation of the second to a certain extent limits possible changes in the whole approach or in the methodology, which is restricted by the availability of data, continuity in allocation and, in general, the time pressure to prepare the allocation plan in time.

Projections of GHG emissions were one of the most problematic and controversial aspects of the allocation plan of the Czech Republic.

This was particularly so in those Member States having no problem in fulfilling the Kyoto quantitative target because specific assumptions in the forecast would determine whether over-allocation, which was the European Commission's main concern, had occurred. National projections of GHG emissions are made for a variety of purposes, with a number of assumptions that might work on a macroeconomic basis but can prove to be a problem when applied at the sector or installation level. Also, projections made for a much longer timescale may involve inaccuracies when interpreted in a short time period. Some assumptions may prove to be wrong or changes of unexpected external influences (such as a rise in oil prices) might outweigh the effect of other trends in the economy so that the projections become more an econometric exercise than realistic visions of the future. Finally, projections might also be influenced by political concerns to portray future developments as being optimistic or pessimistic depending on the context for which the projection is used.

An important issue is the role of the other targets that influence emissions indirectly, such as the share of renewable sources on primary energy consumption. The attitude towards these types of targets is sensitive to the difficulties faced in their achievement once serious problems are faced; the target is termed 'indicative' with a description why the original plan must be revised. This situation is common in a number of European countries, including the EU as a whole, and interpreting these targets as fixed, unbreakable barriers in the process of assessment of allocation plans might be source of confusion and criticism.

Generally speaking, the situation is complicated by the fact that even if the total amount of allowances to be allocated is consistent with the needs of the economy, allocation on the installation level – which must be simplified to stay at least partially transparent – might be facing problems in distributing enough allowances to all installations. Of course, 'enough' allowances is often interpreted differently by the ministry and companies, but still, one company being really under-allocated is a subject of criticism of the whole allocation plan, despite the accuracy used in defining the allocation formula. These 'micro' problems are probably unavoidable – allocation can never be perfect – but are logically a bit difficult to explain to a representative of this particular firm.

From the procedural point of view, being the opposite number of the Commission in the latest stage of negotiations was an extremely hard job. The importance of the issue, which was discussed with real money seen behind each allowance, was enough to be a nightmare for those who were sent to Brussels to obtain as many allowances as possible. The Commission was a tough but fair partner. It was clearly visible that experience in negotiations and negotiating tactics lies on the side of the Commission, with expert background available nearby. It could be interesting to analyse whether those countries which have more experience in working with European institutions and its processes were able to be more efficient in defending their positions in relation to allocation. As new Member States were newcomers, their chance to make the most of the negotiation potential was probably partially limited.

Another element of the scrutiny process, which was relatively unclear, is the role of other Member States. Drafts of NAPs were discussed both at Working Group 3 and finally by the Climate Change Committee, but the question is what was the real impact of the comments raised by Member States on the decision of the Commission, because one can expect that they would be raised anyway. The role of national experts in the process was discussed several times on both platforms, with some reservations presented by some Member States. More detailed guidance about what should be the role of the Member States would be helpful. On the other hand, the discussions served as a good indicator of potential points of interest and provided the only chance of other Member States to comment officially other NAPs. However, it is difficult to be completely objective in this role as everybody's NAP is discussed as being a 'tough' criticism of the NAP of other Member States and could result in the same position of that Member State in return. It could be interesting to further analyse whether Member States were at least partially consistent in their opposition to the Commission and tended to feel that they were 'in the same boat' (which was maybe the case at the beginning), but it seems that at a later stage each Member State was more focused on its own problems with the allocation plan.

5 Issues deserving special mention

Due to the relatively detailed description of the process given in Chapters 2 and 3, there is only a little else to be discussed separately here.

5.1 Auction

From the beginning of the preparatory work on the allocation plan, it was clear that all the allowances would be distributed free of charge and that the auction would not be used. Also, the absence of experience with auctioning at the EU level was one of the elements for this decision. However, auction was back on the scene later on as an option for using unused allowances from various reserves as an alternative to their cancellation. Finally, the CHP and CZT reserves would be fully used as early action reserve; the only type of reserve where auction was theoretically possible was the reserve for new entrants, and this approach was subsequently given up as part of a compromise with the European Commission (all allowances will be grandfathered for free).

5.2 Definition of installation

One of the most intensive factual problems of the EU ETS implementation was the absence of a more strictly harmonised approach to the definition of installation (mainly combustion installation) and the issues of new entrants and closure provisions. Despite the fact that the degree of flexibility might be useful and therefore important for some Member States, a greater harmonisation or at least a more detailed guidance would be useful, because the multinational companies were pointing out that the different approaches used by different Member States were a source of misunderstanding and misinterpretation and were indicated as potential sources of discrimination on the single market.

5.3 New entrants and closures

As *ex-post* changes were strictly prohibited by the European Commission, the Czech Republic uses a relatively common approach to new entrants and closures. In case of new entrants the allowances from the new entrants reserve are used; however, due to the availability of data (benchmarks) for the determination of allocation (i.e. original policy followed for the new entrants), a specific approach was used. Allocation is decided *ex post* after the end of the year when the installation enters the scheme, but before the obligation to surrender allowances for that year. This approach allows the use of real installation-level data for allocation, avoiding problems with the absence of available data for the determination of allocation on an *ex-ante* basis. Allocations will be methodically in line with the allocations for existing installations, thus avoiding the distortion of competition.

The allowances from the new entrants reserve can also be used for those installations that underwent such a change within the trading period as to influence the need of allowances. This particular element was included in the plan to serve as an option for those who intend to invest in further development of the installation, and who would have felt discriminated against should the allocation be insufficient to cover the increased needs. A necessary condition is the change of the permit, i.e the change in the installation must be a change in terms of investment or it must affect the monitoring method, and only the difference between the existing allocation and the revised need is allocated.

In case of closures, the relevant point is the withdrawal of the permit. Once the permit is withdrawn, the installation physically disappears from the scheme and the allowances unallocated so far are cancelled (in fact there is nobody to allocate to). Lawyers are still unclear in their interpretation of the Czech emission trading act in terms of obligations to surrender allowances, especially in case the operator of the installation legally disappears (the company ceases its existence from the legal point of view). There is nobody to shift the obligation to surrender allowances to (in the case where the company ends before April), while the allowances that were already allocated (if it happens between February and April) can be sold. However, even if it is high time that this question is pointed out, such a situation is expected to be very rare, and speculations about closing and reopening the installation with potential double gain in allowances are probably rather theoretical.

5.4 The role of NGOs and the wider public

A relatively surprising point is the involvement of environmental NGOs and the public in general. While the public remained relatively silent – the implementation of the scheme was probably exciting for a limited number of those who were affected on both sides – green NGOs served as a strong partner in discussions, being able both to understand the scheme and to be in contact with the development in other parts of Europe, and therefore proved to be more informed or active than some of the industrial associations. Their influence on the final decision was in fact minimal, but their voices are expected to be raised again at the time of preparation of the second allocation plan.

6 Concluding comments

Implementing the EU ETS in the Czech Republic was an extremely interesting experience. The absence of any experience with emissions trading meant that all the infrastructures needed had to be built from scratch. A number of difficulties were faced; however, the scheme was put into operation successfully. Allocation was, if not the most important part of implementing the EU ETS, then at least the most sensitive element. Even if it is difficult to make any judgements at this stage, looking back at the preparation of the Czech NAP can be an instructive experience.

An important feature of the total allocation for some Member States – including the Czech Republic – is that the final decisions were based on GHG projections rather than on installation-level data. This required a change in the nature of the arguments that could be used for defending the level of allocation. In the case of the Czech Republic, national projections were taken into account only to a limited extent during the preparation of the NAP. Instead, a more or less political decision was made that the needs of industry to support the projected economic growth would prevail over any potential conflict with environmental goals. The willingness of the European Commission to challenge this decision was underrated. As a result, the decision about the total allocation was suddenly changed in mid-process from bottom-up to top-down.

The involvement of the major stakeholders was not the same from beginning to end. In case of the government, the initial antagonism of the ministries responsible for implementation (MoE and MIT) changed to cooperation later on. Except for some minor details, the position of both ministries was uniform at the latest – and key – stage of developing the NAP. The role of the industrial associations varied. Some interest groups were able both to defend their own particular interests and to communicate effectively in the preparatory phases by providing data and arguments. Others served only to present negative reactions to anything that was published, sometimes even contradicting previous positions. In general, the larger the interest group, the more difficult it seemed to find a common position. This was exemplified by the biggest group, the Association of Industry and Transport, which at the final stage officially admitted that it could not form a consensus among its members because of the completely opposite positions to the final variants of the NAP.

Reference

European Commission 2003. Guidance to assist Member States in the implementation of the criteria listed in Annex III to Directive 2003/87/EC and on the circumstances under which force majeure is demonstrated; COM (2003) 830, January 2004.

12 | Poland

BOLESLAW JANKOWSKI

1 Introductory background and context

1.1 Poland's participation in climate protection

Poland signed the Climate Convention on 26 July 1994 and ratified the Kyoto Protocol on 13 December 2002, making the commitment to reduce greenhouse gas (GHG) emissions by 6% within the period 2008–2012 compared to the 1988 emissions. These decisions were taken in Poland after a lot of hesitation and discussions, which often expressed fears as to whether the policy of reducing CO_2 emission would impose too much burden on Poland, because of the heavy domination of coal in its fuel consumption. Many concerns were expressed regarding the introduction of a 'coal tax', which at that time was expected to be the most realistic instrument of coal emission control. However, the political will to support the efforts of the international community on climate protection prevailed.

1.2 General economic, social and environmental context

Until the fall of the communist political regime in 1989, the centralised economic system dominated in Poland, largely based on development of heavy industry. Special environmental policy was not carried out.

The economic transition period started in 1989, after important political changes aimed at creating a market-oriented economy. Transition costs were very high. The GDP dropped drastically during the first years and in 1991 it was 18% lower than in 1989. Unemployment rose and is in constant growth. Part of the detrimental effects have been overcome as illustrated by the systematic GDP increase since 1992. One of the effects of economic transition is the significant decrease of GHG emissions, including CO_2, evidenced by the following data:

- Total CO_2 emission in 1988 – 476.6 Mt
- Total CO_2 emission in 2001 – 317.8 Mt (66.8% of 1988).

These changes were accompanied by huge social cost. A lot of problems are still unsolved. Regarding the possibilities of further CO_2 reduction in Poland, it is necessary to consider the following issues:

- large gap in GDP and electricity consumption per capita between Poland and the EU15 average
- difficult substitution of coal-dominated electricity production (natural gas, nuclear?)
- highest unemployment rate in EU25
- high costs of compliance with EU environmental requirements including emission standards (SO_2, NO_x, particulate matter (PM)).

The most painful issue is unemployment, which exceeded the 20% level in 2003 and is now fluctuating around 17–18%. Reducing unemployment is possible only in the case of rapid GDP growth. It is not possible without a significant increase in energy consumption and related CO_2 emission. The possibilities of using primary fuels with lower emission factors are limited by very high supply costs, mainly from import.

Accession to the EU created new challenges for Poland in the field of environmental protection. The costs of implementing the Integrated Pollution Prevention and Control (IPPC) Directive in Poland was estimated at the level of about €25 billion. The Large Combustion Plant (LCP) Directive (2001/80/EC) sets more stringent requirements concerning SO_2, NO_x and PM emissions for power plants with rated thermal capacity over 50 MW. To comply with new emission standards the operators of existing coal-fired plants must spend about €5 billion before the year 2008 for DESOX equipment. New environmental requirements accompanied by governmental plans of long-term contracts liquidation create a very difficult situation for large power producers in Poland. The implementation of the ETS is an equally important issue. Given the accumulation of important problems the task of implementing the EU ETS in Poland becomes particularly difficult.

1.3 Emission trading as an ecological policy instrument

Because of its high economic efficiency, emission trading has been considered in Poland for years as a realistic alternative to environmental policy dominated by the command-and-control approach. Extensive investigations regarding the economic efficiency of implementing emission trading in Poland were carried out in cooperation with the World Bank in the 1990s with pilot studies in the Opole region. They confirmed that significant savings were possible from applying this instrument.

During the transformation process in the 1990s Poland was looking for new ecological policy patterns and discovered the USA's positive experiences with emission trading systems on the one hand, and the command-and-control approach in the EU and Germany on the other hand. Due to an EU-oriented policy approach the Polish government has decided not to implement emission trading for SO_2 or NO_x, where the biggest savings were expected.

In the years 2003–2004 an extensive and detailed study was prepared again to determine the possible and justified scope of emission trading implementation in Poland. It was based on the current environmental legislation in Poland and the EU. General conclusions have shown three possible levels for emission trading:

(1) Supra-national or national level – as an instrument of solving global problems (e.g. climate protection).
(2) National level – against regional, supra-national problems (e.g. acidification, eutrophication, ground ozone – processes covered by the Geneva Convention).
(3) Local level – against national or sub-national problems (to reduce local pollution concentration, to control emission from means of transport in city zones, after removal of technical barriers that do not allow emissions to be connected with location).

The implementation of emission trading in the *cap-and-trade* formula requires the definition of a longer-run cap for certain polluters and groups of emission sources. The current legal regulations in Poland and the EU make it possible to implement emission trading at the national level with respect to the following pollutants:

• CO_2 – based on Kyoto Protocol caps,
• SO_2, NO_x and PM – based on the National Emission Ceiling Directive (2001/81/EC) and provisions concerning the National Emission Reduction Plan in the LCP Directive.

Within the study serious legal barriers were identified, rendering the implementation of emission trading in Poland with respect to SO_2 and

NO_x difficult (due to conflict with regulation of the IPPC Directive). Moreover, emission trading was not supported by provisions related to the preparation of the National Emission Reduction Plan within the LCP Directive.

As a consequence, some doubts arose regarding the EU policy with respect to the wider use of emission trading as an instrument of environmental protection. The EU was going to implement emission trading of GHG, but was simultaneously keeping barriers for using this instrument in other areas. Such a situation makes it difficult to create a long-term policy of implementing emission trading in areas where it could increase economic efficiency of environmental protection. A wider use of this instrument was considered more effective, because the same legal and institutional infrastructure could also be used to create and manage different emission trading sub-systems.

Due to the above considerations, there were great expectations for the implementation of the EU ETS in new Member States. From a legal perspective the situation of the EU15 is quite different from that of the new Member States. The old Member States have a common emission cap under the Kyoto Protocol. This first legally binding supra-national cap was the basis for the creation of the EU ETS, meant primarily to cover only the EU15. The new EU countries have got their own limits based on the Kyoto Protocol, and they all had conditions to implement emissions trading on the national level. The design of one unified system covering both groups of EU members was thus expected to be a very interesting issue both in theoretical and practical terms.

1.4 Poland and the EU ETS

Integration of the new EU members within the EU ETS was carried out in the simplest way. The rules prepared for the EU15 have been automatically extended to the new members. This was done in spite of the fact that under Article 30 the Commission was obliged to consider and to outline 'how to adapt the Community scheme to an enlarged European Union'.

The situation and specific features of the new EU members were not considered during the design of the EU ETS. As a result, doubts regarding legal fundamentals and economic efficiency of this system have emerged. From the new Member States' perspective, the current version of emission trading could have serious negative consequences for their development. The EU ETS creates significant implementation costs for companies (emission monitoring, changes in accounting, market behaviour, operational planning, strategy planning etc.) but gives neither direct nor cost-saving benefits for the economy. This is due to the fact that in the absence of the ETS Poland does not need to implement any carbon restriction programme as it currently meets its Kyoto target.

The rules regarding the preparation of a National Allocation Plan (NAP) were also prepared without considering the new Member States' situation. In effect these rules were perceived by new Member States as not quite fair, particularly in respect to issues such as:
- current emission level as a basis for the further allocation
- no early action influence on the total number of allowances in NAP
- allocation based on projection, not on Kyoto limits (cap)
- rules regarding the method of verification projections not clearly stated
- controversial interpretation of non-discrimination rule within the competition policy
- no instruments for risk compensation (regarding uncertainty of projections and changes in monitoring methodology)
- limitation of *ex-post* adjustment
- no clear long-term perspective (rules, limits).

Critical assessments of the whole system had a significant influence on the process of the creation of the Polish NAP.

As a consequence of the above-mentioned doubts, work on the Polish NAP was an attempt to combine both rules accepted by the European Commission and rules derived from the Directive as well as Poland's economic and social situation. Questions emerged regarding the legitimacy of the attempt to adapt to disadvantageous EU ETS rules, in a situation where Poland had put in a lot of effort earlier, considerably reducing its emissions of CO_2. Poland being on the path to meet the Kyoto Protocol limits seems to fulfil the goals of the ETS Directive. Because of that, the following rule was accepted as a fundamental guideline for work on the Polish NAP: 'Implementation of EU ETS in Poland should not worsen the situation of the Polish economy compared to the situation of individual realisation of the Kyoto Protocol targets.'

2 The NAP preparation process

2.1 *Organisation, main phases and actors*

The Ministry of Environment, representatives of industrial sectors covered by the ETS and independent experts (consultants) were the most important actors in the NAP preparation. The Ministry of Environment was responsible within the government for the NAP preparation and implementation of the ETS Directive. The Ministry of Economy played a less important role, which was widely criticised by representatives of industry, particularly at later stages of the NAP preparation. The passive position of this ministry could be partly explained by:

- strong involvement in other issues of high importance for the economy: implementation of LCP and IPCC Directives, implementation of the policy for the development of energy production from renewable sources, creating legal solutions for long-term contract liquidation,
- common conviction that CO_2 emissions and ETS are a big issue for the old EU countries, not for Poland.

Work on the NAP preparation was launched relatively late, in December 2003. The Ministry of Environment decided to engage the consulting company EnergSys with support of CMS Cameron McKenna and Ecofys Poland to organise and assist the process.

In a relatively short time the technical and informational framework was created to enable data collection, efficient information distribution and consultation with sectors' representatives and public opinion. Information on NAP preparation progress and related documents were distributed through a special Internet portal. Additional direct communication with sectors' representatives was established by means of branch organisations. Almost all sectors were represented in that way. During the NAP preparation almost twenty meetings with individual branch organisations were organised, devoted to explaining crucial aspects of ETS and NAP. After each phase of work the Ministry of Environment organised general meetings with participation of representatives from all sectors and other departments. During the meetings the results of work were presented and key issues discussed.

The main technical and IT work was performed by consultants. Representatives of each company were directly involved in data preparation, and in assessing the results presented. Representatives of branch organisations were the main partners in discussing the approaches proposed to each sector (assumptions on production growth, technological changes, allocation methods etc.). The Ministry of Environment played the role of supervisor and decision-maker.

Due to the limited commitment of the Ministry of Economy, the economic perspective was taken into consideration by the Ministry of Environment which was not interested in creating additional development barriers for the Polish companies. Therefore, at a relatively early phase of the work, the Ministry of Environment decided to propose additional bonuses above emission projections in order to acknowledge early action. Industry representatives were trying to lobby for the highest possible allocation. The consultants' role was to structure the process and organise constructive discussions based on the ETS Directive and the Commission documents, as well as objective data and clear proposals for calculation limits and their allocation. Overall, this phase of work was quite successful, even though industrial representatives tried to enforce their opinions and expectations.

The NAP document was prepared by the Ministry of Environment, accepted by other government departments and submitted to the European Commission on 22 September 2005. This document proposed to allocate allowances for 286.2 Mt average emissions per year within the period 2005–2007. On 8 March 2005 the European Commission decided that it would assign an allocation at an average level of 239 Mt per year, about 16.5% less than the Polish proposal.

The Polish government, after a lot of hesitation, decided not to appeal and began to work on the NAP adjustments to the limit imposed by the Commission. After the Commission's decision, all the work was carried on directly by the Ministry of Environment without the support of consultants. Since June 2005, the ministry has prepared several proposals for the allocation plan based on the new total limit. These proposals have been discussed with representatives of branch organisations and large market players. The ministry has tried to achieve consensus mainly by negotiating.

On 27 October 2005 the new NAP was published as a Ministers' Board decree. A group of large power plants protested against the decree. Despite this fact, the ministry tried to obtain the Commission's acceptance of that document. On 31 October 2005 the new government was established. It will be responsible for taking a final decision on the Polish NAP.

2.2 General methodology

The NAP was prepared using a methodology that consolidated two types of approaches: top-down and bottom-up.

Top-down analysis is focused on the research of key relations in the whole national economy and on the calculation of CO_2 emissions at the country level with the use of models. The starting point for this type of analysis is a set of statistical data regarding macroeconomic quantities, fuel consumption and CO_2 emissions. The results focus on quantities describing changes in GDP, energy consumption and CO_2 emissions as the country total and in sectoral division.

Bottom-up research is focused on the analysis of data and relations at the installation level and by aggregating individual figures to create results at the sector level.

The calculation of the allocation plan was made in three phases:

- *Phase A* – determination of the total number of allowances within the NAP
- *Phase B* – allocation of allowances at sector level
- *Phase C* – allocation of allowances at installation level.

The top-down approach dominated the first phase, whereas the bottom-up approach dominated the remaining two phases.

At the beginning of the work it was concluded that the dynamics of production and CO_2 emissions are different in different sectors. Therefore, all the installations covered by the ETS system were split into fourteen sectors. In the 'combustion installations' group three sectors were identified: public power plants, combined heat and power (CHP) plants and heating plants. Industrial heating and CHP plants were classified in the industrial sectors (e.g. steel, cement), following the assumption that energy production may depend on the main sector's output.

3 The macro decision concerning the aggregate total

3.1 Methodology of top-down analysis

The purpose of top-down analysis is to define the number of allowances that are to be allocated to the installations covered by the NAP.

The modelling set for medium-term forecasting was used, consisting of the three models: (1) macroeconomic CGE-PL model, (2) final energy demand model (PROSK-E) and (3) energy system model (EFOM-PL).

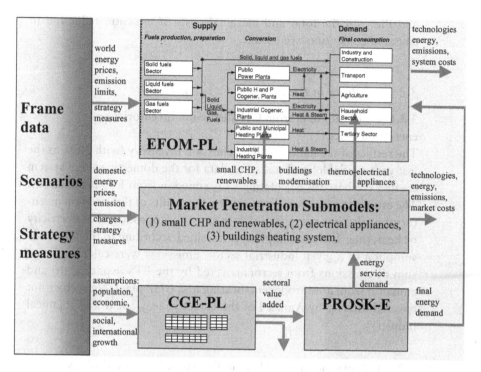

Figure 12.1. The modelling set used to develop CO_2 emission projections in Poland.

The typical configuration of the model set is shown in Figure 12.1. This modelling set was used in several studies for the Polish government in which official energy or emission projections were developed. It was also used to calculate the potential and costs of CO_2, SO_2 or NO_x emission reduction.

The CGE-PL model generates macroeconomic scenarios up to 2020. Based on these results the PROSK-E model generates projection for energy service and final energy demand. Market penetration sub-models simulate the effects of the market behaviour of an individual energy consumer or producer in the most important areas of potential energy efficiency improvement. The results of the market simulation are transferred into the EFOM-PL model, configured as an energy supply model. It covers all processes of fuel production, preparation, conversion as well as fuel and energy transportation. The model calculates energy production, energy supply, emission levels of SO_2, NO_x, CO_2

and PM and the total and marginal costs of emission reduction (if emission constraints are used).

3.2 ETS share in the national CO_2 emissions

3.2.1 Collection data from GHG emission inventories and energy statistics

The results of the national GHG emission inventory (with 2001 as the last year available) and statistical data for the domestic energy system created the base for calculating CO_2 emission from ETS installations. The estimation was based not only on the results of the emission inventory, but also on energy statistics. In case of autoproducers of electricity or heat, their fuel consumption is classified according to the final consumption of a given industrial sector. Emissions were calculated as a sum of emissions from sectors covered by the ETS (e.g. cement) and emission from autoproducers (heating and CHP plants) in sectors not covered directly by Annex I of the Directive (e.g. sugar and chemical industry).

3.2.2 Calculation of ETS share in the national CO_2 emissions

Relatively reliable estimates of CO_2 emission from ETS sectors were achieved, based on official statistics. Emissions were estimated in two types of aggregation: by sectors and by Annex I categories. Top-down results were compared with findings from data collection at installation level. The comparison is presented in Table 12.1.

Total emissions from ETS sectors calculated based on official statistics (top-down approach) and based on the bottom-up approach differ by about 2%. The result of this comparison was taken as a confirmation that ETS emission estimation was correct. It was assumed that lower questionnaire emissions can result from the incomplete list of identified ETS installations and from differences in emission factors used in the inventory and in questionnaires.

Based on the above calculations, the share of ETS sectors in total domestic CO_2 emissions in 2001 was defined at the level of 68.0% (217.2 Mt divided by the corrected national emissions of 319.3 Mt). Table 12.1 also includes the base emissions from the period 1999–2002 by sectors, that were used as the base for sectoral allocation (see Section 4.2).

Table 12.1 *Sectoral emissions based on top-down and bottom-up analyses (kt/year)*

Sector	CO$_2$ emissions based on official statistical data 2001	CO$_2$ emissions calculated on surveys related to the NAP[a] 2001	Base CO$_2$ emissions calculated on surveys related to the NAP[a] 1999–2002
Public power plants	152,120	113,423	116,552
Public heat and power cogeneration plants		37,201	37,327
Public heat plants	13,541	12,396	12,690
Mineral oil refineries	6,210	6,793	6,834
Coking plants	4,491	3,465	3,548
Iron and steel industry	12,132	12,482	13,670
Cement plants	9,227	8,551	9,406
Lime production plants	2,272	1,763	1,939
Glass industry	1,780	1,272	1,400
Ceramic industry	753	1,102	1,253
Paper industry	1,921	2,185	2,284
Sugar production industry	1,828	1,978	2,176
Chemical industry	6,188	5,763	6,086
Other industries	4,728	4,390	4,605
Total	217,191	212,763	219,769

[a] Based on results of data collection May 2004.

3.3 Scenario modelling: main assumptions and results

When the scenario calculations were performed (the first half of 2004) the Polish economy was characterised by dynamic changes in GDP and industry production growth. This fact is illustrated in Figures 12.2 and 12.3.

General macroeconomic assumptions (see Table 12.2) were based on official governmental projections from the year 2003. GDP growth was assumed at the average level of 4.1% despite the very optimistic dynamics of GDP in 2004. Main growth was expected in services, with slightly slower growth in transport and industry. Compared to GDP

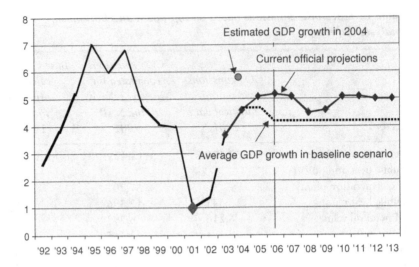

Figure 12.2. Historical GDP growth and current governmental plans (%) in Poland.
Source: GUS and Projection to the National Development Plan 2007–2013.

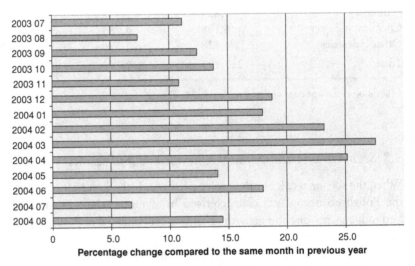

Figure 12.3. Growth of sold production in industry in Poland in 2004.
Source: GUS Central Statistical Office.

Table 12.2 *Main macroeconomic assumptions in the Polish scenario modelling*

Specification	2002 Billion PLN	Structure	2005	2010	2015	2015 Structure
			Growth rate 2002 = 100%			
Agriculture and fisheries	21	3.2%	103.3	119.2	131.9	2.5%
Industry and construction	205	30.4%	112.6	129.2	142.5	25.5%
Transport and communications	53	7.9%	103.4	131.1	165.2	7.7%
Services	394	58.5%	117.0	147.2	187.4	64.4%
Total added value	674	100.0%	114.1	139.6	170.2	100.0%
Gross GDP	771		114.8	140.7	172.1	
			2002–2005	2006–2010	2011–2015	
Mean annual GDP growth over the period			4.7%	4.1%	4.1%	

Source: Polish National Allocation Plan for 2005–2007.

and industry production growth figures in 2004, these assumptions were considered rather conservative.

Other important assumptions used for emission projection include:
- production of energy-intensive products (steel, cement, chemicals etc.) or services (e.g. transport activities)
- changes in energy intensity of production processes
- energy and fuel prices
- energy supply options and their characteristics.

Model calculations were performed for the following two scenarios:
- *BLN (Baseline) scenario* – according to the official governmental documents, GDP growth: 4.7% per year till 2005, 4.1% per year average till 2015
- *BAU'88 (Business as Usual '88)* – the same macroeconomic growth as BLN, CO_2 emission factors as of 1988.

BAU'88 is the reference scenario. In comparison to the BLN scenario it shows the total reduction of CO_2 emission achieved because of structural and technological changes in the Polish economy after 1989.

The main results of the model calculations consist of macroeconomics quantities, fuel and energy consumption and emission of pollutants including CO_2. The changes of GDP emission intensity are shown in Table 12.3. They were calculated from statistical data within the

Table 12.3 CO_2 *emission per unit of GDP in Poland*

Year	CO_2 emissions (kt/year)	GDP (Million PLN'99)	Emission intensity of GDP (t CO_2/1000 PLN'99)	Change on previous year
1988	476,625[a]	532,922[b]	0.894	–
1989	457,412[a]	531,856[b]	0.860	−0.038
1990	380,697[a]	449,951[b]	0.846	−0.016
1991	366,959[a]	415,754[b]	0.883	0.043
1992	371,591[a]	426,564[b]	0.871	−0.013
1993	363,133[a]	442,773[b]	0.820	−0.059
1994	371,588[a]	465,798[b]	0.798	−0.027
1995	348,172[a]	498,403[b]	0.699	−0.124
1996	372,530[a]	528,308[b]	0.705	0.009
1997	361,626[a]	564,232[b]	0.641	−0.091
1998	337,448[a]	591,316[b]	0.571	−0.110
1999	329,697[a]	615,560[b]	0.536	−0.061
2000	314,812[a]	640,182[b]	0.492	−0.082
2001	317,844[a]	646,584[b]	0.492	0.000
2002	323,712[c]	655,636[b]	0.494	0.004
2003	337,909[c]	680,550[b]	0.497	0.006
2004	352,310[c]	715,939[c]	0.492	−0.009
2005	363,299[c]	752,452[c]	0.483	−0.019
2006	369,511[c]	785,966[c]	0.470	−0.026
2007	375,723[c]	819,480[c]	0.458	−0.025

[a] Inventories of the emissions and sinks of greenhouse gases and their precursors in Poland in 2001, Institute for Environmental Protection, Warsaw, 2001.
[b] Statistical Yearbooks of the Main Statistical Office (GUS).
[c] Calculations by EnergSys, BLN scenario.

period 1988–2003 and from results calculated for the period 2004–2007. In the BLN scenario the unit emission decreases by about 1% in 2004 and by about 2–2.5% each year since 2005.

CO_2 emission projections in the BLN scenario split into ETS sectors as one aggregate and main sectors that remain outside the system are shown in Table 12.4.

Results of CO_2 projection as well as the Kyoto Protocol limit within 2008–2012 are shown in Figure 12.4. The limit for 2005–2007 related

Table 12.4 CO$_2$ *emissions in the Polish BLN scenario from trading and non-trading sectors (in Mt)*

Emission sources	2001	2005	2005–2007	2010	2015
NAP sectors	217.1	255.3	259.9	273.9	286.2
Other industries	20.0	19.7	19.8	20.0	20.2
Transport	30.1	35.0	36.1	45.0	54.0
Agriculture	9.1	9.7	9.8	10.1	10.0
Households	39.8	40.5	40.8	42.0	43.0
Commerce and services	3.0	3.1	3.1	3.3	3.5
Total	319.1	363.3	369.5	394.3	416.9
Share of NAP sectors in the total	68.0%	70.3%	70.3%	69.5%	68.6%

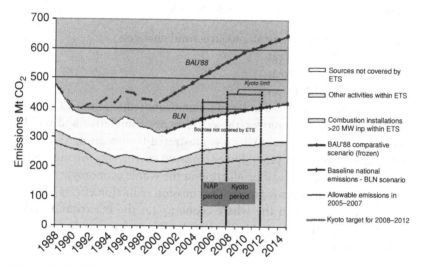

Figure 12.4. CO$_2$ emissions in both Polish scenarios versus the Kyoto Protocol limit.

to the Kyoto obligation was calculated based on emission projection trends in the BLN scenario.

Comparing emission results in scenarios BLN and BAU'88 reveals the following CO$_2$ emission decrease as a result of previous

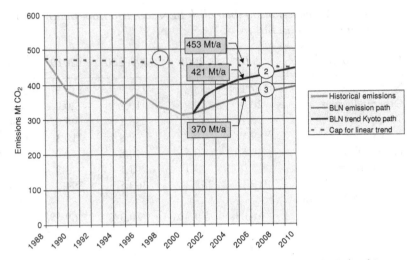

Figure 12.5. Different paths to achieve the Kyoto CO_2 target in Poland (numbers represent average emissions in 2005–2007).

technological improvement and structural changes:

- in 2005 – 142 Mt/year
- in 2010 – 196 Mt/year
- in 2015 – 230 Mt/year.

Economic changes had induced a significant reduction of CO_2 emissions in the past. Moreover, these effects are cumulative and will be prolonged into the future which is illustrated by increasing the calculated effects in future years (lower emission coefficients are multiplied by higher future levels of production in the whole economy).

The above numbers represent the emission reduction effects of early actions undertaken in the whole economy (at the macro and micro levels). It is worth noting that these early CO_2 emission reductions are giving significant cumulative climate protection benefits each year.

3.4 ETS emission limit

Comparison of the BLN scenario projection with the Kyoto limit has shown that emission projections are located under the limit. But, because of significant uncertainty of assumptions, other future emission paths are considered possible. Figure 12.5 shows three different paths starting from 1989 emissions: linear decrease from the 1988 emission to the Kyoto limit (1), the emission path of the BLN scenario (3) and path (2), which secures the trends of BLN while aiming at the Kyoto Protocol limit in 2008–2012.

Because of uncertainties related to the BLN projections, real emissions were expected to lie between path (3) and path (2). Projections regarding emissions from ETS sectors are characterised by uncertainty related to main assumptions such as GDP growth, changes in production and emission intensity. Additional uncertainty is a result of the changes of emission monitoring methodology, issued with the ETS Directive. The great uncertainty surrounding BLN emission projections led to the decision to calculate the emission cap for the ETS sectors based on path (2), which describes maximum allowable emissions when considering the limit established in the Kyoto Protocol and emission projection trends in the BLN scenario. In terms of quantity of emissions, path (2) is about 30 Mt per year lower than the linear reduction path (1).

The maximum level of allocation for ETS sectors was calculated by multiplying the emissions from the chosen path of Kyoto Protocol fulfilment by the ETS sectors' share. The last value was accepted conservatively at the 2001 level.

Total allocation of allowances in ETS sectors = 420.9 Mt × 0.68 = 286.2 Mt/year (as average in period 2005–2007).

3.5 EC decision on the Polish cap

The emission projection presented by Poland was not accepted by the Commission according to its decision dated 8 March 2005: 'total quantity ... is inconsistent with the Commission's most recent assessment made pursuant to Decision 280/2004/EC.' The Commission used the following arguments:

This additional information records the actual carbon dioxide emissions of Poland in 2002 as being 308.3 Mt CO_2eq, the total emissions intensity of gross domestic product (GDP), expressed in units of carbon dioxide emissions per unit of GDP, as decreasing by 4.8% between 2002 and the period 2005–2007, and the total GDP growth in Poland as increasing by 20% between 2002 and the period 2005–2007. Multiplying the 2002 carbon dioxide emissions of Poland by projected GDP growth and projected emissions intensity improvements for each of the years 2005, 2006 and 2007, and then averaging the results, means that the average annual total carbon dioxide emissions of Poland in the period 2005–2007 would be 351.6 Mt

CO_2eq. Applying the share of the trading sector's carbon dioxide emissions in the total carbon dioxide emissions as provided In the plan (68%), the maximum average annual emissions of the trading sector during the same period should be 239.1 Mt CO_2eq.

This decision did not take into consideration that: (a) the year 2002 was characterised by a temporary drop in electricity production (99.1% compared to 2001) while the next year showed dynamic growth (105.2% in 2003 compared to 2002); (b) share of ETS sectors in total CO_2 emission in years 2005–2007 was expected to reach 70.3% according to BLN scenario; (c) uncertainty related to GDP growth and effects of changes in emission monitoring made possible an emission increase above BLN projection.

After the Commission's decision, two factors were crucial in making Poland accept the modifications. A key role was played by updated data, which proved to be lower than the assumed level of production in several sectors, including the steel and power industries. Another reason was of a political nature and was related to the fact that the government was not interested in creating conflicts with the Commission ahead of the new parliamentary election.

4 The micro decision concerning distribution to installations

4.1 Collection and verification of data on ETS installations

The process of data collection was organised very carefully with special focus on completeness and reliability of the data:

- data collected on special forms
- mainly via the Internet (direct input into database)
- the process supported by special WWW portal and consultation team
- more than twenty consultation meetings with industry branch organisations
- data on early actions and their effects collected
- wide process of data validation and refining
- all data stored and structured in special multi-access Internet database
- installations grouped according to different classifications (Annex I, sectoral classification)
- sum of sectoral emissions compared with results of top-down analysis.

As discussed above, Table 12.1 compares the final results of bottom-up data collection with top-down analysis. The difference between the emissions estimated with these two approaches was about 2%, indicating that the data collected at the installation level were reliable and could be used to calculate allocation at the sectoral and installation levels.

4.2 Calculation of CO_2 base emission needs

The procedure applied to calculate expected CO_2 emission needs in 2005–2007 was the same in each of the fourteen sectors analysed:

- Step 1 – base emission in each sector was calculated, defined as a sum of the installations' average emissions over 1999–2002, excluding the year with the lowest emission. This definition was taken in order to filter out situations of temporal drop in an installation's production.
- Step 2 – activity change and CO_2 intensity change indexes were estimated.
- Step 3 – these were multiplied to give the CO_2 emission change index of each sector that was to reflect expected changes between base emissions and the first ETS commitment period.
- Step 4 – the CO_2 emission change index was multiplied by the base CO_2 emission to give the expected CO_2 emission of each sector in 2005–2007.

Activity change indexes were calculated separately for each sector (Table 12.5). Due to the fact that energy production trends are more stable in time and depend on the development of the whole economy, the model set described in Figure 12.1 was used for the electricity and heat production sectors.

Activity changes for non-energy production sectors were defined based on results of bottom-up, market-oriented analysis. Production growth in these sectors is more dynamic and could reach yearly rates higher than 20% (see Figure 12.3). Future activities estimates were based on the current sector situation, market conditions and assessment of development potential. Wide consultations with sectors were performed.

Activity change indexes are within a relatively wide range of values, reflecting real differences in the development potential of different sectors. For example, the expected dynamics of centralised heat production is much smaller than that of electricity generation. The

Table 12.5 *Main indexes that influence CO_2 emissions in the Polish ETS sectors in 2005–2007*

		Change in sectoral quantities in 2005–2007 with respect to 2001 (2001 = 100%)		
Number	Sector	Activity change index (%)	CO_2 intensity change index (%)	CO_2 emission change index (%)
1	Public power plants	118.5	99.7	118.1
2	Public heat and power cogeneration plants	112.1	99.7	111.8
3	Public heat plants	109.0	99.7	108.7
4	Mineral oil refineries	129.0	101.6[a]	131.1
5	Coking plants	133.7	100.0	133.7
6	Iron and steel industry			132.0
	Sinter	150.7	100.0	150.7
	Pig iron	129.4	100.0	129.4
	Steel	140.8	101.1[b]	142.3
7	Cement plants	137.8	100.0	137.8
8	Lime production plants	130.8	105.4[c]	137.8
9	Glass industry	135.0	100.0	135.0
10	Ceramic industry	124.4	100.0	124.4
11	Paper industry			122.7
	Pulp	115.5	100.0	115.5
	Paper	132.3	100.0	132.3
12	Sugar industry	139.3	100.0	139.3
13	Chemical industry	126.7	99.5	127.3
14	Other industries	117.9	99.5	117.3

[a] This increase is caused by deeper mineral oil processing due to new environmental requirements.
[b] This increase is caused by change of technology and product lines.
[c] This increase is caused by higher utilisation of existing installations which are less efficient and are set in operation only with higher demand for products.
Source: Polish National Allocation Plan 2005–2007.

relatively high growth in the coking plants or in the iron and steel industry reflect the advantageous situation of those sectors due to the currently observed demand increase in these markets.

As concerns CO_2 emission intensity indicators, their values are close to 100%. It reflects the fact that existing production technologies were

Table 12.6 *Projected emission needs in the Polish ETS sectors in 2005–2007*

Sector	Baseline CO_2 emission (kt/year)	CO_2 emission change index (%)	CO_2 emission (2005–2007) (kt/year)
Public power plants	116,552	118.1	137,648
Public heat and power cogeneration plants	37,327	111.8	41,731
Public heat plants	12,690	108.7	13,794
Mineral oil refineries	6,834	131.1	8,959
Coking plants	3,548	133.7	4,744
Iron and steel industry	13,670	132.0	18,045
Cement plants	9,406	137.8	12,962
Lime production plants	1,939	137.8	2,673
Glass industry	1,400	135.0	1,890
Ceramic industry	1,253	124.4	1,559
Paper industry	2,284	122.7	2,803
Sugar industry	2,176	139.3	3,031
Chemical industry	6,086	127.3	7,749
Other industries	4,605	117.3	5,403
Total	219,769	119.7	262,991

Source: Polish National Allocation Plan for 2005–2007; calculations by EnergSys.

thoroughly modernised in previous years and that there is only limited time available before 2005–2007.

In some sectors (power generation and refineries) new environmental requirements will stimulate the growth of energy consumption and CO_2 emission (e.g. new DESOX installations).

Multiplying the CO_2 emission change indexes by base emissions created the CO_2 emission needs in each of the sectors analysed. The results are shown in Table 12.6. Emission needs in 2005–2007 are about 20% higher than base emissions from 1999–2002.

4.3 Early action and cogeneration bonuses

The Polish NAP also proposed early action and cogeneration bonuses according to Annex III criteria 7 and 8. They were calculated based

Table 12.7 *Results of sectoral allocation: annual average (kt/year) for 2005–2007*

Sectors	Total allocation	Including				
		Base allocation	New entrants reserve	Early action bonus	Cogeneration bonus	Reserve for remaining purposes
Public power plants	143,627	136,774	874	5,963	16	
Public heat and power cogeneration plants	47,935	40,985	747	3,141	3,063	
Public heat plants	14,345	13,540	254	549	2	
Mineral oil refineries	10,204	8,822	137	735	511	
Coking plants	5,255	4,602	142	483	28	
Iron and steel industry	18,967	17,772	273	917	4	
Cement plants	13,704	12,680	282	742	0	
Lime production plants	2,789	2,615	58	116	0	
Glass industry	2,165	1,848	42	276	0	
Ceramic industry	1,683	1,521	38	125	0	
Paper industry	3,072	2,757	46	144	126	
Sugar industry	3,142	2,988	44	92	19	
Chemical industry	8,070	7,567	183	78	242	
Other industries	5,713	5,265	138	310	0	
Total	280,672	259,736	3,256	13,669	4,012	5,513
Total allocation		286,185				

Source: Polish National Allocation Plan for 2005–2007; calculations by EnergSys.

on real and permanent CO_2 emission reduction effects achieved in installations through early actions or cogeneration production.

4.3.1 Early action (EA) bonus

- Early action bonus is equal to 75% of the yearly CO_2 emission reduction effect achieved through permanent improvements in production technology, proved by the operator and positively verified by the comparison of CO_2 emission intensity factors in 1999–2002 and in 1989 (or the first year of operation).
- Only investment projects were accepted as early action measures.

4.3.2 Cogeneration (CHP) bonus

- CHP bonus is equal to 50% of the yearly CO_2 emission reduction, achieved by the production of electricity and heat in cogeneration.
- CO_2 emission reduction effects in cogeneration production were calculated by adapting formulas from Directive 2004/8/EC.

4.4 Allocation at sectoral level

Calculated emission needs for years 2005–2007 in sectors were split into:

(1) *New entrants budget* – for installations entering ETS, the emission needs were estimated as a share of base CO_2 emissions in 1999–2001 (values from 0.75% in the electricity sector to 4% in coking plants).

(2) *Base allocation budget* – for installations covered by the NAP list, equal to calculated emission in 2005–2007 minus new entrants budget.

Sectoral budgets were then increased by bonus budgets:

- *Early action bonus budget* – sum of early action bonuses for all installations in the sector.
- *Cogeneration (CHP) bonus budget* – Sum of cogeneration bonuses for all installations in the sector.

Table 12.7 summarises the results of the sectoral allocation. The new entrants reserve was calculated in two steps: (1) calculation in each sector, (2) aggregation into one budget for all sectors. The *reserve for remaining purposes* will be described later. The bottom-up allowance budget is equal to the limit calculated through the top-down approach.

This consistency was achieved by justifying factors for early action and cogeneration bonuses.

4.5 *Allocation at installation level*

The allocation at installation level in each sector consisted of three main elements:
- Allocation of *base budget*
- Allocation of *early action budget*
- Allocation of *cogeneration budget.*

Bonus budgets were allocated in the same way in all sectors. Allocation was proportional to CO_2 emission reduction effects. The method of base budget allocation differs across sectors as a result of their specifics and consultation process.

4.5.1 Allocation of base budget

The allocation of the base budget was split into two steps:

(1) *Individual allocation* – to 'developing' installations that require individual approaches to calculate their quantity of allowances.

(2) *Standard allocation* – to all other installations.

The standard allocation was based on data describing the situation of each installation until 2002 (2003 in case of power plants). The allowances from the new entrants reserve were to be granted for installations that increase their capacity after the start of ETS in 2005. During the data collection and consultation process a group of installations reported that after 2002 they either undertook investments leading to a permanent increase of production capacity or took actions leading to a much higher utilisation of production capacity (e.g. new contracts for centralised heat delivery). It was decided to treat such installations on an individual basis:

- the quantities of allowances to be allocated were calculated on a base of verified production and emission plans (the same way as planned for new entrants)
- the allowances granted during individual allocation were extracted from the base budget and were not used in standard allocation.

Individual allocation covered nearly forty installations. It is worth noting that higher growth rates under individual allocation meant fewer allowances to other installations in standard allocation. That is why results of individual allocation were controlled by operators of other installations and were a subject of discussion and acceptation during the consultation process with the sectors.

4.5.2 Grandfathering and other methods

All the installations, with the exception of nearly forty installations that were treated individually, were subject to standard allocation. Given the different situations in the different sectors the allocation methods were allowed to differ across sectors. The grandfathering method used as a reference was applied in ten sectors. It allocated allowances proportional to the base emission of each installation.

In four sectors different methods were approved after consultation with the sectors concerned:

- cement sector – allocation proportional to projected production
- sugar sector – proportional to production in 2003
- coke sector – proportional to projected production capacities
- public power plants – allocation proportional to weighted values of three factors: base emission, electrical capacity in 2003 and maximum utilisation rate in years 1999–2002.

4.6 *Summary of methodology and results*

Let us briefly summarise the main steps of the NAP preparation:

- *Top-down analysis*
    Step 1 – Total quantity of allowances in ETS
- *Bottom-up analysis* (separately in each sector):
    Step 2 – Total CO_2 emission needs in 2005–2007
    Step 3 – Split emissions between existing and new installations
    Step 4 – Early action and cogeneration bonuses
    Step 5 – Budget for unidentified installations and other needs.

Figure 12.6 represents Poland's NAP preparation process and highlights the main budgets aggregated at the level of all ETS sectors.

After the European Commission's decision the two stages of allocation (sector and installation) were preserved in Poland's NAP preparation. The ministry tried to achieve an agreement first on new sector limits and then on new allocations at installation level. The main modifications at sector level compared to the first NAP related to:

- changes in base emissions (additional data from 2003 and 2004 were collected)
- changes in growth rates (in some sectors, e.g. steel industry)

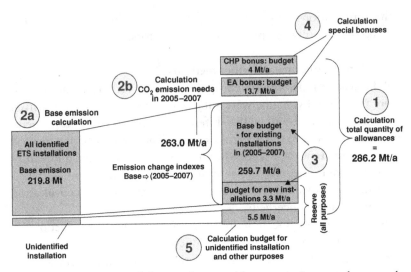

Figure 12.6. Summary of the top-down and bottom-up integrated approach to the Polish NAP preparation.

- liquidation of early action and cogeneration bonuses
- liquidation of reserve for other purposes.

The general rules for allocation at the installation level were the same as in the first NAP. In some sectors, however, including the power plant sector, the new methods based on new data collected in 2003 and 2004 were applied.

In the opinion of the interested parties the proposals of the ministry after the Commission's decision were not based on a clear methodology. New proposals were the result of a negotiation process and the final solution was the result of a (still not complete) consensus among the different actors enforced by the fear that not implementing the NAP would be worse than accepting the current version.

5 Issues of coordination and harmonisation

In the Polish view the most essential point is the harmonisation of the EU ETS design and the NAP preparation rules with economic and social goals, environmental policy and international obligations. The main criterion of proper harmonisation is the ability to generate a positive balance of benefits and costs in each country as a result of ETS implementation. In the Polish case, the basic expectation was that the EU ETS as a tool supporting the fulfilment of the Kyoto obligations would not make the domestic economy and employment situation worse than the situation resulting from the implementation of the Kyoto obligations using alternative instruments. During the NAP preparation several serious doubts were raised. Some of them are described below.

5.1 Harmonisation of EU ETS with social and economic goals

5.1.1 Increase of economic competitiveness and effectiveness

The economies of new EU Member States are characterised by the dynamic changes of sectoral GDP structure and by the market shares of the individual producers. To allow economic improvement the best companies should have the opportunity to increase their production and market share and to displace the less competitive ones. The implementation of the EU ETS according to present rules hinders this process. The allocation of allowances based on historical production or emission creates an impediment for the development of the most effective and competitive companies with large development potential. Within ETS the weaker, less competitive companies gain an extra chance of defending their previous market position. Such an effect could hinder improvement of the whole economy, particularly in countries with high dynamics of economic changes. New EU members belong to that group, because of the high intensity of modernisation and investment processes, rapid effectiveness improvement and dynamic increase in exports. These processes are additionally stimulated by significant changes in legal and market conditions due to EU accession.

5.1.2 Energy market liberalisation

According to an EU policy recommendation, Poland is strengthening the liberalisation of its power market. Advanced work on the elimination of long-term power supply contracts is being carried out, with the aim of increasing competition among power producers. The limitation of long-term contracts and other market processes stimulates changes in the market shares of different producers. Administrative allowance allocation without the possibility of *ex-post* adjustment would lead to the situation where market forces are replaced by administration decisions. In the case of CO_2 emission, as opposed to the emission of SO_2, power plants have very limited possibilities for rapid emission reduction by changes in technology.

In case of power-generating installations, the allocation of allowances at a level lower than their market development potential creates in practice a strong barrier to development. The purchase of additional allowances to enable higher production in coal-fired power plants means the growth of marginal production costs by more than 50% and cannot be compensated by the higher efficiency of that producer.

5.1.3 Stimulation of new investment
The effectiveness of an emission trading system is mainly based on four factors: (1) long-term emission targets, (2) clear and stable rules, (3) efficient enforcement of rules and (4) availability of emission reduction technologies. The fulfilment of the above conditions contributes to increase investment in emission reduction.

The EU ETS seems to have some weak points in the above-mentioned aspects. The current legally binding emission limits expire in 2012. How to transform the national GHG emission limits into ETS caps is not clearly defined, particularly in countries with emissions under the Kyoto limits. Even some most important rules of system functioning are not clearly defined for the long term. Major changes (e.g. linking directive (European Community 2004)) are being implemented after the launch of the system. The availability of reduction technologies is limited, and using new ones often has serious consequences for the whole domestic economy. In Poland more than 70% of ETS emissions comes from power plants. Significant emission reduction is possible mainly by replacing newly modernised coal-fired power plants with new ones fired with natural gas. Such an option is very costly and intensifies Polish dependence on Russian fuel.

The current implementation of the EU ETS seems to complicate rational long-term investment decisions. The consequences of any current decision on technology choice for a new power plant would be felt in years 2030–2040. In situations where the longer perspective of emission trading and its rules are not fully clarified, the result could be the slowing down of new investments instead of a revival, while more precise 'rules of the game' are awaited.

5.2 Harmonisation of the NAP with technical, market and environmental conditions

5.2.1 Methodology of monitoring CO_2 emissions

Practically each NAP was prepared based on emission data, which were not in accordance with EU ETS rules (European Commission 2004). The implementation of a new monitoring method requires more than one year, while official guidelines were issued a few months before the compulsory date for NAP submission.

During the work on the Polish NAP, the previous inventories were found to have significantly underestimated CO_2 emissions from coking plants compared with the results of the application of the mass balance method. The feedback from companies (not only in the coking sector) indicates that following the implementation of the methodology consistent with the Commission guidelines, higher CO_2 emission factors may emerge, compared with those used during the NAP preparation. This means that the changes in the monitoring methods may cause changes in the emission figures regardless of changes in real physical volumes.

In the Polish NAP, the possible increase of emissions following changes in the monitoring methodology was estimated at 10–12 Mt per year. It was one of the reasons for proposing allocation above the projection level. The Commission did not agree to this risk-compensation method, and moreover rejected the proposal of using *ex-post* allocation adjustment.

5.2.2 Competition policy: non-discrimination between companies

Non-discrimination between companies and sectors was an important criterion of NAP acceptance defined in Annex III of the Directive. The interpretation of this note in the Commission's guidance (European Commission 2003) was very general, however, and quoted Treaty notations (Articles 87 and 88 of the Treaty).

The interpretation of fair competition rules used by the Commission during NAP assessment seems not to have considered the legal nature of the national GHG emission limits. Within the framework of the Kyoto Protocol agreement and later the Burden Sharing Agreement (BSA) each EU country has a national cap of GHG emission defined for the period 2008–2012. The legal implementation of these requirements in the EU could have been easily done by applying new GHG emission limits in the so-called NEC Directive (European Community 2001). Fair competition under such obligations relies on each country being obliged to fulfil its obligations. The cost of fulfilling the obligations should be charged in particular to the operators of installations emitting GHGs according to the 'polluter pays' principle.

The EU ETS as an instrument supporting the fulfilment of the Kyoto commitments should not disturb the competitive relations created by the introduction of a hypothetical national CO_2 emissions ceiling. In the Polish case, however, the implementation of the EU ETS worsens the competitiveness of the Polish sectors, compared to the situation where each country itself fulfils its GHG emission target.

5.2.3 Accordance with climate protection policy

The Kyoto Protocol was signed years before the implementation of the EU ETS Directive. For years some countries have carried out their activities with the purpose of limiting CO_2 emissions. From the climate protection viewpoint, it is preferable to reduce GHG emissions as soon as possible, because the effects of such reductions are cumulative over longer periods of time. Climate protection policy should thus encourage earlier emission reduction. The harmonisation of the Directive with rational climate policy rules should prefer early emission reduction. This was not fully respected in the case of the EU ETS. The Commission did not take into account the effects of earlier activities during the determination of the national emission caps. The installations and sectors that took early action are thus treated in the same way as the installations and sectors which did not make the effort or whose efforts were insufficient for the fulfilment of national obligations.

5.3 Harmonisation with the EC and other Member States

In spite of critical remarks concerning the ETS Directive and the NAP preparation rules, the Directive and EC documents were important elements of the NAP preparation in Poland. When several sectors tried to clarify the definition of 'installation', the official EC documents and the opinions of experts from the Commission helped to explain the problem.

Beyond the official EC documents an important role in NAP preparation was played by information from other Member States and

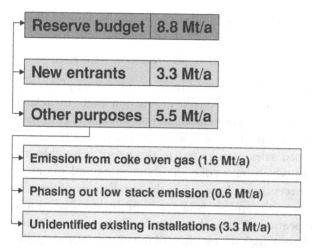

Figure 12.7. The structure of the reserve in the Polish NAP proposal.

outcomes from expert groups working under the EC auspices, mainly from the Working Group 3 (WG3) 'emission trading' group.

Reports from WG3 were very useful for taking decisions on the definition of 'combustion installation' (level 2 was chosen). Information from other Member States was also analysed and considered in deciding the general approach. After submitting a first group of NAPs it was clear that other countries were trying to avoid high pressure on emission reduction under ETS.

6 Issues deserving special mention

6.1 Reserve for new installations

The total reserve consists of two main parts (see Figure 12.7):
- New entrants reserve (3.3 Mt/year)
- Reserve for other purposes. (5.5 Mt/year).

The reserve for other purposes consists of three parts, which are described below.

6.1.1 The reserve for emissions from coke oven gas

Emissions of CO_2 from coke plants come mainly from the combustion of coke gas, which is a by-product of the coking process. Part of this gas is sold and burned by external users off the site of the coking plant. CO_2 emissions in coking plants depend on the amount of gas sold:

gas used in the coking plant = gas produced − gas sold.

The allocation of CO_2 emission allowances for coking plants was calculated on the basis of projected coke production levels, the related projected amounts of coke gas produced and the amount of coke oven gas sold, which was assumed at the average level of years 1999–2002. Lower than assumed sales of coke gas in the future (e.g. when the cogeneration plant which buys coke gas shifts to natural gas as its fuel) will make it necessary to burn gas in the installations of the coking plant, generating CO_2 emissions in excess of the levels taken into account in the allocation. A special reserve of allowances was therefore established, from which the existing coking plants were to receive additional allowances when the sales of coke gas fall below the level in 1999–2002. The quantity of the reserve (1.6 Mt/year) reflects the average coke gas sales in 1999–2002 recalculated to CO_2 emissions from its combustion.

6.1.2 The reserve for the liquidation of 'low' emission sources

This reserve was earmarked for large district heat producers. It was created to cover their additional CO_2 emissions caused by the takeover of district heat supplies in the place of small and inefficient boiler-houses. Such actions are highly beneficial for the local environment and climate protection and this is why they were supported by the ETS.

6.1.3 The reserve for unidentified installations

The reasons for the establishment of the reserve for unidentified installations were as follows:

- the quantity of allowances allocated to each sector depends on the base emission levels, calculated from bottom-up data
- the process of collecting data was carried out without the relevant legal basis, on a voluntary basis.

For these reasons, the government expected that some installations that were not included in the NAP list might join the ETS in the future, particularly after the adoption of relevant legal obligations. The applications from such installations will cause the growth of base emissions and will thereby increase the allocation to a given sector.

The government proposed that this reserve be maintained only until the middle of the first annual reporting period.

The unused allowances from all parts of the reserve for other purposes were planned to be cancelled before the end of the period 2005–2007, thus not increasing the total emission cap.

7 Concluding comments

7.1 Main conclusions

(1) The main difficulties in preparation of the Polish NAP were caused by the fact that the EU ETS and NAP guidance were designed to support fulfilment of the common Kyoto limit of EU15, an unsuitable situation for new accession countries. These countries have a different economic situation and are below the Kyoto limit due to significant emission reduction in the past. From the new Member States' perspective some elements of the EU ETS seem to be not quite fair or too risky, particularly with respect to such issues as:

- current emission level as a basis for further allocation
- no influence of early action on the total number of allowances in NAP
- allocation based on projection, not on Kyoto limits (cap)
- rules regarding the verification projections method not clearly stated
- controversial interpretation of the non-discrimination rule within the competition policy
- no instruments for risk compensation (regarding uncertainty of projections and changes in monitoring methodology).

(2) Preparing the NAP proposal in Poland could be compared to wearing an ill-fitted suit. It was necessary to choose between strict adjustment to the European Commission's rules (with possible adverse consequences for the Polish economy) and their reinterpretation taking into consideration the different situation of Poland.

(3) One of the most important issues is that implementing the ETS in Poland generates no cost savings in CO_2 emission reduction, because no carbon restriction programme is needed to comply with

the Kyoto Protocol in the absence of ETS. Once the high transaction costs of implementing the ETS (emission monitoring and changes in accounting rules, market behaviour, operational planning, strategy planning etc.) are taken into account, a cap that is equal to or lower than projected emissions will have the effect of creating a negative balance of benefits and costs.

(4) In September 2004 the complete NAP proposal covering allocation at sectoral and installation level was prepared and submitted officially to the European Commission. The total number of allowances of 286.2 Mt/year was built from emission projection and additional allocation reflecting early action and cogeneration emission reduction effects.

(5) In a decision taken in March 2005 the Commission accepted the cap level of 239 Mt/year which is 16.5% below the Polish proposal. This decision was based on the Commission's revision of emission projections and its refusal to accept the allocation of additional allowances above the projections to assign benefits for significant early action and cogeneration production and to compensate main risk factors.

(6) A decision on reducing the quantity of allowances by 16.5% was difficult to accept because of the common conviction of industry representatives in Poland regarding the restrictiveness of this decision. Poland's participation in the EU ETS under such a cap seems to lead to a negative balance of benefits and costs compared to the situation of separate compliance with the Kyoto Protocol.

(7) Based on the Commission's decision, the Ministry of Environment prepared several NAP versions and adopted the final decision by issuing the NAP decree on 27 December 2005. It has still not been accepted by the Commission and the new Polish government. The expected negative economic balance of implementing ETS makes the further work on NAP and the implementation of EU ETS harder.

7.2 Technical and political aspects

(1) Most of the Commission's recommendations had a positive influence on the technical feasibility of NAP preparation. The

interpretations regarding the definitions of installation and combustion installation (information published after the meetings of the WG3 group) were the most useful.

(2) The process of data collection at the installation level could be positively evaluated. It reached 2% concordance with evaluation at the macro level. The software tools used, formal procedures, and wide support of the process by detailed instructions, consultations, meetings and the WWW proved their usefulness.

(3) The integrated top-down and bottom-up approach allowed the development of a good picture of the issue from both macro and micro perspectives at the same time. It caused a better understanding of the problem and made discussions between different actors easier.

(4) The individual approach to each sector in respect to emission projections and the allocation method could be positively assessed. Significant differences in production and emission dynamics made implementing the same factor for different sectors controversial. It was also useful to have the possibility of using different allocation methods in different sectors. The power sector is a good example to illustrate this need. Testing a grandfathering method in this sector showed that some installations had been allocated above their real production capacity, while others were allocated at a very low level of capacity utilisation.

(5) After the Commission's decision on the total cap the additional work on the NAP developed in Poland were less clear in their methodological aspects and more focused on negotiating. The large number of iterations suggests that such an approach is not quite effective, particularly in a situation where a large group of the companies involved is not satisfied with the general level of allocation.

References

European Commission 2004. 'Decision of 29/01/2004 establishing guidelines for the monitoring and reporting of greenhouse gas emissions pursuant to Directive 2003/87/EC of the European Parliament and of the Council'.

European Community 2001. 'Directive 2001/81/EC of the European Parliament and of the Council of 23 October 2001 on national emission ceilings for certain atmospheric Pollutants', *OJ* L 309, 22, 27 November 2001.

European Community 2004. 'Directive 2004/01/EC of the European Parliament and of the Council of 27 October 2004 amending Directive 2003/87/EC establishing a scheme for greenhouse gas emission allowance trading within the Community, in respect of the Kyoto Protocol's project mechanisms', *OJ* L 338, 18, 13 November 2004.

PART III

Concluding remarks and background material

13 | *Unifying themes*

A. DENNY ELLERMAN, BARBARA K. BUCHNER AND CARLO CARRARO

The preceding chapters describe and analyse the NAP process in ten representative Member States of the European Union. It is now time to pull together the unifying themes that emerge from these individual accounts. In this final chapter, the editors discuss the lessons that emerge from this experience and make some concluding comments on what seem to be more general principles informing the allocation process and on what are the global implications of the EU ETS. As any reader will have noticed, the diversity of experience among the Member States is considerable, so that it must be understood that these lessons and unifying themes are drawn from the experience of most of the Member States, not necessarily from all. For every lesson and each general principle, there is typically at least one Member State for which it does not apply, or it does so only weakly. Accordingly, as is true of all lessons and concluding comments, these will need to be applied carefully and some may not apply to future allocations in the EU ETS or to the circumstances surrounding future allocations of CO_2 rights by other countries. Still, it seems likely that most of the problems experienced by the Member States of the European Union will be encountered by others who follow this example and that the lessons and general principles drawn from them will be helpful.

These lessons and unifying observations are grouped in three categories: those concerning the conditions encountered, the processes employed and the actual choices.

1 Conditions encountered

1.1 Data availability limits allocation choices

The lack of data at the level of the installation was perhaps the biggest problem confronted in the allocation process by nearly all Member States. This came as a surprise to most people since all countries had developed reasonably good inventories of CO_2 emissions data. The problem was that the inventory data were developed from statistics of aggregate energy use and they did not extend to the level of the installation, which was the mandated recipient of the allowance allocations by the EU Directive. Since all Member States were operating under very tight time constraints in submitting NAPs, obtaining installation-level data became the first major hurdle that had to be cleared and the final allocation choices were strongly influenced by considerations of what data could be obtained within the available time.

The data problem existed regardless of the history of data collection or the extent of pre-existing energy or environmental regulation. Member States with a long history of energy and environmental regulation such as Germany and Sweden faced as large a problem as those with less, such as Spain and Italy, not to mention the accession countries of Eastern Europe. Some Member States had collected installation-level data for many of the facilities covered by the EU ETS, but the discrepancies between the earlier data and those submitted in the allocation process could be as much as 20% (see Zetterberg, 'Sweden'). The only Member State that did not face a significant data problem was Denmark, which had already established a CO_2 emissions trading scheme that included most of the emissions to be included in the EU ETS (see Lauge Pedersen, 'Denmark').

The problem of data availability was compounded by the lack of legal authority to collect the relevant data. When combined with the pressing deadlines for NAP submission, governments had little choice but to rely heavily on voluntary submissions from industry, while they also initiated action to acquire the requisite legal authority. The surprising thing is that the affected firms cooperated as fully and in as good faith as appears to have been the case. This cooperation may have reflected recognition of the ultimate power of the government to compel performance, but it is also true that the production of the requested data established a claim on the allowances being distributed and a failure to produce data would have resulted in no allocation to the installation, as well as other sanctions.

The limitations imposed by data availability had important consequences in ruling out certain baselines and types of allocation for which an a-priori preference might have existed. For instance, Germany had advocated that allocations be based on 1990 emissions (see Zapfel, 'European Commission'). This would have been in keeping with the

Kyoto Protocol and with the EU Burden Sharing Agreement (BSA) and it would have recognised 'early action.' It soon became evident, however, that data on installation-level emissions in 1990 were non-existent and, in 2003, irretrievable in any reliable or meaningful form. Some Member States with better data could choose baselines that extended as far back as 1998 (United Kingdom, Sweden, Denmark), but for most countries, the baseline or reference periods for allocation included only the most recent few years because these were the only years for which installation-level data could be easily retrieved. Consequently, baselines that would automatically recognise early action were infeasible. If any recognition was given to early action, it was the subject of special provisions for those who had the data and could make a convincing case. Among the ten Member States examined in this volume, only Germany, the Czech Republic, Hungary and arguably the United Kingdom (in the baseline and rationalisation rules) made such provision. The more general pattern was to disregard early action not only because of the data problems but also on account of the conceptual problem of distinguishing early action from emission reductions taken for other reasons (see Barry, 'Ireland').

While data availability limited allocation choices, a much noted by-product of the need to acquire installation-level data for allowance allocation was the resulting significant improvement in the quality of the data on emissions and energy use, as is stressed also by Atle Christer Christiansen (see Box 13.1).

1.2 Inclusion of small facilities is not worth it

The EU Emissions Trading Directive established a very low level of heat input (20 MW_{th}) as the threshold for inclusion in the ETS. If there is one refrain that arises from virtually every one of the ten NAP processes included in this volume, it is that the inclusion of small installations was not worth it. As noted in one contribution after another, a large proportion of CO_2 emissions originate from a small number of installations, while a very large number of the installations contribute only a small percentage of emissions. For instance, in the United Kingdom, 20% of the sites account for 94% of emissions and 80% of the sites contribute 6% of the emissions (see Harrison and Radov, 'United Kingdom'). Similar statistics are found in every Member State.

Box 13.1 The challenge of data availability during the allocation process

The establishment of the EU Emissions Trading Scheme is one of the most interesting chapters in the history of EU environmental law and policy. The size of the EU ETS on its own makes it the largest emissions trading system seen to date – covering over 11,000 installations in twenty-five countries. In addition, the links to reduction projects in developing countries ensure that the price of carbon in Europe in reflected in investment decisions around the world. The carbon market has also proved to be fairly liquid and prices have to a large extent responded to fundamentals and policy signals. Point Carbon's assessment of the system halfway through Phase 1 indicates that it is a qualified success.

However, there are some apparent flaws in the system, which hopefully will be rectified for Phase 2, starting in 2008. The verified emissions data for 2005 have shown that the market actually had a surplus of allowances in its first year of operation. While some of this surplus might have come through actual reductions, it can to a large extent be explained by generous allocations – going against the intention of the scheme.

This has occurred in part due to a lack of reliable information on historical emissions, intended to provide a starting point for allocations at 'the right level'. The gathering and processing of emissions data was an enormous task for most Member States when designing their Phase 1 National Allocation Plans, and there were limited possibilities to check the quality of the submitted data. The EU Commission faced the same problem in its assessment of the NAPs, even though it was able to cut off almost 300 Mt from the initial allocation plans.

The good news is that we now have reliable emissions data from installations in the scheme, providing an improved factual basis for the Phase 2 allocations. As some of the 'teething problems' can now be left behind, the EU ETS might – if the political will is there – be given the bite needed to ensure long-term CO_2 reductions in Europe.

Atle Christer Christiansen

The problems relating to the size threshold are two-fold. First, and most evident in this first allocation cycle, data requirements are installation-specific. Therefore, much of the data problem discussed above was created by the small size threshold. Second, the reporting and verification requirements will impose costs on small installations that are disproportionate to their emissions or the abatement that could be expected from them.

While the inclusion of small installations required more time and effort than would appear to be justified by their emissions or abatement potential, the alternatives are not obvious. The problem with any size threshold is that it has the potential to create a competitive disadvantage for installations covered and a perverse incentive to downsize in order to avoid regulation. And the higher the threshold, the greater these problems are likely to be. Perhaps a staged approach, whereby small installations were brought in later, would have reduced the initial data problems, but those problems are now solved and no longer at issue.

Reporting and verification has just started so that the extent of the burden is not fully known. Similar size thresholds in US systems have not resulted in transaction costs that have created a noticeable problem despite similar reporting procedures.[1] The only way around this burden would seem to be an upstream point of regulatory obligation – at the refinery, gas terminal or coal mine – that would result in a fuel price that included the price of carbon. This would have the same effects on abatement by small installations without the transaction costs involved in a downstream monitoring and reporting requirement.

Without a doubt, small installations will need to be part of a successful emissions trading scheme in the longer run. In the medium term, the obvious burden for small installations deriving from the high transaction costs in relation to relatively low environmental benefits needs to be eased. Yet, the ultimate solution presented above presumes that the system eventually becomes more comprehensive, covering other sectors also. In order to partially resolve the problem of small installations in the shorter term, a judicious opt-out provision might be a promising

---

[1] Small installations in the US SO_2 program are generally exempt from the requirement to install a Continuous Emissions Monitoring System and instead report emissions based on fuel use and engineering calculations.

way. Currently, the Directive foresees a temporary exclusion of installations only for the Phase 1 of the EU ETS. In order to avoid excessive transaction costs for small installations, the continuation of the opt-out possibility would be an attractive way to increase the efficiency of the EU ETS. As is the case with the Phase 1 opt-out provision, the exercise of this option could be coupled with a requirement of equivalent regulatory measures.

As a matter of fact, a number of small installations appear to be interested in continuing in the EU ETS for reasons related to the investments made for Phase 1, allowance allocations, or circumstances related to their advanced technologies or low energy intensity. Allowing installations below a certain threshold to opt out appears to be a better way to lower the burden for these facilities and to increase the scheme's positive participation incentives and its overall efficiency than attempting to set new thresholds, given the potential danger of creating some discriminatory effects.

2 The processes employed

2.1 Emitters are involved in allocation decisions

In all the Member States examined in this volume, except one, the allocation process can best be described as an extended dialogue between the government and industry. The involvement of industry in the process is not surprising given the data problem we have just described, but there were other factors as well. The Emissions Trading Directive mandated that at least 95% of the allowances in Member States be allocated to the installations that would be included in the scheme. Even had there been no data problems at the installation level, any democratically elected government would have considered it prudent to consult with the recipients, who were in addition well aware of the value of the endowments they were to receive. These two factors worked together to create an intense iterative process between the relevant parts of the Member State governments and the affected industry whereby data were collected, cross-checked and refined at the same time that distribution proposals were made, evaluated against the data and modified until a final NAP emerged. As noted in the comment by Hans Warmenhoven (see Box 13.2), this interactive process was a key factor in successfully completing the NAP process.

Box 13.2 The importance of an interactive process for allocation

The interesting thing about the allocation processes was that most of the countries entered into it as a technical exercise, even though they realised that creating an equitable system would be impossible. No one seemed willing to acknowledge that at the end of the day no technical solution would suffice and that a political solution would be necessary. This is strange because it should have been clear that subdividing billions of euros among a large number of companies would necessitate strong, high-level management of the process.

Most of the countries started off with a clear plan for developing the NAP; however this plan was typically frustrated by the frequent interactions with stakeholders. These interactions, mostly governed by different lobby groups, made the process more chaotic and less transparent.

A difference can be observed between those countries that planned the NAP process to be highly interactive with the stakeholders and those countries that took a more technocratic approach. For the first group, the process was slow because all the stakeholders had to be brought along. The second group could make faster progress in the beginning, but later their method had the disadvantage that, when the stakeholders realised what was going on, they where not well educated, more distrustful and more likely to resist the process fiercely.

Looking back, one can conclude that there are three success factors that should be taken into account in this kind of process:
- make sure that the process is actively governed on a high level
- develop a explicit process plan and communicate it to all stakeholders (and change it if necessary but then communicate the new plan)
- make sure there is a strong interaction with all stakeholders throughout the process.

Hans Warmenhoven

The government's role in the process was as much one of managing a process by which conflicting claims could be resolved as it was one of imposing any preconceived idea of how allowances should be distributed. The government was always the final arbiter of conflicts, but the actual exercise of this role was more the exception than the rule and it was always a last resort. On the part of industry, there was of course much lobbying, but the fixed total forced all players into a zero-sum game where a defensive concern about what competitors would receive became as important as offensive attempts to gain more for themselves.

Evidence of the government's role as organiser and arbiter of the process can be seen in several choices not taken. The 'pooling' option, which would have effectively delegated installation-level allocation to some industrial association, was never chosen. Similarly, in Spain, an early idea to have sector associations make installation-level allocations was set aside at an early date in favour of a process managed by the government (see del Río, 'Spain'). More generally, the frequency with which the word 'fair' appears in this volume in describing industry concerns indicates the extent to which the government's role was one of finding a reasonably equitable resolution of the conflicting claims that would permit a final NAP to emerge. The process was inevitably contentious, but it never broke down and there are at least as many comments on the cooperative aspects of the process as there are to the evident conflicts.

The government participants in this process were nearly always the environmental ministry in the lead with the ministry charged with economy or trade heavily involved. Sometimes the process started out as a more or less technical exercise within the environmental ministry, but the economics/trade ministry became heavily involved either as a means of obtaining the necessary data or at the instigation of industry. The relations between the environmental and economics/trade ministries could be contentious and even require resolution by the head of the government (see Matthes and Schafhausen, 'Germany'), but the relation was as often a cooperative one especially towards the end of the process when the prospect of confronting and persuading Brussels loomed. On the industry side, sector associations played a vital role in nearly every Member State both in obtaining the necessary historical data and in negotiating for the sector, although where few firms were involved and towards the end of the process the role of individual firms became greater (see Chmelik, 'Czech Republic').

Ireland is one exception to the active involvement of industry in the allocation process. Here, an ongoing investigation of an earlier scandal involving similar endowments led to the delegation of the task to the independent expert agency that issued environmental permits and was responsible for monitoring compliance with environmental regulations. The government retained the power to decide the total and the basic distribution principles upon the agency's recommendation after public consultation, but all the technical work was done within the agency with the help of consultants and some advisory groups that included industry representatives.

Two parties were noticeably absent from the distributive part of the NAP process, environmental non-governmental organisations (NGOs) and Brussels. Depending on the country, NGOs were either absent in any meaningful sense or they tended to focus on the total number of allowances. To the extent that they were concerned about how the total was to be split, it was to ensure that favoured activities, such as cogeneration or district heating, received favourable treatment. Perhaps the most notable absence from the debate on internal allocation, given the frequent calls for 'harmonisation', was the European Commission. Aside from suggesting how the internal distribution might be done in an 'informal' non-paper and performing a perfunctory review for state aid, the Commission stayed out of the controversies about allocating allowances within Member States in keeping with the subsidiarity principle. In its NAP decisions, the Commission fixed the overall amount but explicitly allowed for redistribution within that envelope in case of data improvements.

2.2 Projections played a major role despite their unreliability

In all Member States, projections of CO_2 emissions and the associated modelling played a large role in determining national and sector totals. Although the use of predictions is sure to involve some error and be subject to subtle gaming, their use was unavoidable and they had the merit of narrowing debate about projected emissions to the underlying assumptions and imposing some top-down discipline on expansive bottom-up claims.

At the national level, projections became necessary because no Member State wished to deviate far from expected emissions in deciding the total to allocate to installations. Business-as-usual (BAU) emissions were explicitly the constraint for Member States not facing a Kyoto constraint, as was the case for most of the accession countries in Eastern

Europe. But even for the EU15, for whom compliance with the targets of the Kyoto Protocol or the EU BSA pose more of a problem and for whom the 'Path to Kyoto' provided an alternative criterion, only a gently constraining total was chosen, as will be discussed shortly, and that criterion necessitated the use of projections to determine what emissions could be expected to be.

The second major use of projections in the EU ETS was in establishing sector totals. Most Member States chose to allocate the national total in a two-step process whereby the national total was broken down into sector totals, which were then split among the installations in each sector. For these sector allocations, the use of projections followed from the decision to endow (non-electric) industrial sectors with as many allowances as 'needed', as will be discussed more fully in a subsequent section. Moreover, since all sectors were not expected to grow at the same rate, it became necessary to develop projections for each sector.

The use of projections also led to modelling problems. At the national level, no model captured the trading sector exactly so that every Member State had to revise existing sector models to approximate more closely to the ETS sectors, and this in turn had to wait on the availability of data to define the baselines for those sectors. The heterogeneity of non-electric industrial production also led to a phenomenon of hyper-differentiation of sectors. Just as cement, steel or pulp and paper would not be expected to experience the same rates of growth and therefore of 'need', so it was that groups of installations within the each of the broad sectors did not expect to have similar rates of emissions growth because of the somewhat different products that they produced and other intra-sector differences among firms. This process of differentiation was carried furthest in the United Kingdom where the originally proposed fourteen sectors became fifty-two (see Harrison and Radov, 'United Kingdom').

The problem with the use of projections is that no projection will be accurate because of errors in expectation concerning important determinants of CO_2 emissions such as the rate of economic growth, relative energy prices (especially that between coal and natural gas), the ongoing rate of improvement in energy efficiency and other structural transformations in the economy that will either increase or decrease CO_2 emissions. When the totals are at or close to the projected total, prediction errors will have a much greater effect on the expected tightness or slackness in the constraint and on allowance prices. At best, projections provided a range of estimates of BAU emissions and the choice of a total implicitly involved estimates of the probability of over-allocation. Agreeing on a central value or even a range would have been hard enough for any of the EU15, but it was even harder for the East European economies that are undergoing a fundamental structural transformation. From the standpoint of the Commission, the problem was one of avoiding a national total that had a high probability of creating surplus allowances. Although Brussels reduced a number of the proposed caps by significant amounts, the likelihood of some excess allowances was not eliminated, especially in Eastern Europe.

Notwithstanding these problems, projections did serve some useful purposes. Most generally, they provided a form of top-down discipline by constraining aggregated, bottom-up estimates of 'need'. As noted by Istvan Bart ('Hungary'), 'a reasonable emissions forecast for a sector is by definition lower than the sum of the safety points that would satisfy the expectations of all the sector's players'. Projections also served to channel the debate concerning totals into arguments about the reasonableness of underlying assumptions and consistency with projections used for other purposes. The contributions concerning the Czech Republic and Poland provide especially good accounts of the use of projections for both of these more beneficial uses (see Chmelik, 'Czech Republic' and Jankowski, 'Poland').

2.3 Central coordination is important

A distinguishing feature of the allocation process in the EU ETS is the highly decentralised manner in which it was done. This characteristic is what could be expected of a multilateral system in which the constituent members retain significant elements of national sovereignty, not to mention one in which the principle of subsidiarity is enshrined in principle and practice. Nevertheless, the role of the centre was critical in arriving at the result that can be observed today. Indeed, it is hard to imagine how twenty-five nations could have succeeded in such a multinational enterprise without the central coordinating role played by the European Commission. Three aspects of this role are especially important.

The first and most visible role of the Commission was as agent for the whole in implementing a commonly agreed-upon policy. As such, it found itself in the unenviable, and somewhat unexpected, role of being the enforcer of scarcity, as well as the agent insisting upon certain rules (such as no *ex-post* adjustments) that promote an effective trading regime (see Zapfel, 'European Commission'). The Commission could insist upon these conditions because the Emissions Trading Directive granted it the power to review and to reject individual NAPs, but this power had to be exercised judiciously. This delegation of considerable power to the central agent also allowed Member State governments, perhaps disingenuously, to shift the blame for unpopular decisions to an external authority that represented some greater good and thereby to make it easier for individual Member States to take unpopular decisions.

A less visible but probably equally important role of the Commission was as educator and facilitator of the decisions that Member States had to take. The degree of familiarity with emissions trading varied greatly among the Member States and for most a quick learning process was required. Studies funded by the Commission and guidance documents served to share the Commission's technical expertise in emissions trading and to inform Member States of the options available to them. A number of mechanisms were set up, such as Working Group 3, to facilitate the exchange of information at the technical level and to allow those charged with implementation in the Member States to share experiences among themselves. Had the Commission not taken this active role as educator and facilitator, it is doubtful that the EU ETS could have succeeded given the ambitious schedule for implementation and the inexperience of most Member States with this regulatory instrument. One other result is a degree of 'soft' harmonisation that is obscured by the not infrequent calls for still greater harmonisation of one provision or another.

A final aspect of the role played by the Commission is the technical competence and political capability generally displayed in bringing the scheme to fruition (see Chmelik, 'Czech Republic'). Technical competence in understanding what trading systems required was evident as early as the Green Paper which in March 2000 first publicly suggested that emissions trading might be one of the instruments to be included in the European Climate Change Programme (European Commission 2000), and it continued to be displayed in later proposals, guidance documents and directives. A politically sensible approach was evident in the choice of an instrument that would not fall foul of the Community's unanimity rule (as had the earlier carbon tax proposal) but more generally in the minimalist approach in exercising its power to approve NAPs by focusing on only two issues, a total that was not overly generous and no *ex-post* adjustments. These two conditions ensured some degree of scarcity and that trading would be necessary for compliance. Finally, low-key, back-channel, informal consultations were heavily used to avoid confrontations and to allow the process to move forward. For instance, conditional approval and subsequent, technical changes to the NAPs avoided outright rejection, thereby sparing Member States the unwelcome news coverage concerning the widely supported European endeavour to comply with the Kyoto Protocol. Yet, the Commission's assessment process has also been criticised by some Member States as being too 'high level' and not involving enough technical expertise in the sense that decisions on the evaluation of the allocation plans were not always made by those who were familiar with the technical details of the different countries (see Barry, 'Ireland').

In a broader perspective, the Commission exercised the central coordination that the theoretical literature emphasises as necessary to correct for undersupplied goods and services in a decentralised market and to address distributional problems that can arise from uncoordinated, decentralised decisions. In this instance, the centre provided a large part of the educational services needed to obtain a smooth and timely implementation and it acted to ensure a reasonable degree of equity among Member States with respect to the burden that would be placed upon national industries included in the EU ETS. In this latter role, aspects of the BSA were effectively renegotiated to allow Member States with a greater deviation from the Path to Kyoto to adopt totals that were similar to those of other Member States not facing these problems. Finally, in their review of NAPs, the Commission also acted to ensure that individual provisions did not constitute unwarranted subsidy (i.e. state aid).

3 The choices made

3.1 Benchmarking is little used

In no aspect of the allocation process for the EU ETS was the disparity between advocacy and practice greater than for benchmarking.

Although not always well defined, benchmarking refers to an allocation in which allowances are distributed according to some common emission rate multiplied usually by historical output. The emission rate is often one associated with best available technology, but it could also be an average emission rate for the sector. The common feature is that installations having an emission rate higher than the standard will not receive more allowances, and those having an emission rate lower than the standard will not receive fewer allowances. As will be discussed in the next section, the basis for allocation was almost always recent emissions so that installations emitting more or less per unit of output received commensurately more or fewer allowances.

The failure to adopt benchmarking more widely was not because of a lack of trying (see especially Matthes and Schafhausen, 'Germany', Bart, 'Hungary' and Zapfel, 'European Commission'). Many benchmarks were proposed; but, every time one was tried, the resulting deviations of allocations from recent emissions at the installation level were too great to gain wide acceptance. This points to what is the biggest problem in applying benchmarks: source heterogeneity. If all sources were more or less alike, benchmarking would be easy; but in practice installations differ greatly even within the same sector. Moreover, these differences lie not so much in energy or emissions efficiency, but in the specific output produced by the installation. For instance, the energy and emissions associated with producing steel slab is not the same as what is required for finished rolls. Thus, two facilities that may seem alike in producing similar quantities of output measured in some common denominator, such as tonnes, may have very different emissions, not because one is producing more efficiently than the other, as is often implied in arguments for benchmarking, but because the products are different. An example of the extent to which output heterogeneity led to differentiation is provided by the Netherlands where 120 benchmarks were developed before the concept was abandoned (see Zapfel, 'European Commission').

Heterogeneity is not restricted to output; it can also affect inputs into a highly homogeneous product, such as electricity. A single, fuel-blind benchmark emission rate proved impossible to choose for existing facilities whenever the fuels used to generate electricity differed significantly in CO_2 content. Still other sources of heterogeneity exist as illustrated by the twenty-six separate benchmarks that were requested for the electric utility industry in Germany (see Matthes and Schafhausen, 'Germany').

Heterogeneity in emission sources can be overcome if there is a widely accepted, pre-existing standard that can be applied. A good example is provided by the US SO_2 cap-and-trade program to which the EU ETS is often compared. Installations in the SO_2 program received an allocation that was benchmarked to 1.2 pounds of SO_2 per million Btu (approximately 500 g/GJ) despite heterogeneity among affected sources that was comparable to that for CO_2 in the EU ETS. This emission rate is the same as the New Source Performance Standard (NSPS), adopted in the early 1970s as the best available control technology then available, and it has been applied since then with some modifications to all new generating units. By 1990, when the SO_2 program was adopted, the NSPS applied to about 60% of generation and it was an obvious benchmark. By comparison, nothing like this well-established standard for SO_2 emissions in the US exists for CO_2 emissions in Europe or elsewhere. Thus, although plausible CO_2 benchmarks could be and were proposed, none had the institutional precedent and legal force that made adoption of the NSPS benchmark feasible in the US SO_2 trading program.

A final explanation for the absence of benchmarking is the problem of data availability. The informational requirements for a benchmarked allocation are more demanding than an allocation based on past emissions. CO_2 molecules are uniform and it was a lot easier to collect this common single data point across the great heterogeneity of affected facilities than it would have been to classify the facilities according to product (and sub-product) and decide upon a benchmark for each. When time was short and voluntary cooperation was required to produce the necessary data, this consideration became important. In addition, competitive considerations might also make firms less willing to reveal output or input data than emissions.

Despite the limited use of benchmarking, it was used in some cases and these cases illustrate the reasons for the limited use of benchmarking. The most common use is for allocations to new entrants for whom there is, by definition, no prior emissions history. Although the criteria applied for allocations from the new entrants reserves are often imprecise, the usual practice is to scale the allocation to capacity or expected output and to apply some benchmark generally reflecting best available technology, which is an obvious if ill-defined norm for new installations. Benchmarking is also applied to some existing installations. In both Denmark and Italy, benchmarks were used for the electricity sector. In both cases, the benchmark was the number of allowances

ETS Totals

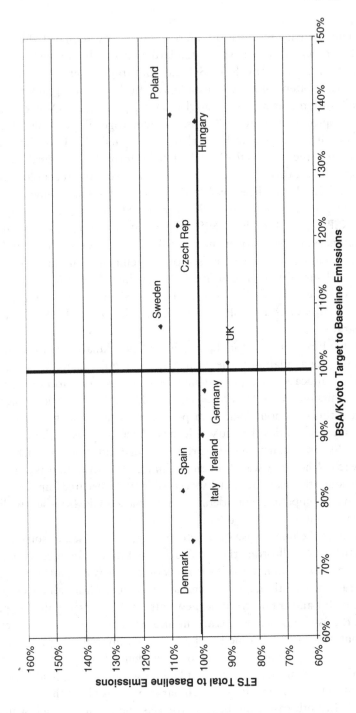

Figure 13.1. The relation of recent emissions to the ETS allocation and the Kyoto target.

remaining for the electric utility sector after allocation to the non-electric industrial sectors divided by baseline output of the electricity sector in Denmark and by expected output in Italy. In Denmark, the use of a benchmark for existing generation facilities was mandatory but it was made easier by the similarity of the generating fuel profile of the two main electric utilities and by the circumstance that the use of prior emissions would have punished facilities that reduced emissions under the earlier emissions trading program (see Lauge Pedersen, 'Denmark'). In Spain, combined cycle gas turbine (CCGT) units received a benchmarked allocation based on good practice, but these units were all recently built or under construction and therefore without complete baseline data. Also, certain industrial sectors received allocations proportional to capacity with an implicit benchmark of the average sector emission and utilisation rate because installation-level data on emissions were lacking. In Poland and Italy, industrial sectors were given a choice and in each country several opted for benchmarking. In all of these cases, benchmarking was chosen either because historical data were not available, sources were homogeneous or the sources in the affected industry could agree on a benchmark.

3.2 The main reference point for allocation is recent emissions

In the absence of an obvious or practical benchmark, recent emission levels became the basic reference point for the allocation process in the EU ETS. This was true both for the macro decision on the level of the cap and on the micro distribution of allowances to existing installations within individual Member States.

The macro level adherence to recent emissions is shown in Figure 13.1. The horizontal axis represents the relationship between the Kyoto or BSA target (for the EU15) for national emissions and 2003 emissions. The spread among the ten Member States is considerable. Poland and Hungary have a target that is about 40% higher than recent emissions, while Denmark has a BSA target that is about 26% below recent emissions. The vertical axis represents the relation between total allowances and recent or baseline emissions for the trading sectors of each economy. Here the spread is much reduced. Poland, the Czech Republic, Sweden, Spain and Denmark created totals that were from 3% to 13% above recent emissions, while Italy, Ireland, Germany and

Hungary were very close to recent emissions, and the United Kingdom was 10% below.[2]

The criteria for deciding total allocations by each Member State would suggest that countries for which the Kyoto/BSA target is greater than recent emissions could adopt a cap that would be somewhat higher, depending on projections of 2005–2007 emissions. However, for countries with a Kyoto/BSA target below recent emission levels, as is the case for five of the countries in this sample of ten, an ETS total lower than recent emissions (although not necessarily lower than projected emissions) would seem to be implied. As can be readily seen, there is no relation between the Path to Kyoto and the ETS totals. All are clustered close to recent emissions. Significant gaps from the Path to Kyoto are to be bridged by the intended purchase of Clean Development Mechanism (CDM) and Joint Implementation (JI) credits, as indicated in the NAPs of those Member States with Kyoto/BSA targets below recent emission levels.

While national and sector totals were typically based on recent or projected emissions, the distribution of the national total to installations could have been done on another basis such as capacity or past activity levels. This was done in a few instances, as noted in the section on benchmarking, but the nearly universal pattern was to allocate to installations on the basis of their share of baseline emissions. Where sector emissions were projected to increase, this could mean an allocation larger than baseline emissions, but it could as easily imply a smaller allocation often due to the subtraction of certain quantities from the national total for new entrants or special bonus provisions, such as for central heat and power (CHP), early action or auctioning. Basing the micro-level distribution on emissions was also dictated by data limitations, especially in the industrial sector where output or input data were either not available or not easy to collect due to the heterogeneity

---

[2] These relationships cannot be taken to indicate the relative stringency facing each of these countries. This depends upon expected growth which varies among the member states of the EU. Spain for instance is widely regarded as short in the EU ETS and it was so revealed in the 2005 emissions data (−6%), as were also the United Kingdom (−18%) and Ireland (−16%). Also, Denmark has highly variable annual emissions due to coal-fired electricity exports to the rest of Scandinavia, which depend on rainfall in Norway and Sweden. For instance, CO_2 emissions in 2002 and 2003 were 30.9 and 36.6 Mt, respectively. Denmark's ratio of 1.03 is calculated from the 1998–2003 average emissions for the ETS sectors.

of output. More generally, emission shares had the merit of recognising the heterogeneity of emissions sources whether the causes were differing products or earlier investments that had not anticipated a price being imposed on CO_2 emissions.

While recent emissions constituted the reference point for allocations, the baseline used to define recent emissions was not uniform among Member States. Data limitations dictated some choices, but variations in the definition of the baseline usually reflected other factors. Virtually all Member States chose an annual average of a multi-year baseline to avoid the idiosyncracies of any single year; however, that multi-year baseline ranged from three years in Germany and Spain to six years in the United Kingdom. Another variation was to adopt a drop-minimum rule, such as in the United Kingdom, to allow firms to eliminate an unrepresentative year from the baseline. In Spain, Denmark or the Czech Republic, the standard baseline average could be set aside if the most recent year (or average of the most recent two years in Spain) was greater than the standard baseline, and the more recent year(s) used as the allocation baseline for that installation. The United Kingdom also adopted a set of Baseline, Commissioning and Rationalisation Rules which had the effect of allowing further adjustments of the baseline for installations to reflect conditions that would more accurately reflect recent emissions (see Harrison and Radov, 'United Kingdom') The end result of all these variations is that the baseline shares do not represent the actual shares of any single recent year or years, but shares of an artificial baseline consisting of what could be considered an appropriate average level of recent emissions. The extreme version of such an artificial baseline occurred in Hungary where the baseline was negotiated individually with all large emitters and with groups of small emitters to create an aggregate that comprised each installation's negotiated level of recent CO_2 emissions (see Bart, 'Hungary').

3.3 Shortage was allocated to electricity generation

Most of the Member States constituting the EU15 adopted a total allocation for the trading sectors that was less than predicted BAU emissions, although often slightly more than recent emissions. This total implied an expected shortage that had to be allocated somehow. One approach would have been to distribute the shortage equally among all sectors and sources. Instead a clear distinction was usually made between electricity generation and industrial sources and the shortage was allocated to the electricity sector. This approach was adopted by the United Kingdom, which was the first Member State to publish a draft NAP (in January 2004) and many subsequent NAPs made the same choice when confronted with an expected shortage.

The reasons for allocating the shortage to the electricity is two-fold. First, electricity generation did not face international, non-EU competition as did the products of many of the industrial installations. As noted in the comment from Christine Cros (see Box 13.3), all governments were in a quandary concerning their climate change policy commitment and the feared competitive effects of that commitment. Second, power plants are commonly believed to have the ability to abate emissions at less cost than others, typically by switching to natural gas instead of the continued use of coal.

The main exceptions to the allocation of the shortage to electricity generation are Italy and Germany. Italy has little coal-fired generation so that the ability to switch to gas is limited and an expected significant turn-over of the electricity generating plant was a prominent feature of the preparation of the Italian NAP (see Agostini, 'Italy'). In Germany, the distinction took another form, but the rationale was similar. Here the distinction was between combustion and process emissions. Since electricity generation is entirely combustion and process emissions arise only from industrial processes, the difference concerns industrial combustion. In the German NAP, process emissions were given a preferential allocation equal to baseline emissions, while industrial combustion sources were treated like electricity generation in receiving the same compliance factor, or ratio of allowances to baseline emissions.

The general method of determining the shortage to be allocated to the electricity sector was: industrial sources would receive what they could reasonably be expected to need and the shortage would be allocated to the electricity sector by means of a uniform reduction from installation baselines. The method of determining what the industrial sectors would need was typically determined by sector-specific projections, but some countries, notably Denmark and Germany, gave industrial facilities (or process emissions in Germany) their historic baseline amounts without involving projections. Most other Member States allocated the remaining power sector allowances according to installation shares of recent sector emissions.

Box 13.3 Competitiveness concerns as a major force behind allocation decisions

It has been very difficult for all the EU governments to elaborate their National Allocation Plans because of two conflicts between policy objectives that had to be resolved.

The first was between the commitment to climate change mitigation and the competitive effects of that commitment. In committing themselves in a binding quantitative objective, even if developing countries and the US did not, the EU as a whole and all Member States agreed to pay for climate change control, but that does not mean that they agreed to lose competitive position internationally or to suffer unemployment in the industrial sector. This is a politically very sensitive issue that was emphasised repeatedly in public consultations and usually led to an increase in the national totals. This dilemma could be solved by establishing some kind of 'competition arrangement', perhaps through the use of World Trade Organisation rules, to compensate the industrial sector for the competitive burden vis-à-vis extra EU companies. Since there was no way to implement this option during the first NAP process, governments often allowed the industrial sector to bear less of the effort than other domestic sectors because of its exposure to international competition.

The second conflict was between using the EU ETS as a strong instrument in meeting the country's Kyoto or BSA objectives and not disadvantaging the country's own industry in EU-wide competition. These two objectives are not compatible because of the significant differences in the gaps between emissions and targets among EU countries. This is a typical free-riding problem: everyone has an interest that real reductions are made but also that others make the reductions. The effective political choices were:

- to consider climate change policy a domestic issue so that any country can choose its own ETS target so long as it can prove that it will meet its Kyoto or BSA objectives
- to adopt more strongly harmonised rules, based on very simple and objective elements, such as technological benchmarks by sector, auctioning, or installation-level allocations that would be an equal percentage below some baseline for each sector.

Christine Cros

3.4 New entrant and closure rules are a common feature

Article 11(3) of the Emissions Trading Directive required Member States to take into account the need to provide access to allowances for new entrants, but it provided no specific guidance and did not direct free allocation. Nevertheless, all twenty-five Member States set up reserves to provide free allowances to new entrants and most require closed facilities to forfeit post-closure allowances. This feature is the more remarkable in that it is not found in other cap-and-trade systems (in the US) where with few exceptions new entrants must purchase whatever allowances they may need and the owners of closed facilities are able to keep the allowances distributed to those facilities.

The motivation for these provisions is invariably explained by a desire not to be placed at a disadvantage in the competition for new investment and a complementary concern to avoid an incentive to shut down facilities in the Member State and to move production elsewhere. The argument concerning the closure provisions produced a particularly effective slogan in Germany where the absence of a closure rule was said to be equivalent to creating a 'shut-down premium'. In addition to these arguments based on employment concerns, comments are often heard that it doesn't seem fair to award allowances to incumbents and deny them to new entrants, or to continue the endowment of allowances to facilities where there is no ongoing need for them. New entrant provisions were also seen as required by pre-existing energy or industrial policy (see Matthes and Schafhausen, 'Germany').

A number of observers have remarked on the distortionary effects of these provisions either as a subsidy to production or as biasing technology choices in a more CO_2 emitting manner. As noted in a number of the contributions to this volume, officials in the Member States and at the European Commission were well aware of these effects; but they were unable to resist the political demands that such provisions be included in NAPs. These political demands did not come from incumbents who favoured retention of allowances upon closure and who, by definition, did not represent new entrants. A good example of the political importance of these provisions is provided by Ireland where one of the few technical recommendations overridden by the government was the one recommending against the provision of new entrant endowments and the forfeiture of allowances upon closure of a facility (see Barry, 'Ireland').

While new entrant provisions are common, their specific characteristics are not. The reserves established to provide free allowances to new entrants vary greatly in size. Among the ten countries included in this volume, new entrants reserves range from 6.5% of the national total in the United Kingdom and Italy to as little as 0.5% in Germany and Poland. Distribution is generally by a 'first-come–first served' rule, but countries vary according to what happens if the reserve is exhausted. For most, late-comers will have to resort to the market but Italy and some other countries have stated that the government will purchase allowances on the market to provide for all new entrants. Provisions also differ if the new entrants reserve is not all claimed. Most have stated an intention to sell the unclaimed surplus either by auction or on the market, generally in 2007, but Germany and Spain will annul remaining allowances. The criteria for determining the number of allowances to award to new entrants also differ considerably. As already noted, all employ some variation of a benchmark based on some definition of best practice or technology multiplied by expected production or by new capacity; however, these benchmarks can differ by fuel or technology used, especially in the electricity sector. For instance, the United Kingdom, Denmark and Spain use a common benchmark for all fuels, while most other countries, notably Germany and Italy, differentiate by fuel. Among the ten countries analysed in this volume, Sweden is unusual in allowing closed facilities to keep their allocations, at least for the rest of the allocation period, and Sweden also restricts access to the new entrants reserve to industrial and district heating facilities, implicitly excluding fossil-fuel-fired generating units (of which none are planned to be built) (see Zetterberg, 'Sweden').

A final variation that is worth noting in this respect is the transfer rule that was pioneered by Germany and adopted by some other Member States. A transfer rule allows the owner of a closed facility to transfer the allowances of the closed facility to a new facility, which is thereby not eligible for allowances from the new entrants reserve. Transfer rules can be very complex. In Germany, for instance, the new facility must be put in operation within eighteen months before or after the closing of an existing facility. The transferred allowances are good for four years after which time the new facility receives allowances from the new entrants reserve. Also, the transfer is proportionate to the productive capacity of the closed facility. The key point in all of these transfer rules is that they operate only for new facilities within the

Member State. Thus, while the 'shut-down premium' was not avoided for specific facilities, the social costs thereby incurred were offset by the benefits of the compensating new investment in the same Member State.

Given the widespread inclusion of special rules for new entrants, the general objective of avoiding investment disincentives through the allocation process was not accomplished. An endowment of allowances to new entrants reduces the cost of an investment and if it varies among Member States, a potential further distortion to the common economic market is introduced. The resulting differences have led to a lot of dispute (primarily from academic quarters), emphasising that harmonisation would seem especially appropriate for new facilities. Harmonisation could be accomplished by introducing harmonised provisions, a unified central EU new entrants reserve, or by prohibiting any new entrant allocations, in which case all new entrants would have to buy allowances in the unified market. The main argument for harmonising new entrant provisions is to avoid adding to the already existing differences in investment incentives across Member States for CO_2 emitting facilities, and the corresponding case for harmonised conditions is strong. Moreover, if an eventual harmonisation of all allocations, to existing and new installations alike is desired, harmonising new entrant provisions is a first practicable step.

3.5 Auctioning is little used

One of the most striking features of the EU ETS allocation process is the extent to which auctioning was not chosen, despite the option provided by the Emissions Trading Directive to auction up to 5% of the Member State's total. Only four Member States (Denmark, Ireland, Hungary and Lithuania) decided to set aside an explicit amount for auctioning and only Denmark opted for the full 5%.[3] The total amount to be auctioned is 8.4 Mt annually out of a total EU ETS allocation of about 2.2 billion tons, or 0.13% of the total. This figure might be augmented by any unclaimed allowances in new entrant reserves.

Auctioning was strongly opposed by the owners of existing facilities in almost all Member States since the amount set aside meant fewer allowances for incumbents. The motivation for auctioning among the

[3] The percentages for Hungary, Ireland and Lithuania are 2.5%, 0.75% and 1.5%.

four Member States choosing to auction varied. In Ireland, the motivation is explicitly budgetary: proceeds will fund the agency set up to administer the trading system. In both Denmark and Hungary, auction proceeds go to the general treasury. In Hungary, the finance minister was active in promoting auctioning but the initially proposed 5% was cut back to 2.5% under pressure from incumbents (see Bart, 'Hungary'). Denmark is unusual also in that the electricity industry favours full or at least harmonised auctioning throughout the EU ETS in subsequent periods in order to reduce the competitive disadvantage the industry believes it suffers when allowances are distributed gratis (see Lauge Pedersen, 'Denmark'). The reasoning is that the BSA obligation assumed by Denmark (-21% from 1990 emissions) implies fewer allowances per unit of BAU output for Danish firms than for competing generators in other Member States. If allowances were auctioned, the differences in the distribution of free allowances and the consequent endowment effects would be wiped out.

4 Concluding remarks

4.1 A more general principle?

Entwined with these ten lessons and unifying themes are two more general issues that are raised by the European experience. To what extent is the allocation of emission rights for CO_2 different from that for other conventional pollutants, such as SO_2 or NO_x? Is there some more general consideration that influences the choices when emission rights are distributed? Or, to rephrase this second question more negatively, why are the many welfare-enhancing choices universally advocated by economists not chosen?

There are many ways in which CO_2 is different from SO_2 and NO_x, but the one that seems to matter for allocation is the perceived potential for abatement. Put simply, the perception is that, with few exceptions, CO_2 emissions cannot be reduced in relation to production other than by carbon capture and storage (CC&S), which is available only at costs that are higher than any society is now willing to bear. At lower costs, this perception maintains that the only way of reducing CO_2 emissions is to reduce output. This is a very different view of abatement potential from what characterises SO_2 and NO_x. In both cases, deep reduction technologies achieving 90% or better removal efficiency have been technically demonstrated and are available at costs that do not imply significantly higher product prices. In addition, less expensive abatement methods effecting smaller reductions are available, such as switching to lower-sulphur fuels or low NO_x firing. With this panoply of abatement options, the effect on production is expected to be slight and the only question is which options will be used to effect the significant emission reductions ($>50\%$) mandated by the SO_2 or NO_x caps. For CO_2, the perception is quite different and the result is manifest in the unwillingness to adopt more than a gently constraining cap initially, to furnish most participants with as many allowances as they will probably need in order not to curtail production, and to allocate the modest shortage to the one sector, electricity, where some abatement is acknowledged as possible by switching from coal to natural gas.

Whether this perception of the abatement potential for CO_2 is correct is not the issue. There is much to suggest that it is not. If there is one lesson from the US experience with cap-and-trade systems, it is that unexpected forms of abatement appear when a price is imposed on emissions. Moreover, it is commonly asserted in Europe that further gains in energy efficiency can be achieved a relatively low cost, in which case equal reductions in CO_2 emissions logically follow at low prices. While it is too early to know the degree of abatement that has occurred in Europe in response to CO_2 prices, the question is whether the perception of limited abatement potential was accepted sufficiently to be politically important in determining allocation choices. That appears to have been the case.

There is hardly an economist who does not deplore the limited use of auctioning and the concomitant extensive use of free allocation in the EU ETS (as well as in other cap-and-trade systems). The choice in the EU ETS is the more puzzling in that the economic arguments for auctioning (i.e. the so-called 'double dividend') are highly applicable to the European Union. In brief, if allowances are auctioned, the revenues received can be used to reduce taxes on labour and capital thereby reducing the disincentive effect of such taxes on the supply of labour and capital, which would then result in greater output than would otherwise occur (and thus less loss of welfare from the environmental measure). Such 'recycling' of auction revenues would seem to have great appeal anywhere, but particularly in Europe where high social charges on labour are commonly seen as the cause of persistently high unemployment levels. That auctioning was so little chosen suggests that more is involved than arguments based on economic welfare.

The usual explanation is lobbying or a version of public choice theory whereby some perversion has entered the system. Yet this is a strangely incomplete explanation. While lobbying is no doubt present, the distinction between this form of advocacy in what are demonstrably democratic systems and other forms of pleading that are considered legitimate is never clearly made. In the case of the EU ETS, it must also be remembered that the decision to allocate at least 95% of the EUAs for free was taken by duly constituted political authorities, namely, the European Council of Ministers and the European Parliament. At the Member State level, industry was by necessity heavily involved in the process of allocation, but the government role is nowhere described as one of awarding allowances to the highest bidder. As noted above, the government role was largely one of managing a process whereby competing claims could be reconciled, being the final arbiter, and imposing some top-down discipline on the process. Notably, concerns for fairness are at least as prominent as demands for more. Finally, lobbying notably fails to explain the phenomenon of the new entrants and closure provisions. Incumbents favoured retaining allowances if facilities are closed (see Harrison and Radov, 'United Kingdom') and they generally did not advocate new entrants reserves, which implied fewer allowances for incumbents.

A different and perhaps broader perspective comes from political science where the argument is made that these distributions of private rights in public resources, whether air, grazing land, fisheries etc., express social norms that often grant prior use a strong claim (Raymond 2003). The argument starts with the dual recognition that the rights to emit now being limited were previously freely exercised and that there will be continuing use after the constraint has been imposed. The question very quickly becomes whether the entitlement to the continuing rights should have any relation to the exercise of the implicit rights that existed prior to the constraint. A prior-use norm implies that a strong relationship between the two is legitimate and appropriate.

This relationship is made stronger by the difference in regulatory obligation imposed on firms by the cap-and-trade form of regulation. In conventional, prescriptive environmental regulation, often pejoratively termed 'command-and-control', continuing use incurs no charge so long as the installation is deemed in compliance with the relevant 'command'. In contrast, market-based approaches such as the EU ETS impose a charge on continuing use while also explicitly recognising some aggregate level of continuing emissions as allowable. Since those who freely exercised the implicit right before the constraint and those who will exercise the continuing right afterwards are very largely the same, imposition of the charge without some offsetting mechanism implies a drastic redistribution of those rights and one that would not occur with more conventional means of regulation. The simple way to solve the dilemma is to offset the liability imposed on continuing emissions with an endowment of assets conveying the rights to the newly created scarcity rent. Thus, free allocation to regulated entities becomes the means by which the new form of liability is imposed. The compensating endowment may or may not be fully compensating, but whatever the balance, the incentive is clearly created to reduce emissions to the new aggregate limit as cost-effectively as possible.

In this perspective, the newly regulated emissions are not so much a 'bad' as a heretofore fully authorised by-product of useful economic activity which is expected (and indeed fervently hoped) to continue. This view also helps explain the new entrant and closure provisions, which are not implied by a strict application of a prior-use norm, and it suggests a modification of that norm. If useful economic activity is desired and CO_2 emissions are regarded as being to some extent unavoidable, then it is ongoing production that conveys a right to emit. The amounts provided to installations may differ according to vintage or industry, but the basic principle remains. This too seems to be a distinctive feature of CO_2 in contrast to SO_2 and NO_x, as well as other public resources, where new entrants are generally obliged to buy rights from those who implicitly exercised them prior to the imposition of the constraint.

4.2 Global implications

Notwithstanding the increasingly common and convenient reference to the European Union as a single entity, it is well to remember that the Union consists of twenty-five sovereign states, all of whom jealously guard national prerogative. Accordingly, it is appropriate to look at the EU ETS not just as an unusual example of a common undertaking within the European Union, but also as an exercise in implementing a multinational climate change regime in which not all participants are equally committed to taking effective action to restrict CO_2 emissions.

It is not surprising that EU Member States that are highly committed to meaningful action on climate change within the ambit of the Kyoto Protocol, such as Germany, the United Kingdom, Sweden, Denmark and the Netherlands, should develop an international trading scheme, but the same cannot be said for a number of the other Member States. Eight of the ten accession states have national limits under the Kyoto Protocol, but for seven of these states those limits are slack so that nothing is required of them over the 2005–2012 horizon by the Kyoto Protocol. Moreover, two of the accession states, Cyprus and Malta, are not Annex I countries and therefore do not have any caps under the Kyoto Protocol. Yet all of these states have adopted national caps and are participating in this multilateral trading scheme. The three contributions to this volume concerning Poland, Hungary and the Czech Republic make it clear that none regard the rules as especially appropriate for their circumstances, and yet they belong.

A more interesting case is presented by Italy and Spain. These countries participated fully in the European BSA and accepted targets that are in both cases considerably below likely BAU emissions. No serious measures appear to have been contemplated to ensure compliance with the BSA targets apparently in the belief that the targets were aspirational goals to be used to justify measures that would be taken anyway, but certainly not the basis for imposing a significant price on a significant share of national emissions. In these countries, the EU ETS was not easily accepted, but in the end it was (see del Río, 'Spain' and Agostini, 'Italy').

Given the highly different circumstances of the EU Member States and their equally varying commitments to adopting meaningful measures to restrict CO_2 emissions, one must ask: what caused the reluctant followers to adopt caps and to enter into the multinational trading regime? The short answer is that participation was a requirement of membership in the European Union.[4] But this easy answer relying on legal formalities evades the more serious issue of what caused highly sovereign nations to accept a measure that they did not seek, would not

[4] The issue of voluntary participation was actually discussed in various shades during the negotiations of the Emissions Trading Directive, and could be settled only through the introduction of the time-limited opt-out clause by installation, being subject to stringent conditions as well as Commission scrutiny.

otherwise have chosen, and that involved costs that they would have preferred to avoid. The answer would appear to be the broader benefits of participation in the European Union – not so much the European identity, however that may be defined, but the more concrete benefits of freer trade, access to larger markets, freer movement of labour and capital, and of other benefits (including aid) that come with becoming part of a broader community. As stated in the contribution concerning Hungary, accepting the EU ETS was 'just another obligation on the long march to the EU' (see Bart, 'Hungary'). In Poland, there was strong opposition from industry to accepting the European Commission's reduction of the total, but the Polish government was unwilling to challenge the Commission because of these broader interests (see Jankowski, 'Poland'). In these instances and others, the European idea served as the glue that both attracted reluctant participants and fastened them to undertakings that they would not otherwise have accepted. The European idea will not serve beyond Europe, but some similar combination of desirable community and practical advantage will need to be found.

The importance of the EU ETS extends well beyond its usefulness as a laboratory in which twenty-five experiments in the allocation of carbon rights can be studied or as an example of a multinational endeavour that has successfully navigated the shoals of differing circumstance and motivation. It has cast the die concerning the nature of a future global climate regime if there is to be one. Europe's choice of emissions trading has created a fact on the ground that will be as difficult to ignore in the future as it is to imagine an effective global regime without the United States.

This influence on the nature of a future global climate regime is easiest to imagine in the event of failure of the EU ETS. Had the indispensable first step of allocation descended into a cacophony of conflicting interests and political mayhem, CO_2 trading as a national, not to mention global instrument of climate policy, would have been badly set back, perhaps irretrievably. The success of the European CO_2 trading experiment cannot yet be taken for granted, but the grounds for optimism are much greater than they were some years ago when the opinion quoted from Point Carbon in the introduction was a commonplace (Point Carbon 2001). Assuming that the whole experiment succeeds, and not just this first step of allocating carbon rights, the EU ETS will set the standard for a global regime and provide an unexpected but propitious example that others seeking effective measures to limit GHG emissions will find increasingly hard to resist.

References

European Commission 2000. 'Green paper on greenhouse gas emissions trading within the European Union', COM(2000) 87, March 2000.

Point Carbon 2001. 'Towards EU-wide emissions trading? Politics, design and prices', *The Carbon Market Analyst*, 25 September 2001.

Raymond, L. 2003. *Private Rights in Public Resources: Equity and Property Allocation in Market-Based Environmental Policy*. Washington, DC: Resources for the Future Press.

I | *Participants*

Daniele Agostini, Adviser, Italian Ministry for the Environment. Daniele Agostini holds a degree in environmental engineering and economics from the Massachusetts Institute of Technology (USA), University of California, Berkeley (USA) and University College London (UK). He has been advising government and private sector companies on environmental issues for more than ten years. Since 1999 he has been serving as expert adviser for the Italian Ministry for the Environment and Territory coordinating the Kyoto Flexible Mechanisms team within the Climate Change Unit. In this role he has participated in international negotiations as well as in the national implementation of emissions trading, joint implementation and clean development mechanism activities.

Conor Barry, B.Comm., M.Sc., is a Programme Officer in the CDM Section of the UNFCCC Secretariat. At the time of writing his chapter he was Senior Manager in the Emissions Trading Unit of the Irish Environmental Protection Agency, where his responsibilities included the preparation of the Irish National Allocation Plan. Prior to this he was an Environmental Economist in the Irish Department (Ministry) of Environment, Heritage and Local Government with responsibility for the implementation of economic instruments to support the National Climate Change Strategy, including negotiation of the EU ETS Directive.

Istvan Bart has been working in the Environment Directorate General of the European Commission since March 2005. As part of the team responsible for the EU ETS, he specialises in the operation of the Community Independent Transaction Log, and he also works with JI/CDM on the issues related to new Member States and Accession Countries. Before joining the Commission, he worked for Hungary's Ministry of Economy, where he participated in the drafting of Hungary's National Allocation Plan for 2005–2007. A law graduate from ELTE University,

370

Budapest, he worked at Baker & McKenzie's Budapest office before entering public service.

Barbara K. Buchner, Ph.D. in economics, University of Graz, Masters Degree in economics within the Economics/Environmental Sciences Joint Program, University of Graz and University of Technology of Graz, has been a Senior Researcher at the Fondazione Eni Enrico Mattei (FEEM) since 2000. She is involved in a number of activities related to FEEM's Climate Change Modelling and Policy Unit in the field of environmental economics. In particular, she works on the economic evaluation of climate policies, trying to analyse the incentives that are caused by different strategies. Special attention is given to the ancillary benefits of abatement measures and to effects concerning endogenous technical change induced by various policies. Within the FEEM research activities she is furthermore working on the analysis of international negotiations and the formation of international economic coalitions, focusing on climate negotiations.

Carlo Carraro, Ph.D. in economics, Princeton University, is currently Professor of Econometrics and Environmental Economics at the University of Venice. He is Research Director of the Fondazione Eni Enrico Mattei as well as Chairman of the Department of Economics and Vice-Provost for Research Management and Policy of the University of Venice. His research activities include the econometric evaluation of environmental policies to control global warming; the microanalysis of environmental policies and of their impact on market structure, the analysis of international negotiations and the formation of international economic coalitions (with particular emphasis on environmental negotiations). Since 1992, he has been working in several OECD and EU projects. The most important of these dealt with modelling environment–economy linkages, studying environmental innovation, analysing climate policies, market-based policy instruments and in particular voluntary agreements.

Tomas Chmelik, M.Sc. in economics (University of Economics, Prague) is Director of the Climate Change Department, Ministry of Environment of the Czech Republic. After completing his M.Sc. degree, with major specialisation in public administration and minor specialisation in environmental economics he started to work for the Ministry of Environment, Department of Environmental Economy, Unit of Economic Instruments with responsibility for economic instruments in air, water and waste policy and environmental tax reform. Since 2003 he has been head of the newly established Climate Change Unit (since 2005 Climate Change Department), with responsibility, among other issues, for the EU ETS and Kyoto flexible instruments. He is a member of various working groups on climate change and energy policy issues, of the national focal point for UNFCCC and actively cooperating with academia. He is currently working on his Ph.D. at the Czech Technical University with a thesis on the implications of the EU ETS and allocation plans in particular on the energy sector.

Atle Christer Christiansen, Ph.D in chemical engineering, is Director of Point Carbon, a global provider of independent analysis, news, market intelligence and forecasting for the emerging carbon emission markets. His main research areas include climate policy, emissions trading, mathematical modelling, numerical and statistical methods, technological change and innovation, new renewable energy, and energy policy. His knowledge of the European environmental and emissions trading circles and of the emerging carbon market was an important contribution to the project.

Christine Cros has been head of the unit on Global Public Goods in the Economic Studies and Environmental Evaluation Department of the French Ministry of Ecology and Sustainable Development since 2004. This unit deals with the economic aspects of international issues such as climate change, biodiversity, international trade and environment and regulation on international environmental issues. The unit was involved in the elaboration of the French national allocation plans. She has worked on trading systems for a long time since she began a Ph.D. in 1994 on the comparative analysis of the US SO_2 tradable system and the French SO_2 regulation, with a view to assessing the possibility of using such trading systems in the French regulatory context. She had worked for the French agency for environment and energy management (ADEME) from 1998 to 2001 as a climate change economic expert. During this period she had been involved in the French climate change delegation and in the negotiation of the Marrakech Accords. From 2002 to 2004, she worked for the French department that coordinates French views on EU negotiation issues. In this job she had coordinated the French positions during the negotiation of the Directive 2003/87/EC implementing an emission trading system for the European community.

Frank Convery, B.Agr.Sc., M.Agr.Sc., M.S., Ph.D., is Heritage Trust Professor of Environmental Policy at University College Dublin. He is active on a number of EU-wide investigations and bodies. He has written extensively on resource and environmental economics issues with particular reference to agriculture, forestry, energy, minerals, land use, urbanisation, environment and development in developing countries. At present, his research relates to EU environmental policy with particular reference to the use, potential and effectiveness of market-based instruments.

Pablo del Río, Ph.D., University of Castilla–La Mancha. Pablo del Río is Associate Professor of Environmental Economics at the University of Castilla–La Mancha, Toledo. He has worked and written extensively on climate change mitigation measures and renewable energy support schemes, particularly on emission trading and tradable green certificates schemes as well as their respective interaction. His research also focuses on the factors influencing environmental technology change in firms. He has been involved in several EU and national projects and has published his work in international journals.

A. Denny Ellerman, Ph.D. in political economy and government, Harvard University, is Senior Lecturer at the Sloan School of Management at the Massachusetts Institute of Technology and the former Executive Director of Center for Energy and Environmental Policy Research and the Joint Program on the Science and Policy of Global Change. Before coming to MIT, Dr Ellerman worked in government and industry as an energy economist. He was president of the International Association for Energy Economics in 1990 and received that organisation's Award for Outstanding Contributions in 2001. His research interests focus on emissions trading, climate change policy and the economics of fuel choice, especially concerning coal and natural gas. He is co-author, with colleagues from MIT, of *Markets for Clean Air: The US Acid Rain Program.*

David Harrison, B.A., M.Sc., Ph.D., is Head of the Global Environmental Practice of NERA Economic Consulting, an international consultancy with offices in Europe, the United States, Australia and Asia. He directs projects in environmental economics and policy, climate change, natural resource damage assessment, energy policy, economic impact assessment, and transportation. Dr Harrison has participated actively for more than twenty-five years in the development of emissions trading programmes and other innovative means of increasing the flexibility and reducing the costs of environmental regulation. He was a member of the advisory committee for the RECLAIM program, an innovative emissions trading programme in the Los Angeles air basin, and has advised on numerous other programmes, including the acid rain trading programme and the averaging, banking, and trading programmes developed for mobile sources in the United States. Most recently, Dr Harrison has co-led NERA efforts to assist the European Commission and the UK government with regard to the EU ETS. He has consulted frequently to various public groups and private firms on the use of emissions trading to deal with climate change in Europe, the United States and Asia. Before joining NERA, Dr Harrison was an Associate Professor at the John F. Kennedy School of Government at Harvard University, where he taught among other subjects energy and environmental economics and policy, microeconomics, and regional development. He also served as a Senior Staff Economist on the US government's President's Council of Economic Advisors, where he was responsible for environment, energy and transportation regulatory and policy issues.

Boleslaw Jankowski, M.Sc. Technical University of Warsaw, Ph.D. in energy planning from Silesian Technical University. From 1991 to 1997 he worked in the Department of Energy Problems in the Polish Academy of Sciences. Since 1997, he has been vice-president and leader of the development strategies team in the research and consulting company EnergSys. His activity concentrates on complex energy and environmental problems tackled with the wide use of a systems approach. He has led several complex studies on energy system development at the national and local level using a set of simulation and optimisation models, including studies for the Polish government, the World Bank and EU projects. He is co-author of energy policy of Poland in the years 1991–1995 and main author of the national strategies on SO_2, NO_x and CO_2 emissions accepted by the Ministry of Environment. He led the team that in 2002–2003 developed the Polish Programme for Flexible Environmental Protection assuming implementation of emissions trading in several areas of environmental protection. In 2004, he led the team preparing the draft NAP for Poland. He is also active at the micro level promoting and implementing the systems approach and tools in strategic management in large infrastructure enterprises.

Sigurd Lauge Pedersen, M.Sc., Ph.D., Danish Energy Authority. He acquired an M.Sc. degree in physics from Copenhagen University in 1981. After five years in research at Copenhagen Technical University, including a Ph.D. degree in energy planning, he joined the Danish Energy Authority, a government body currently under the Ministry of Transport and Energy. His main responsibilities are: emissions trading (legislation, allocation, auctioning and the 2001–2004 domestic emissions trading system), electricity regulation and energy planning (including energy systems modelling). His present position is Senior Advisor.

Per Lekander is equity analyst at the UBS Investment Research. UBS is a premier global financial services firm offering wealth management, investment banking, asset management and business banking services to clients. Lekander's main research interests are concentrated on the emerging carbon market and the related impact of the EU ETS, with a particular focus on the role of the power sector.

Felix Christian Matthes, Dr.rer.pol., worked in industry as a graduate engineer for electrotechnology, having successfully completed his degree. He has been working at Öko-Institut (Institute for Applied Ecology) since 1991; in 1997 he was made Coordinator of the Energy and Climate Division. In 1999 he received a doctorate in political science from the Freie Universität Berlin for a dissertation on energy policy. He has published a number of studies on national and international energy and climate policy and is engaged in political consultation on a national, European and international level. The main focus of his work lies in energy and emissions projections and the development, analysis and valuation of political instruments of energy and climate protection policy. He has been working as a Consultant to the German federal government in several fields of climate policy, since 2003 particularly in the area of emissions trading. From 2000 to 2002 he was an Expert Member of the Study Commission of the 14th German Bundestag for 'Sustainable Energy under the Conditions of Globalisation and Liberalisation'. Alongside his position at Öko-Institut, he teaches energy policy analysis at the Freie Universität Berlin.

Daniel Radov, B.A., M.Phil., M.Sc., is a Senior Consultant in the Energy, Environment, Transport, and Water Global Practice of NERA Economic Consulting. His work concentrates on environmental economics, with a particular focus on emissions trading, climate change and economic issues associated with energy use. His projects have spanned a range of industries, including electric power, automobile and engine manufacturing, forest and paper products, cement, agrochemicals and refining. Mr Radov has led NERA's work to assist the UK government in its development of its NAP for the EU ETS. He also assisted the European Commission in evaluating options for the initial allocation of allowances in the EU ETS and co-wrote NERA's report to the Commission on the prospects for using market-based instruments to help reduce emissions from marine shipping sources. He has studied and evaluated the design, implementation and effects on different industries of other emissions trading programmes, including existing and proposed greenhouse gas trading programmes in the US and Europe.

Franzjosef Schafhausen graduated in economics and public finance at the University of Cologne after having been trained as a banker in Düsseldorf. Since 1990, he has been Head of the Division of National Climate Change Programme, Energy and Environment at the Federal Ministry for the Environment, Nature Conservation and Nuclear Safety in Germany. He is also Chairman of the Interministerial Working Group on CO_2 Reduction established by the Federal Cabinet in 1990 to develop and implement the National Climate Change Programme as well as Chairman of the Working Group on Emissions Trading to Combat Climate Change (Arbeitsgruppe Emissionshandel zur Bekämpfung des Treibhauseffekts – AGE) established by the Federal Cabinet in 2000 to assist the development and implementation of emissions trading in Germany. Before that, he was responsible for the thematic 'Energy and Environment' at the Federal Ministry for the Environment, Nature Conservation and Nuclear Safety and for the area 'Environment and Economics' at the Federal Ministry of the Interior. Previous to joining the Ministry, he had also worked for the Environmental Protection Agency in Berlin.

Hans Warmenhoven is an environmental engineer and has been active in the field of energy-saving policy and energy-saving technology and climate change issues since 1989. Since 1999 he has been active as a consultant first within PricewaterhouseCoopers and later within his own firm Spin Consult. As consultant he has been involved in several projects related to allocation: implementation of emission trading in the

Czech Republic; review of allocation proposals for cogen installations in the Netherlands; study for the EU on how to allocate allowances within the EU ETS system; evaluation of the Dutch allocation process; a study on the allocation of CO_2 emission rights in the post-Kyoto period.

Peter Zapfel, Ph.D., European Commission, holds academic degrees from the University of Business and Economics in Vienna and the John F. Kennedy School of Government at Harvard University. He has been with the European Commission since 1998, first for two years in the DG for Economic and Financial Affairs, and since 2000 with DG Environment. He has represented the Commission as a delegation member in UN climate negotiation sessions. He is responsible for the economic assessment of climate policy. This comprises the development and quantitative assessment of cost-effective strategies and instruments to implement the EU climate policy objectives. He has been involved in the Commission's work on emissions allowance trading since 1998 and the work on national allocation plans in the implementation of the Emission Allowance Trading Directive. Since May 2005 he has coordinated the DG Environment's EU ETS team.

Lars Zetterberg is head of the climate change unit at the Swedish Environmental Research Institute (IVL). He is project leader of the Mistra-financed project 'ETIC, the role of emission trading in climate policy'. In addition, he has worked (and is currently working) with the Swedish government, its agencies and Swedish industry on a wide range of assignments in connection with the implementation of the emission trading system. These assignments include: analysing allocation principles; benchmarking; data collection from participating installations; developing the Swedish NAP; analysis of carbon reduction technologies for the Swedish energy sector; analysis of other Member States' NAP; implications of including the transport sector in the EU ETS; implications of including the energy sector in the ETS. He chaired the workshop 'Allocation in the Emission Trading System – Lessons Learned' in Göteborg 2005 and chaired the workshop session 'European and American Business perspectives on Emission Trading' in New York City 2005.

Length: 7,000 to 10,000 words (about 21–30 pages with $1\frac{1}{2}$ line spacing)

General: This Member State outline is intended as a means for providing unity in presentation to facilitate comparison while allowing room for individual authors to tailor their contributions to the story of their respective Member States. Accordingly, the section headings (Sections 1 to 6 below) should be respected, but authors should feel free to use subheadings within each section as appropriate, and as will probably be inevitable in Sections 4 and 5. The percentages in parentheses following each section heading are suggestions concerning the length of that section. In general, about half the paper should be devoted to an explanation of how the total quantity of allowances to be allocated was decided and distributed and the other half to aspects deserving emphasis for understanding the process in the particular Member State. Some sections will be more descriptive than analytical and each author will have to decide on the mix that serves best to tell the story of the particular Member State's experience in allocating carbon rights in an accessible and understandable manner.

Introductory background and context (10%)

This section should be used to inform the reader of important country-specific background and contextual features of the Member State that will help in understanding the allocation decisions that were made and the problems and issues that presented themselves in the process. Among these features, also the political capacity of each Member State should be tackled. These insights are therefore expected to facilitate the understanding of the influence of the political aspects on the process.

The macro decision concerning the aggregate total (25%)

Member States were confronted with two analytically distinct decisions: (1) what was to be the total number of EUAs to be allocated during the first period, and (2) how to distribute that total to included installations. This section and the next are to be devoted to these two issues, with this section devoted to the first of these decisions. The roles played by the 'Path to Kyoto' and forecasts of BAU emissions will figure largely here; however, the underlying issue is the degree of scarcity that the regulator sought to create and how it was determined. Various Member States were in very different positions with respect to the Kyoto or BSA criteria and all Member States were concerned not to place too great a burden on their industries. An important issue in this section is the extent of interdependence between the macro and micro decisions and how the two were reconciled.

The micro decision concerning distribution to installations (25%)

The section focuses on how the total was split among installations. Usually, some basic principle (such as historical emissions or benchmarking) is applied with a greater or lesser degree of deviation to deal with specific circumstances. Frequently, there will be a distinction made between sectors or groups of sectors, such as between electricity and general industry, so that each may have to be described separately. Of course, whether such a distinction was made and why it was made is important. Also, the selection of baseline periods for determining historical emissions or activity levels (for benchmarks) is not obvious and the reasons for choosing a particular baseline period should be discussed.

Issues of coordination and harmonisation (15%)

This section and the next one are the place to elaborate on particular issues that might have been particularly important in the macro or micro decisions or both. For instance, competitiveness concerns are inherently ones of harmonisation; and, if this was a major concern in allocation, it might be worth discussing it in more detail in this section than in the macro or micro sections. The EU ETS is distinctive in being relatively decentralised yet having relatively strong coordination from the centre. Thus, the role that Commission's guidance and its approval played in deciding both the macro and micro aspects of allocation should figure importantly in this section. Issues of coordination may also arise with respect to other EU and Member State policies such as energy market liberalisation, taxation or renewable energy policy. To the extent that these policies figured importantly in allocation decisions at either the macro or micro level, they should be discussed in this section.

Issues deserving special mention (15%)

Inevitably, certain features and issues will deserve more attention than can be given to them in the micro or macro sections. For instance, inadequacies in data constrained some decisions and data availability seems to have been a problem, and to have constituted a major part of the process, in most if not all Member States. If the issue deserves further mention, this is the section for it. Also Member States had several options explicitly left to them that deserve some discussion, such as whether to auction up to 5% of the total allocation. Certain features of the NAPs also stand out, such as new entrants reserves, closure provisions, and the treatment of central heating plants and biomass burning installations. While all of these may be described in Section 3, this section provides the room to develop the rationale and factors influencing the decisions that were made. This is also the place to discuss other factors that don't fit neatly into the previous categories or that were sufficiently important to deserve more discussion, such as the role of rumour, special-interest lobbying, or particular legal issues.

Concluding comments (10%)

This is the place for the broad conclusions or lessons that arise from the experience of the particular Member State in confronting the issues raised by the NAP process. Factors that facilitated or complicated the required decisions are especially important.

III | *The country tables*

1. *NAP summary for Czech Republic*

Cap statistics	Latest year (2004)	Projected BAU 2005–2007	Approved ETS cap
Annual ETS sector emissions (Mt CO_2)	91 (approx)	97.6	97.6

BSA/Kyoto statistics	Latest year (2003)	Projected 2005–2007	BSA/Kyoto target
National emissions (Mt CO_2e)	146.9	150	174.8

NAP timing	Answers
Initial notification to the Commission	October 2004 (incomplete NAP) 25/01/2005 (complete NAP)
Final Commission approval of NAP	12/04/2005
Final MS allocation law/regulation	20/07/2005

NAP coverage	Answers
Number of installations	426

Allocation process	Answers	
Base period years	1999–2001 (adjustments allowed in certain cases), 2004 also a part of allocation formula	
Two-step (sector/installation) process?	Yes	
If so, on what basis?	Historical emissions, sectoral growth projections	
Basis of allocation to installations	Historical emissions determining the 'share' of installation on sectoral emissions	
Are electric utilities allocated less relative to base period or need than others?	Yes	

Other NAP provisions	Brief description if applicable
Auction provisions? (% if yes)	No
New entrants reserve? (size or % if present)	Yes, 345,163 allowances annually, to be used for new entrants (forgotten installations to be allocated extra)
Closure/transfer provisions?	In case of closing down the plant, allowances will not be allocated for further years of the period, no specific transfer provisions
Early action provisions?	Yes, NAP includes recognition of EA in the form of a EA bonus
CHP/cogeneration provisions?	Yes, NAP includes bonification of CHP producers in the form of a CHP bonus
Pooling provisions?	Pooling generally allowed (no interest so far)
Banking provisions?	Banking to be forbidden between 2005–2007 and 2008–2012; is already written in NAP, legislation to be adjusted accordingly
Opt-ins/opt-outs?	No

2. NAP *summary for Denmark*

Cap statistics	Latest year (2002)	Projected BAU 2005–2007	Approved ETS cap
Annual ETS sector emissions (Mt CO_2)	30.65	39.3	33.5

BSA/Kyoto statistics	Latest year (2003)	Projected 2005–2007	BSA/Kyoto target
National emissions (Mt CO_2e)	73.9	78.3	54.9

NAP timing	Answers
Initial notification to the Commission	31/03/2004
Final Commission approval of NAP	07/07/2004
Final MS allocation law/regulation	09/06/2004

NAP coverage	Answers
Number of installations	380

Allocation process	Answers	
Base period years	1998–2002	
Two-step (sector/installation) process?	Yes	
If so, on what basis?	Historical emissions, and historic electricity production (for electricity sector).	
Basis of allocation to installations	Benchmark (allocation per MWh) for electricity sector, historic emissions for other sectors	

Are electric utilities allocated less relative to base period or need than others?	Yes	

Other NAP provisions	**Brief description if applicable**
Auction provisions? (% if yes)	Yes, the full 5%
New entrants reserve? (size or % if present)	Yes, 1 Mt/yr (3%)
Closure/transfer provisions?	On closure: allowances returned to NER. No transfer provisions
Early action provisions?	Only indirectly (through electricity benchmark and long base period)
CHP/cogeneration provisions?	Yes. CHP gets the general electricity benchmark allocation for electricity plus historical CO_2 for heat production
Pooling provisions?	One pooling application received (DH sector)
Banking provisions?	No
Opt-ins/opt-outs?	No

3. NAP summary for Germany

Cap statistics	Latest year (2002)	Projected BAU 2005–2007	Approved ETS cap
Annual ETS sector emissions (Mt CO_2)	506.5	495	499

BSA/Kyoto statistics	Latest year (2003)	Projected 2005–2007	BSA/Kyoto target
National emissions (Mt CO_2e)	984.3	n.a.	962

NAP timing	Answers
Initial notification to the Commission	31/03/2004
Final Commission approval of NAP	07/07/2004
Final MS allocation law/regulation	26/08/2004 (signature) 31/08/2004 (entry into force)

NAP coverage	Answers
Number of installations	1,849

Allocation process	Answers	
Base period years	2000–2002	
Two-step (sector/installation) process?		No
If so, on what basis?	—	
Basis of allocation to installations	Historical emissions as the common rule, option to opt for projections for production and new installation emission benchmark.	

Are electric utilities allocated less relative to base period or need than others?		No

Other NAP provisions	Brief description if applicable
Auction provisions? (% if yes)	No
New entrants reserve? (size or % if present)	9 million EUA for the period 2005–2007 (0.6%)
Closure/transfer provisions?	Yes. Closure provision with ex post adjustment if actual emissions fall below 60% of base year emissions because of production decrease (subject to legal dispute with the Commission), capacity related transfer provision for four years
Early action provisions?	Yes. Based on year of commissioning or efficiency improvement thresholds
CHP/cogeneration provisions?	Yes. Special allocation for existing CHP and double benchmark (allocation based on power and heat production) for new CHP
Pooling provisions?	No pooling provisions
Banking provisions?	No banking allowed
Opt-ins/opt-outs?	No opt-in or opt-out allowed

4. NAP *summary for Hungary*

Cap statistics	Latest year (2002)	Projected BAU 2005–2007	Approved ETS cap
Annual ETS sector emissions (Mt CO$_2$)	30.986	31.266	31.266

BSA/Kyoto statistics	Latest year (2003)	Projected 2005–2007	BSA/Kyoto target
National emissions (Mt CO$_2$e)	83.200	87.200	11.487

NAP timing	Answers
Initial notification to the Commission	08/10/2004
Final Commission approval of NAP	27/12/2004
Final MS allocation law/regulation	Not adopted at time of writing

NAP coverage	Answers
Number of installations	229

Allocation process	Answers	
Base period years	Varying by sector	
Two-step (sector/installation) process?	Yes	
If so, on what basis?	Projection of emissions in 2005–2007	
Basis of allocation to installations	Calculation of installation's share in the base period's sectoral emissions, and the allocation of an equivalent share of the projected emissions	

Are electric utilities allocated less relative to base period or need than others?		No

Other NAP provisions	Brief description if applicable
Auction provisions? (% if yes)	2.5%
New entrants reserve? (size or % if present)	2%
Closure/transfer provisions?	Yes, closing installation is allowed to transfer allocations to another installation in the same sector and same owner
Early action provisions?	0.6% is redistributed on the basis of early action investments
CHP/cogeneration provisions?	No
Pooling provisions?	No
Banking provisions?	No
Opt-ins/opt-outs?	No

5. NAP summary for Ireland

Cap statistics	Latest year (2003)	Projected BAU 2005–2007	Approved ETS cap
Annual ETS sector emissions (Mt CO_2)	22.474	22.984	22.32

BSA/Kyoto statistics	Latest year (2002)	Projected 2005–2007	BSA/Kyoto target
National emissions (Mt CO_2e)	68.985	68.71	60.365

NAP timing	Answers
Initial notification to the Commission	31/03/2004
Final Commission approval of NAP	07/07/2004
Final MS allocation law/ regulation	08/03/2005

NAP coverage	Answers
Number of installations	106

Allocation process	Answers	
Base period years	2000–2003 2000–2003 (installation)	
Two-step (sector/installation) process?	Yes	
If so, on what basis?	Historical emissions	
Basis of allocation to installations	Historical emissions	

Are electric utilities allocated less relative to base period or need than others?	Yes	

Other NAP provisions	Brief description if applicable
Auction provisions? (% if yes)	Yes, 0.75%
New entrants reserve? (size or % if present)	Yes, 1.5% for new entrants
Closure/transfer provisions?	Installations do not receive allocations in years subsequent to closure
Early action provisions?	No
CHP/cogeneration provisions?	Reserve for new CHP
Pooling provisions?	No
Banking provisions?	No
Opt-ins/opt-outs?	No

6. NAP *summary for Italy*

Cap statistics	Latest year (2000)	Projected BAU 2005–2007	Approved ETS cap
Annual ETS sector emissions (Mt CO_2)	224	n.a.	233

BSA/Kyoto statistics	Latest year (2002)	Projected 2005–2007	BSA/Kyoto target
National emissions (Mt CO_2e)	513.04	n.a.	475

NAP timing	Answers
Initial notification to the Commission	21/07/2004
Final Commission approval of NAP	25/05/2005
Final MS allocation law/regulation	February 2006

NAP coverage	Answers
Number of installations	950

Allocation process	Answers	
Base period years	2000–2003	
Two-step (sector/installation) process?	Yes	
If so, on what basis?	Projections	
Basis of allocation to installations	Power sector – projections Non-power sector – share based on emissions or output	

Are electric utilities allocated less relative to base period or need than others?		No

Other NAP provisions	Brief description if applicable
Auction provisions? (% if yes)	No
New entrants reserve? (size or % if present)	46.65 Mt
Closure/transfer provisions?	Yes
Early action provisions?	No
CHP/cogeneration provisions?	Yes
Pooling provisions?	No
Banking provisions?	No
Opt-ins/opt-outs?	No

7. NAP *summary for Poland*

Cap statistics	Latest year (2003)	Projected BAU 2005–2007	Approved ETS cap
Annual ETS sector emissions (Mt CO$_2$)	217.2[a] 219.8[b]	259.9[c] 263[d]	239.1

BSA/Kyoto statistics	Latest year	Projected 2005–2007	BSA/Kyoto target
National emissions (Mt CO$_2$e)	317.8[e] 308.3[f] 319[g]	369.5[h]	448

NAP timing	Answers
Initial notification to the Commission	22/09/2004
Final Commission approval of NAP	08/03/2005
Final MS allocation law/regulation	27/12/2005

NAP coverage	Answers
Number of installations	878

[a] Top-down estimation – based on statistics and emission inventories for the year 2001.

[b] Bottom-up estimation – base emissions from period 1999–2002 (average from three years after exclusion of the lowest emission).

[c] Top-down projection (70.3% of the national emissions in these years).

[d] Sum of sectoral projections, based on average emissions from the period 1999–2002 (219.8 Mt).

[e] 2001 – last year emissions available during NAP preparation by Poland in 2004.

[f] 2002 – last year emissions taken into consideration by EC in decision from March 2005.

[g] 2003 – current latest emission data from the national inventory.

[h] Top-down projection based on emissions from the year 2001.

Allocation process	Answers
Base period years	1999–2002
Two-step (sector/installation) process?	Yes
If so, on what basis?	14 Sectors – allocation based on emission projections Determination of sector budgets = historical emissions in 1999–2002 × sector-specific growth factor × emission intensity change factors
Basis of allocation to installations	10 sectors – grandfathering method; 4 sectors (cement, sugar, coke, power sector) – other methods, (projection, benchmark, historic multi criteria)[i]
Are electric utilities allocated less relative to base period or need than others?	No

Other NAP provisions	Brief description if applicable
Auction provisions? (% if yes)	100% free allocation; auction for remaining part of New entrant reserve before the end of 2007
New entrants reserve? (size or% if present)	NAP proposal 2004: 3.3 Mt/yr (1.2%)[j] Latest proposal in 2005: 0.8 Mt/yr (0.32%)
Closure/transfer provisions?	Yes, transfer of allowances in case when new installation takes over the production from closed installation

[i] Cement sector – allocation proportional to forecast production in 2005–2007; sugar sector – proportional to production in 2003; coke sector – proportional to forecast production capacities in 2005–2007; public power plants – allocation proportional to weighted values of three factors: base emission (50%), electrical capacity in 2003 (25%) and maximum utilisation rate in years 1999–2002 (25%).

[j] Additional reserve of 5.5 Mt (1.9%) proposed for special purposes.

Early action provisions?	NAP proposal 2004: additional 13.7 Mt/yr EA budget proposed to be allocated in NAP Latest proposals in 2005: no EA provisions
CHP/cogeneration provisions?	NAP proposal 2004: additional 4 Mt/yr CHP budget proposed to be allocated in NAP Latest proposals in 2005: no CHP provisions
Pooling provisions?	Pooling allowed
Banking provisions?	Allowed (under condition of emission intensity improvement confirmed by reduction of unit CO_2 emission per product output)
Opt-ins/opt-outs?	NAP proposal 2004: opt-out proposed for 221 installations mainly from ceramic sector with emissions below 5 kt/yr, total emission below 0.2% of total ETS emissions Latest proposals in 2005: no opt-in proposals

8. *NAP summary for Spain*

Cap statistics	Latest year (2002)	Projected BAU 2005–2007	Approved ETS cap
Annual ETS sector emissions (Mt CO_2)	164.32	n.a.[a]	174.4

BSA/Kyoto statistics	Latest year 2002	Projected 2005–2007	BSA/Kyoto target
National emissions (Mt CO_2e)	401.34	400.70	328.9

NAP timing	Answers
Initial notification to the Commission	07/07/2004
Final Commission approval of NAP	27/12/2004
Final MS allocation law/regulation	09/03/2005

NAP coverage	Answers
Number of installations	819

Allocation process	Answers	
Base period years	2000–2002	
Two-step (sector/installation) process?	Yes	
If so, on what basis?	—	
Basis of allocation to installations	Historical emissions and projections	

[a] Only available for the electricity generation sector (94 Mt CO_2/yr).

Are electric utilities allocated less relative to base period or need than others?	Yes	

Other NAP provisions	Brief description if applicable
Auction provisions? (% if yes)	No
New entrants reserve? (size or % if present)	Yes, 2.994 Mt CO_2/yr[b]
Closure/transfer provisions?	No
Early action provisions?	No
CHP/cogeneration provisions?	Yes. CHP given as many allowance as they are likely to need in the future. Distinction between CHP in Annex I sectors and non-Annex I sectors
Pooling provisions?	Yes (banned for the electricity sector)
Banking provisions?	No
Opt-ins/opt-outs?	No

[b] An additional 0.364 Mt CO_2 is reserved for non-Annex I cogeneration installations, making a total of 3.358 Mt CO_2/yr.

9. *NAP summary for Sweden*

Cap statistics	Latest year (1998–2001)	Projected BAU 2005–2007	Approved ETS cap
Annual ETS sector emissions (Mt CO_2)	20.2	n.a.	22.9

BSA/Kyoto statistics	Latest year 1998–2001	Projected 2005–2007	BSA/Kyoto target
National emissions (Mt CO_2e)	69.3	69.4[a]	74.3

NAP timing	Answers
Initial notification to the Commission	22/04/2002
Final Commission approval of NAP	07/07/2004
Final MS allocation law/regulation	01/01/2005

NAP coverage	Answers
Number of installations	Approx. 700

Allocation process	Answers	
Base period years	1998–2001	
Two-step (sector/installation) process?		No
If so, on what basis?	—	

[a] Projections made for the period 2008–2012.

Basis of allocation to installations	Historical emissions and projected increase in process emissions; energy sector scaled down by factor 0.8, except industrial CHP
Are electric utilities allocated less relative to base period or need than others?	Yes

Other NAP provisions	Brief description if applicable
Auction provisions? (% if yes)	No
New entrants reserve? (size or % if present)	2.19 Mt (3.2%)
Closure/transfer provisions?	Closures keep allowances
Early action provisions?	No
CHP/cogeneration provisions?	For New entrants, only CHP receives allocation
Pooling provisions?	No
Banking provisions?	No
Opt-ins/opt-outs?	Sweden has used the possibility to opt in energy installations below 20 MW in district heating nets where total net capacity is above 20 MW. About 60 installations have been opted in to the emission trading system

10. *NAP summary for United Kingdom*

Cap statistics	Latest year (2003)	Projected BAU 2005–2007	Approved ETS cap
Annual ETS sector emissions (Mt CO_2)	271.6	267.1	245.4

BSA/Kyoto statistics	Latest year (2004)	Projected 2005–2007	BSA/Kyoto target
National emissions (Mt CO_2e)	656.3	644.0[a]	671.5[b]

NAP timing	Answers
Initial notification to the Commission	Provisional NAP in April 2004, with proposed amendments in November 2004
Final Commission approval of NAP	July 2004 of Provisional NAP
Final MS allocation law/ regulation	The Greenhouse Gas Emissions Trading Scheme Regulations 2005 (Statutory Instrument 2005 No. 925)

NAP coverage	Answers
Number of installations	1,055

Allocation process	Answers
Base period years	1998–2003
Two-step (sector/installation) process?	Yes

[a] Figure based on interpolation between 2004 and 2010.
[b] Figure based on data from the UK Greenhouse Gas Inventory, 1990 to 2004, Annual Report for submission under the Framework Convention on Climate Change, April 2006.

If so, on what basis?	Sector-level based on projections incorporating other policies and benchmarks (CCAs) and adjustment factors
Basis of allocation to installations	Installation-level based on share of historical emissions where available, otherwise benchmarks
Are electric utilities allocated less relative to base period or need than others?	Yes

Other NAP provisions	Brief description if applicable
Auction provisions? (% if yes)	Yes (but only of unclaimed new entrant reserve)
New entrants reserve? (size or % if present)	Yes, total of 15.6 Mt CO_2 annually, 6.3% of total cap
Closure/transfer provisions?	Yes, forfeit of subsequent years with transfer possibility
Early action provisions?	Only via baseline and rationalisation rule
CHP/cogeneration provisions?	0.7% of total allocation reserved for unknown CHP new entrants
Pooling provisions?	None
Banking provisions?	No
Opt-ins/opt-outs?	Yes, for 329 CCA participants (2005–7) and 59 UK ETS participants (2005–6); 7 installations in both categories

IV | *Background material from the European Commission*

Excerpts from Directive 2003/87/EC establishing a scheme for green-house gas allowance trading within the Community[1]

Article 9 – National allocation plan

1. For each period referred to in Article 11(1) and (2), each Member State shall develop a national plan stating the total quantity of allowances that it intends to allocate for that period and how it proposes to allocate them. The plan shall be based on objective and transparent criteria, including those listed in Annex III, taking due account of comments from the public. The Commission shall, without prejudice to the Treaty, by 31 December 2003 at the latest develop guidance on the implementation of the criteria listed in Annex III.

For the period referred to in Article 11(1), the plan shall be published and notified to the Commission and to the other Member States by 31 March 2004 at the latest. For subsequent periods, the plan shall be published and notified to the Commission and to the other Member States at least 18 months before the beginning of the relevant period.

2. National allocation plans shall be considered within the committee referred to in Article 23(1).

3. Within three months of notification of a national allocation plan by a Member State under paragraph 1, the Commission may reject that plan, or any aspect thereof, on the basis that it is incompatible with the criteria listed in Annex III or with Article 10. The Member State shall only take a decision under Article 11(1) or (2) if proposed amendments are accepted by the Commission. Reasons shall be given for any rejection decision by the Commission.

Article 10 – Method of allocation

For the three-year period beginning 1 January 2005 Member States shall allocate at least 95% of the allowances free of charge. For the five-year

---

[1] European Community (2003), 'Directive 2003/87/EC of the European Parliament and of the Council of 13 October 2003 establishing a scheme for greenhouse gas emission allowance trading within the Community and amending Council Directive 96/61/EC', *OJ* L 275, 32–45, 25 October 2003.

period beginning 1 January 2008, Member States shall allocate at least 90% of the allowances free of charge.

Article 11 – Allocation and issue of allowances

1. For the three-year period beginning 1 January 2005, each Member State shall decide upon the total quantity of allowances it will allocate for that period and the allocation of those allowances to the operator of each installation. This decision shall be taken at least three months before the beginning of the period and be based on its national allocation plan developed pursuant to Article 9 and in accordance with Article 10, taking due account of comments from the public.

2. For the five-year period beginning 1 January 2008, and for each subsequent five-year period, each Member State shall decide upon the total quantity of allowances it will allocate for that period and initiate the process for the allocation of those allowances to the operator of each installation. This decision shall be taken at least 12 months before the beginning of the relevant period and be based on the Member State's national allocation plan developed pursuant to Article 9 and in accordance with Article 10, taking due account of comments from the public.

3. Decisions taken pursuant to paragraph 1 or 2 shall be in accordance with the requirements of the Treaty, in particular Articles 87 and 88 thereof. When deciding upon allocation, Member States shall take into account the need to provide access to allowances for new entrants.

4. The competent authority shall issue a proportion of the total quantity of allowances each year of the period referred to in paragraph 1 or 2, by 28 February of that year.

Annex III – Criteria for National Allocation Plans

1. The total quantity of allowances to be allocated for the relevant period shall be consistent with the Member State's obligation to limit its emissions pursuant to Decision 2002/358/EC and the Kyoto Protocol, taking into account, on the one hand, the proportion of overall emissions that these allowances represent in comparison with emissions from sources not covered by this Directive and, on the other hand, national energy policies, and should be consistent with the national climate change programme. The total quantity of allowances to be allocated shall not be more than is likely to be needed for the strict application of the criteria of this Annex. Prior to 2008, the quantity shall be consistent with a path towards achieving or over-achieving each Member State's target under Decision 2002/358/EC and the Kyoto Protocol.

2. The total quantity of allowances to be allocated shall be consistent with assessments of actual and projected progress towards fulfilling the Member States' contributions to the Community's commitments made pursuant to Decision 93/389/EEC.

3. Quantities of allowances to be allocated shall be consistent with the potential, including the technological potential, of activities covered by this scheme to reduce emissions. Member States may base their distribution of allowances on average emissions of greenhouse gases by product in each activity and achievable progress in each activity.

4. The plan shall be consistent with other Community legislative and policy instruments. Account should be taken of unavoidable increases in emissions resulting from new legislative requirements.

5. The plan shall not discriminate between companies or sectors in such a way as to unduly favour certain undertakings or activities in accordance with the requirements of the Treaty, in particular Articles 87 and 88 thereof.

6. The plan shall contain information on the manner in which new entrants will be able to begin participating in the Community scheme in the Member State concerned.

7. The plan may accommodate early action and shall contain information on the manner in which early action is taken into account. Benchmarks derived from reference documents concerning the best available technologies may be employed by Member States in developing their National Allocation Plans, and these benchmarks can incorporate an element of accommodating early action.

8. The plan shall contain information on the manner in which clean technology, including energy efficient technologies, are taken into account.

9. The plan shall include provisions for comments to be expressed by the public, and contain information on the arrangements by which due account will be taken of these comments before a decision on the allocation of allowances is taken.

10. The plan shall contain a list of the installations covered by this Directive with the quantities of allowances intended to be allocated to each.

11. The plan may contain information on the manner in which the existence of competition from countries or entities outside the Union will be taken into account.

Index

acceptability *see* political acceptability
adjustments to plan 27, 32–3, 34, 93
  Czech Republic 282–3, 285–7
  Germany 93–7
  Netherlands 150–1
  Spain 197–8
  Sweden 145–6, 150–1
  United Kingdom 61
administration costs: United Kingdom
  70
administrative rules: prevailing
  regulatory tendency in some
  plans to substitute market with
  administrative rules 34–5
Agostini, Daniele 370
agriculture: Ireland 158, 165
allocation process 35–7, 363–9
  allocation aimed for non-achievable
    perfection 32–3
  central coordination 349–51
  challenges 16–20
  distributional character of allocation
    of allowances 28–9
    between companies 29
    between Member States 29–31
  emitters involved in allocation
    decisions 344–7
  formulae at sector and installation
    level 18
  global implications 366–9
  indirect beneficiaries 33–4
  limitations on allocation choices
    31–2
  notification of allocation plans 22–3
  periodic nature 36
  projections used in 347–9
  rules in Directive 2003/87 14–15
  second round 37
  soft harmonisation 35

  see also National Allocation Plans
    (NAPs)
allowances 18
  distributional character of allocation
    28–9
assessment of plans 22–8
  Czech Republic 271, 278–80,
    292–3, 350
  Germany 103
  Hungary 263–4
  Ireland 173–5, 179–80
  Italy 225–6
  Poland 317–18, 334–5
  Spain 198
  Sweden 147
  United Kingdom 52, 54, 62, 70
auctions 362–3, 364
  Czech Republic 297
  Denmark 127–8, 363
  Germany 84
  Hungary 260–1, 265–6
  Ireland 178
  Spain 21–211, 202
  Sweden 148–9
  United Kingdom 68
Austria: assessment of plan 23, 24,
  26

Barry, Conor 370
Bart, Istvan 370–1
baseline rules 355–7
  Czech Republic 275
  Denmark 108, 111, 116–17
  Hungary 257–9
  Italy 229–30
  Poland 313, 319–21, 323, 324–5
  Sweden 146–7
  United Kingdom 58
Belgium: assessment of plan 24, 26